Biological Dosimetry

Cytometric Approaches to Mammalian Systems

Edited by
W. G. Eisert and M. L. Mendelsohn

Springer-Verlag
Berlin Heidelberg New York Tokyo

Biological Dosimetry

Cytometric Approaches
to Mammalian Systems

Edited by

W. G. Eisert and M. L. Mendelsohn

With 147 Figures and 26 Tables

Springer-Verlag
Berlin Heidelberg New York Tokyo 1984

Priv.-Doz. Dr. Dr. Wolfgang G. Eisert
Gesellschaft für Strahlen- und Umweltforschung mbH,
Arbeitsgruppe Zytometrie
Herrenhäuser Straße 2, 3000 Hannover 21, F.R.G.

and

Dr. K. Thomae GmbH
Biologische Forschung, 7950 Biberach-Riss, F.R.G.

Dr. Mortimer L. Mendelsohn
Lawrence Livermore National Laboratory, Biomedical Sciences
Division, P.O. Box 5570, Livermore, CA 94550, USA

ISBN 3-540-12790-9 Springer-Verlag Berlin Heidelberg New York Tokyo
ISBN 0-387-12790-9 Springer-Verlag New York Heidelberg Berlin Tokyo

Library of Congress Cataloging in Publication Data. Main entry under title:
Biological dosimetry. Papers presented at a symposium held in Munich at the
Gesellschaft für Strahlen- und Umweltforschung, Oct. 1982 as a satellite meeting
of the 9th International Conference on Analytical Cytology. Bibliography: p.
Includes index. 1. Ionizing radiation–Physiological effect–Congresses. 2. Ionizing
radiation–Dose–effect relationship–Congresses. 3. Cells–Effect of radiation on–
Congresses. 4. Radiation dosimetry–Congresses. 5. Mammals–Physiology–Congres-
ses. I. Eisert, W. G. (Wolfgang G.), 1947-. II. Mendelsohn, Mortimer L. III. Gesell-
schaft für Strahlen- und Umweltforschung. IV. International Conference on
Analytical Cytology (9th : 1982 : Garmisch/Germany)
QP82.2.I53B56. 1984. 599'.01915. 84-1239
ISBN 0-387-12790-9 (U.S.)

Printing and Binding: J. Beltz, Hemsbach
2127/3140-543210

Preface

In October 1982, a small international symposium was held at the Gesellschaft für Strahlen- und Umweltforschung mbH (GSF) in Munich as a satellite meeting of the IX International Conference on Analytical Cytology. The symposium focussed on cytometric approaches to biological dosimetry, and was, to the best of our knowledge, the first meeting on this subject ever held. There was strong encouragement from the 75 attendees and from others to publish a proceedings of the symposium. Hence this book, containing 30 of the 36 presentations, has been assembled.

Dosimetry, the accurate and systematic determination of doses, usually refers to grams of substance administered or rads of ionization or some such measure of exposure of a patient, a victim or an experimental system. The term also can be used to describe the quantity of an ultimate, active agent as delivered to the appropriate target material within a biological system. Thus, for mutagens, one can speak of DNA dosimetry, meaning the number of adducts produced in the DNA of target cells such as bone-marrow stem cells or spermatogonia.

Biological dosimetry carries the concept one step further by describing dose in terms of a defined biological response. For mutagens this might take the form of induced mutations, for clastogens of induced chromosome aberrations, for ionizing radiation of changes in circulating blood cells, for a reproductive toxin of altered sperm or decreased fertility. Such dosimetry has two general functions: a backwards-looking function to reconstruct the effective physical or chemical dose actually received by the subject; and a forwards-looking function to discover the biologically meaningful dose and thereby to predict some later or more important effect or to titrate more delicately the ultimate physical or chemical dose that should be given to achieve a certain effect. Biological dosimetry is particularly valuable in situations where physical or chemical dose is unknown (as in accidental, occupational or environmental exposures), where dose is difficult to standardize (as in irregular scheduling), or where individual biological susceptibility is likely to vary (as in pharmacologic variants or repair defectives).

Biological dosimetry can take many forms, but of particular concern to this symposium are those endpoints, involving cells and subcellular organelles, which can be approached by automated cytometry. The rapidly evolving technologies for flow cytometry, machine-based image analysis, and cytochemistry offer the promise of objective, precise, sensitive, rapid, and inexpensive processing of such biological specimens, and appear to be ideally suited for biological dosimetry in many human and experimental biological systems.

The specific applications to be found in this volume include cytogenic, mutational, reproductive, hematologic, immunologic, morphologic, kinetic and metabolic effects. These dosimetries are covered both by overviews of conventional methods and by state-of-the-art presentations of cytometric methods ranging from first glimmers of possible approaches to well advanced systems on the verge of practical application. Although this could hardly be called a book of recipes for biological dosimetry, it is a broad, exciting glimpse of a fledgling field that has major potential for the future.

Many organizations and people contributed to the success of the Symposium and the appearance of this volume. We are most grateful for the funding supplied in Germany by the Gesellschaft für Strahlen- und Umweltforschung mbH and the Deutsche Forschungsgemeinschaft, and in the United States by the Department of Energy. The help of the GSF Conference Service and Administration was greatly appreciated. The personal support of the staff members of the Biomedical Sciences Division of Livermore and of the GSF Cytometry Group in Hannover is also gratefully acknowledged.

Wolfgang G. Eisert
Mortimer L. Mendelsohn

Contents

Cytogenetic Effects

An Overview of Radiation Dosimetry by Conventional Cytogenetic
Methods (D.C. Lloyd)   3

Cytogenetic Effects in Human Lymphocytes as a Dosimetry System
(M. Bauchinger) 15

Flow Cytometric Detection of Aberrant Chromosomes
(J.W. Gray, J. Lucas, L.C. Yu, and R. Langlois) 25

Flow Cytometric Measurement of Cellular DNA Content Dispersion
Induced by Mutagenic Treatment (F.J. Otto, H. Oldiges, and V.K. Jain) 37

Flow Cytometric Analysis of Chromosome Damage After Irradiation:
Relation to Chromosome Aberrations and Cell Survival (J.A. Aten, M.W. Kooi,
J.Th. Bijman, J.B.A. Kipp, and G.W. Barendsen) 51

Chromosome Analysis of Leukemia and Preleukemia: Preliminary Data
on the Possible Role of Environmental Agents (H. van den Berghe) 61

Radiation Dosimetry Using the Methods of Flow Cytogenetics
(D.K. Green, J.A. Fantes, and G. Spowart) 67

Preliminary Reanalysis of Radiation-Induced Chromosome Aberrations
in Relation to Past and Newly Revised Dose Estimates for Hiroshima and
Nagasaki A-Bomb Survivors (A.A. Awa, T. Sofuni, T. Honda, H.B. Hamilton,
and S. Fujita) 77

Reproductive Effects

Biological Dosimetry of Sperm Analyzed by Conventional Methods:
Cautions and Opportunities (B.L. Gledhill) 85

Detection of Male Reproductive Abnormalities by Flow Cytometry
Measurements of Testicular and Ejaculated Germ Cells
(D.P. Evenson, P.J. Higgins, and M.R. Melamed) 99

Cytometric Analysis of Mammalian Sperm for Induced Morphologic
and DNA Content Errors (D. Pinkel) 111

Mammalian Spermatogenesis as a Biological Dosimeter for Ionizing
Radiation (U. Hacker-Klom, W. Göhde, and J. Schumann) 127

Mutagenic Effects

Biological Dosimetry of Mutagenesis: Principles, Methods, and
Cytometric Prospects (M.L. Mendelsohn) 141

Detection of HGPRT-Variant Lymphocytes Using the FIP High-Speed
Image Processor (P.E. Perry, E.J. Thomson, M.H. Stark, and J.H. Tucker) 149

Somatic Mutations Detected by Immunofluorescence and Flow Cytometry
(R.H. Jensen, W. Bigbee, and E.W. Branscomb) 161

Computer Scoring of Micronuclei in Human Lymphocytes
(H. Callisen, A. Norman, and M. Pincu) . . . .

Hematopoietic and Immunologic Effects

Review of Biological Dosimetry by Conventional Methods in
Haematopoiesis (J.W.M. Visser) 183

Assessment of Environmental Insults on Lymphoid Cells as Detected
by Computer Assisted Morphometric Techniques (G.B. Olson and
P.H. Bartels) 203

Dose Related Changes in Cell Cycle of Bone Marrow and Spleen Cells
Monitored by DNA/RNA Flow Cytometry (W.G. Eisert, H.U. Weier,
and G. Birk) 219

Hoechst 33342 Dye Uptake as a Probe of Membrane Permeability in
Mammalian Cells (R.G. Miller, M.E. Lalande, and E.C. Keystone) 229

Probing Macromolecular Structures by Flow Cytometric Fluorescence
Polarization Measurements (W. Beisker and W.G. Eisert) 235

Physiological and Preparatory Variation of Nuclear Chromatin as a
Limiting Condition for Biological Dosimetry by Means of High Resolution
Image Analysis (W. Giaretti, W. Abmayr, G. Burger, and P. Dörmer) 243

Morphologic, Kinetic, and Metabolic Effects

Automated Autoradiographic Analysis of Tumor Cell Colonies *in vitro*
(R.F. Kallman) 255

Flow Cytometric Quantification of Radiation Responses of Murine
Peritoneal Cells (N. Tokita and M.R. Raju) 265

Cell Cycle Kinetics of Synchronized EAT-Cells After Irradiation in
Various Phases of the Cell Cycle Studied by DNA Distribution Analysis
in Combination with Two-Parameter Flow Cytometry Using Acridine Orange
(M. Nüsse) 273

Nuclear Morphology of Hepatocytes in Rats After Application of
Polychlorinated Biphenyls (W. Abmayr, D. Oesterle, and E. Deml) 285

Computer Graphic Displays for Microscopists Assisting in Evaluating
Radiation-Damaged Cells (P.H. Bartels and G.B. Olson) 295

Flow Cytometric Measurement of Intracellular pH: A Dosimetry Approach
to Metabolic Heterogeneity (O. Alabaster, M. Andreeff, C. Spooner, and
K. Clagett) . . . .

Flow Cytometric Measurement of the Metabolism of Benzo [A] Pyrene by
Mouse Liver Cells in Culture (J.C. Bartholomew, C.G. Wade, and
K.K. Dougherty) 311

High-Affinity Monoclonal Antibodies Specific for Deoxynucleosides
Structurally Modified by Alkylating Agents: Applications for Immunoanalysis
(J. Adamkiewicz, O. Ahrens, and M.F. Rajewsky) 325

List of Contributors

W. Abmayr
Gesellschaft für Strahlen- und Umweltforschung mbH, Institut für Strahlenschutz,
Ingolstädter Landstraße 1, 8042 Neuherberg, Federal Republic of Germany

J. Adamkiewicz
Institut für Zellbiologie (Tumorforschung), Universität Essen (GHS), Hufelandstraße 55,
4300 Essen 1, Federal Republic of Germany

O. Ahrens
Institut für Zellbiologie (Tumorforschung), Universität Essen (GHS), Hufelandstraße 55,
4300 Essen 1, Federal Republic of Germany

O. Alabaster
The George Washington University Medical Center, Washinton, DC, U.S.A.

M. Andreeff
Memorial Sloan-Kettering Cancer Center, 1275 York Avenue, New York, N.Y. 10021,
U.S.A.

J.A. Aten
Laboratory for Radiobiology, University of Amsterdam, Plesmanlaan 121,
1066 CX Amsterdam, The Netherlands

A.A. Awa
Department of Clinical Laboratories, Radiation Effects Research Foundation (RERF),
5-2 Hijiyama Park, Minami-ku, Hiroshima 730, Japan

G.W. Barendsen
Laboratory for Radiobiology, University of Amsterdam, Plesmanlaan 121,
1066 CX Amsterdam, The Netherlands

P.H. Bartels
Department of Microbiology, University of Arizona, Tucson, AZ 85721, U.S.A.

M. Bauchinger
Abteilung für Strahlenbiologie, Zytogenetik, Gesellschaft für Strahlen- und
Umweltforschung mbH, Ingolstädter Landstraße 1, 8042 Neuherberg,
Post Oberschleißheim, Federal Republic of Germany

W. Beisker
Arbeitsgruppe Zytometrie, Gesellschaft für Strahlen- und Umweltforschung mbH,
Herrenhäuser Straße 2, 3000 Hannover 21, Federal Republic of Germany

H. van den Berghe
Center for Human Genetics, University of Leuven, Kapucijnenvoer 33, 3000 Leuven,
Belgium

W. Bigbee
Lawrence Livermore National Laboratory, Biomedical Sciences Division, University of
California, P.O. Box 5507 L-452, Livermore, CA 94550, U.S.A.

J.Th. Bijman
Laboratory for Radiobiology, University of Amsterdam, Plesmanlaan 121,
1066 CX Amsterdam, The Netherlands

G. Birk
Institut für Biophysik, Universität Hannover, Herrenhäuser Straße 2, 3000 Hannover 21,
Federal Republic of Germany

E.W. Branscomb
Lawrence Livermore National Laboratory, Biomedical Sciences Division, University
of California, P.O. Box 5507 L-452, Livermore, CA 94550, U.S.A.

G. Burger
Institut für Strahlenschutz, Gesellschaft für Strahlen- und Umweltforschung mbH,
Ingolstädter Landstraße 1, 8042 Neuherberg, Federal Republic of Germany

H. Callisen
Department of Radiological Sciences, University of California, School of Medicine,
900 Veteran Avenue, Los Angeles, CA 90024, U.S.A.

K. Clagett
The George Washington University Medical Center, Washington, DC, U.S.A.

E. Deml
Institut für Toxikologie und Biochemie, Abteilung für Toxikologie, Gesellschaft
für Strahlen- und Umweltforschung mbH, Ingolstädter Landstraße 1,
8042 Neuherberg, Federal Republic of Germany

P. Dörmer
Institut für Hämatologie, Abteilung für Experimentelle Hämatologie, Gesellschaft für
Strahlen- und Umweltforschung mbH, Landwehrstraße 61, 8000 München 2,
Federal Republic of Germany

K.K. Dougherty
Laboratory of Chemical Biodynamics, Lawrence Berkeley Laboratory, University of
California, Berkeley, CA 94720, U.S.A.

W.G. Eisert
Arbeitsgruppe Zytometrie, Gesellschaft für Strahlen- und Umweltforschung mbH,
Herrenhäuser Straße 2, 3000 Hannover 21, Federal Republic of Germany

D.P. Evenson
Memorial Sloan-Kettering Cancer Center, 1275 York Avenue, New York, NY 10021,
U.S.A.

J.A. Fantes
MRC Clinical and Population Cytogenetics Unit, Western General Hospital, Crewe Road,
Edinburgh EH4 2XU, Great Britain

S. Fujita
Department of Epidemiology and Statistics, Radiation Effects Research Foundation
(RERF), 5-2 Hijiyama Park, Minami-ku, Hiroshima 730, Japan

W. Giaretti
Institut für Hämatologie, Abteilung für Experimentelle Hämatologie, Gesellschaft für
Strahlen- und Umweltforschung mbH, Landwehrstraße 61, 8000 München 2,
Federal Republic of Germany

B.L. Gledhill
Lawrence Livermore National Laboratory, Biomedical Sciences Division, University
of California, P.O. Box 5507 L-452, Livermore, CA 94550, U.S.A.

W. Göhde
Institut für Strahlenbiologie, Universität Münster, Hittorfstraße 17, 4400 Münster,
Federal Republic of Germany

J.W. Gray
Lawrence Livermore National Laboratory, Biomedical Sciences Division, University of
California, P.O. Box 5507 L-452, Livermore, CA 94550, U.S.A.

D.K. Green
MRC Clinical and Population Cytogenetics Unit, Western General Hospital, Crewe Road,
Edinburgh EH4 2XU, Great Britain

U. Hacker-Klom
Institut für Strahlenbiologie, Universität Münster, Hittorfstraße 17, 4400 Münster,
Federal Republic of Germany

H.B. Hamilton
Department of Clinical Laboratories, Radiation Effects Research Foundation (RERF),
5-2 Hijiyama Park, Minami-ku, Hiroshima 730, Japan

P.J. Higgins
Memorial Sloan Kettering Cancer Center, 1275 York Avenue, New York, NY 10021,
U.S.A.

T. Honda
Department of Clinical Laboratories, Radiation Effects Research Foundation (RERF),
5-2 Hijiyama Park, Minami-ku, Hiroshima 730, Japan

V.K. Jain
National Institue of Mental Health and Neuro-Sciences, Bangalore 560029, India

R.H. Jensen
Lawrence Livermore National Laboratory, Biomedical Sciences Division, University of
California, P.O. Box 5507 L-452, Livermore, CA 94550, U.S.A.

R.F. Kallman
Department of Radiology, Stanford University Medical Center, Stanford, CA 94305,
U.S.A.

E.C. Keystone
Rheumatology Unit, Wellesley Hospital, Toronto, Ontario M4X 1K9, Canada

J.B.A. Kipp
Laboratory for Radiobiology, University of Amsterdam, Plesmanlaan 121,
1066 CX Amsterdam, The Netherlands

M.W. Kooi
Laboratory for Radiobiology, University of Amsterdam, Plesmanlaan 121,
1066 CX Amsterdam, The Netherlands

M.E. Lalande
Ontario Cancer Institute, 500 Sherbourne Street, Toronto, Ontario M4X 1K9, Canada

R. Langlois
Lawrence Livermore National Laboratory, Biomedical Sciences Division, University of
California, P.O. Box 5507 L-452, Livermore, CA 94550, U.S.A.

D.C. Lloyd
National Radiological Protection Board, Chilton, Didcot, Oxon, OX11 ORQ,
Great Britain

J. Lucas
Lawrence Livermore National Laboratory, Biomedical Sciences Division, University of
California, P.O. Box 5507 L-452, Livermore, CA 94550, U.S.A.

M.R. Melamed
Memorial Sloan-Kettering Cancer Center, 1275 York Avenue, New York, NY 10021,
U.S.A.

M.L. Mendelsohn
Lawrence Livermore National Laboratory, Biomedical Sciences Division, University of
California, P.O. Box 5507 L-452, Livermore, CA 94550, U.S.A.

R.G. Miller
The Ontario Cancer Institute, 500 Sherbourne Street, Toronto, Ontario, M4X 1K9,
Canada

A. Norman
Department of Radiological Sciences, University of California School of Medicine,
900 Veteran Avenue, Los Angeles, CA 90024, U.S.A.

M. Nüsse
Institut für Biologie, Abteilung für Biophysikalische Strahlenforschung, Gesellschaft
für Strahlen- und Umweltforschung mbH, Paul-Ehrlich-Straße 20, 6000 Frankfurt/Main,
Federal Republic of Germany

D. Oesterle
Institut für Toxikologie und Biochemie, Abteilung für Toxikologie, Gesellschaft für
Strahlen- und Umweltforschung mbH, Ingolstädter Landstraße 1, 8042 Neuherberg,
Federal Republic of Germany

H. Oldiges
Fraunhofer-Institut für Toxikologie und Aerosolforschung, Institutsteil Grafschaft,
5948 Schmallenberg, Federal Republic of Germany

G.B. Olson
Department of Microbiology, University of Arizona, Tucson, AZ 85721, U.S.A.

F.J. Otto
Fraunhofer-Institut für Toxikologie und Aerosolforschung, Institutsteil Grafschaft,
5948 Schmallenberg, Federal Republic of Germany

P.E. Perry
MRC Clinical and Population Cytogenetics Unit, Western General Hospital,
Crewe Road, Edinburgh, EH4 2XU, Great Britain

M. Pincu
Department of Radiological Sciences, University of California School of Medicine,
900 Veteran Avenue, Los Angeles, CA 90024, U.S.A.

D. Pinkel
Lawrence Livermore National Laboratory, Biomedical Sciences Division, University
of California, P.O. Box 5507 L-452, Livermore, CA 94550, U.S.A.

M.F. Rajewsky
Institut für Zellbiologie (Tumorforschung), Universität Essen (GHS), Hufelandstraße 55,
4300 Essen 1, Federal Republic of Germany

M.R. Raju
Live Sciences Division, Los Alamos National Laboratory, University of California,
Los Alamos, NM 87545, U.S.A.

J. Schumann
Fachklinik Hornheide, Universität Münster, Dorbaumstraße 44–48, 4400 Münster,
Federal Republic of Germany

T. Sofuni
Department of Clinical Laboratories, Radiation Effects Research Foundation (RERF),
5-2 Hijiyama Park, Minami-ku, Hiroshima 730, Japan

C. Spooner
The George Washington University Medical Center, Washington, DC, U.S.A.

G. Spowart
MRC Clinical and Population Cytogenetics Unit, Western General Hospital, Crewe Road,
Edinburgh EH4 2XU, Great Britain

M.H. Stark
MRC Clinical and Population Cytogenetics Unit, Western General Hospital, Crewe Road,
Edinburgh EH4 2XU, Great Britain

E.J. Thomson
MRC Clinical and Population Cytogenetics Unit, Western General Hospital, Crewe Road,
Edinburgh, EH4 2XU, Great Britain

N. Tokita
Life Sciences Division, Los Alamos National Laboratory, University of California,
Los Alamos, NM 87545, U.S.A.

J.H. Tucker
MRC Clinical and Population Cytogenetics Unit, Western General Hospital, Crewe Road,
Edinburgh EH4 2XU, Great Britain

J.W.M. Visser
Radiobiological Institute, REPGO-TNO, 151 Lange Kleiweg, 2288 GJ Rijswijk,
The Netherlands

C.G. Wade
Laboratory of Chemical Biodynamics, Lawrence Berkeley Laboratory, University of
California, Berkeley, CA 94720, U.S.A.

H.U. Weier
Arbeitsgruppe Zytometrie, Gesellschaft für Strahlen- und Umweltforschung mbH,
Herrenhäuser Straße 2, 3000 Hannover 21, Federal Republic of Germany

L.C. Yu
Lawrence Livermore National Laboratory, Biomedical Sciences Division, University of
California, P.O. Box 5507 L-452, Livermore, CA 94550, U.S.A.

Cytogenetic Effects

An Overview of Radiation Dosimetry by Conventional Cytogenetic Methods

David C. Lloyd

National Radiological Protection Board, Chilton, Didcot, Oxon, OX11 ORQ, Great Britain

Summary

A brief summary is given of the development of chromosome aberration dosimetry and its present status as a routine technique in radiological protection. The dose effect relationships for dicentric aberrations are outlined and the statistical uncertainties on dose estimates made from these calibration data are discussed. The experience of approximately 600 investigations over 15 years in the UK is summarised. Speculations are made on the likely effect of the introduction of a metaphase finder into a laboratory undertaking routine cytogenetic dosimetry.

Introduction

It is now twenty years since the Recuplex accident at Hanford, USA, when the first attempt at quantitative biological dosimetry based on chromosome damage in peripheral blood lymphocytes was made on three men who were exposed to a mixed field of gamma and neutron radiation. [1] Improvements during the intervening years have established chromosome dosimetry as a reliable technique in radiological protection where it plays an important role in conjunction with physical methods of dosimetry. Among the more recent advances has been research on automated methods for chromosome analysis. These have concentrated along two lines: metaphase finding with operator-assisted karyotyping and flow cytometry. Many people would probably agreed that at times the development of these systems has been frustratingly slow but automation now seems to be poised ready to make a significant reduction in the labour intensive and fatiguing aspects of chromosome analysis.

In several centres around the world, chromosome dosimetry is now frequently employed when someone is known, or suspected, of having been overexposed. However, to my knowledge, no laboratory currently uses an automated system for the routine workload. There is a strong possibility that the NRPB laboratory will shortly acquire a metaphase finder. In this paper, therefore, it is intended to present an overview of the routine application of chromosome dosimetry drawing mainly on personal experience and then to conclude by speculating on some possible effects from the introduction of the metaphase finder.

Biological Dosimetry. Edited by W. G. Eisert and M. L. Mendelsohn
© Springer-Verlag Berlin Heidelberg 1984

Specimen Handling

Blood samples are obtained usually through the post from people who are known or suspected of having been overexposed. Transit times of a few days can be tolerated without detriment to the lymphocytes so that in principle the laboratory could handle samples from anywhere in the world. On receipt of the blood, the lymphocytes are cultured for about two days in order to obtain metaphase spreads and then the cultures are fixed and the cells dispensed onto microscope slides. These are simple routine procedures which have been fully described elsewhere. [2] A recent innovation, however, which most laboratories now employ, is to include bromodeoxyuridine in the cultures so that fluorescence plus Giemsa (FPG) staining may be used to ensure that the analysis is confined to first division cells. Second and subsequent division cells have an increasing probability of unstable aberration loss and some laboratories have reported that even after only 48 h in culture an appreciable number of such cells may be present. [3] From this one must infer that earlier data from overdose cases and *in vitro* experimental work which was carried out before the introduction of the FPG technique, could have been subject to some uncertainty. This is particularly so at lower radiation doses where the effect of mitotic delay would be minimal and for data derived from lymphocytes cultured for longer than 48 h.

Dose Calculation

It is possible to provide a dose estimate three days after receipt of the specimen. As most of this time comprises the culturing period the introduction of automated microscopy will do little to reduce the delay. Although the microscope analysis is time consuming, in an emergency several people can collaborate in scoring replicate slides. This three day delay means in practice that, if there are no other reliable means of quantifying the dose, such as by a film badge, and a serious over-exposure seems possible, the patient must be kept under close medical surveillance. In these cases it is advisable to make frequent differential white cell counts to check for any rapid decline in numbers of circulating leucocytes. [4]

The cytogenetic estimate of dose is based for a number of reasons on the incidence of dicentric aberrations. The dicentric is an easily identified aberration, which occurs with a low background frequency in unirradiated controls as it is produced by few other clastogens. However, following exposure of cells in G_0, such as circulating human lymphocytes, the dicentric is the most frequently produced chromosome aberration. In addition a major reason for counting the dicentric is that reliable *in vitro* dose response curves can be prepared. Within a laboratory, these data are reproducible over several years and there is sufficient evidence to show that the dose effect relationships for *in vitro* and *in vivo* radiation are comparable. [eg 5]

In order to produce a dose estimate with a statistical uncertainty small enough to be of value, a large number of cells need to be scored. This requirement is in marked contrast to other uses of cytogenetic analyses, such as for genetic counselling, where relatively small numbers of good quality metaphases are examined and fully karyotyped. The decision on how many cells to score must be a compromise based on the importance of the case and the available labour. At NRPB we occasionally score 200 cells, more usually 500, and in a few instances up to 1000. The decision to analyse 1000 metaphases depends on whether there is evidence of a serious overexposure, in which case there is a scientific reason, or if the continued employment of a radiation worker is in jeopardy. Uncertainty in the dose estimation is usually expressed as 95% confidence limits and in Table 1 these are shown assuming Poisson statistics for acute exposures to gamma radiation where several dose estimates have been made from the analysis of 200, 500 or 1,000 cells.

Table 1. The effect of increasing the number of cells examined on the lower and upper 95% confidence limits for 4 estimates of acute gamma radiation dose

		Confidence limits (Gy)		
		No. of cells examined		
Dose estimate (Gy)		200	500	1000
0.1	Upper	—	.34	.26
	Lower		.01	⟨.02
0.25	Upper	.64	.48	.41
	Lower	.04	.08	.13
0.5	Upper	.87	.72	.65
	Lower	.20	.31	.37
1.0	Upper	1.35	1.21	1.13
	Lower	.70	.80	.85

In order to provide a biological dosimetry service a laboratory needs a family of *in vitro* curves relating dicentric yield to dose. This is because the relationship is LET dependent and, for low LET radiations, also dose-rate dependent. Curves for X- and gamma radiation are essential. The need for neutron curves is less pressing, although neutron sources are becoming more common, especially for medical purposes, and the possibility of accidental overexposures from these sources is increasing. It is possible, as was shown in the Recuplex accident, [1] to perform cytogenetic dosimetry following a criticality accident, so it is worthwhile for the laboratory to have a calibration for fission spectrum neutrons.

Figure 1 shows a selection of curves produced several years ago [6, 7] by NRPB and their coefficients are given in Table 2. Generally the curves conform to the well documented radiobiological expression of a linear relationship $(Y= aD)$ for the high LET radiations and linear-quadratic $(Y= aD + \beta D^2)$ for low LET. An exception, however, is for protracted exposures to X- or gamma rays where the data are not linear, but the fit to the quadratic expression is poor. This is due to the dose rate effect which influences the dose squared term. When one applied a time related function to this co-

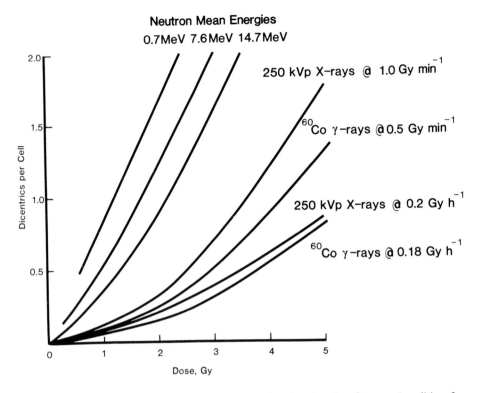

Fig. 1. *In vitro* curves for dicentric aberration yields plotted against dose for several qualities of high and low LET radiation

Table 2. Values of the coefficients α and β in the equation $Y = \alpha D + \beta D^2$ which give the dose response curves for dicentric aberrations in Fig. 1

Radiation type	$\alpha \pm$ SE $\times 10^{-2}$	$\beta \pm$ SE $\times 10^{-2}$	(ratio of α coefficients to acute Co-60 γ rays)
0.7 MeV neutrons	83.5 ± 1.0		53.2
7.6 MeV neutrons	47.8 ± 3.3	6.4 ± 2.0	30.4
14.7 MeV neutrons	26.2 ± 4.0	8.8 ± 2.8	16.7
250 kVp X-rays, acute	4.8 ± 0.5	6.2 ± 0.3	3.0
250 kVP X-rays, chronic	4.1 ± 0.5	2.6 ± 0.4	2.6
Co-60 γ rays, acute	1.6 ± 0.3	5.0 ± 0.2	1.0
Co-60 γ rays, chronic	1.8 ± 0.8	2.9 ± 0.5	$\sim$1.0

efficient as was originally proposed by Catcheside et al., [8] a plausible relationship between acute and chronic dose rates emerges. This requires knowledge of the break repair time for chromosome lesions and there is a considerable amount of published data indicating a time of around 2 h. The dose response curves illustrated here were produced before FPG staining became routine and may now be open to criticism because of the inclusion of second cycle metaphases in the analysis. However, several of the curves have been repeated (to be published) and give essentially similar coefficients suggesting that the errors from this source in the old data were probably minor. Other recent experimental work [9] suggests that the particular culture conditions which have been employed in the laboratory over 15 years, such as the choice of medium, generally result even at zero dose in less than 10% second cycle metaphases at 48 h.

A considerable number of dose effect curves have been published for X-, gamma and neutron radiation. Marked variations in the yield coefficients for comparable radiations are reported between laboratories and a comprehensive review of this is in press. [10] One consistent feature, which does emerge is a significant difference in the relative biological effectiveness (RBE) of neutron radiations relative to X or γ rays which is most marked at low doses. This is illustrated in the data in Table 2, column 4, where the ratio of the alpha coefficients normalised to the value for acute cobalt-60 gamma rays is given. The value for 0.7 MeV fission spectrum neutrons is 53 and for 250 kVp X-rays is 3. As the majority of accidents in our experience fortunately involve low doses (⟨ 0.5 Gy) this LET dependence is of considerable practical significance. It implies that the technique is highly sensitive to fission spectrum neutrons with a lower limit of detection of about 0.01 Gy whole body exposure. For acute X-irradiation a lower limit of perhaps 0.05 Gy is possible, but for gamma rays the limit is nearer to 0.1 Gy. However, there are considerable statistical uncertainties associated with these lower limits (Table 1).

Chromosome Dosimetry in the UK

Several laboratories have presented résumés of the overdose investigations which they have undertaken. [eg. 11] The largest published series is that from the United Kingdom which since 1972 have been described in a sequence of annual reports [eg.12]. To the end of 1981, 580 cases had been referred to the laboratory and Table 3 shows that the majority resulted from industrial uses of radiation. These were particularly associated with the use of iridium-192 for non-destructive testing. The remainder were approximately equally divided between the major nuclear organisations and research education and health institutions. Table 3 also shows that there were 368 cases in which no dicentric aberrations were observed. This indicates that the doses received were either zero or trivial. These are, of course, scientifically unremarkable but they represent a large number of people who may have derived considerable reassurance from the knowledge that they had not been overexposed. Allaying anxiety, particularly of people who may have been exposed but were not wearing a dosemeter is an important aspect of cytogenetic dosimetry.

Table 3. The origins of the cases up to December 1981 and the number of zero dose estimates

Case origins	No. of cases	No. of zero dose estimates
Industrial radiography	371 (64%)	244
Major nuclear organisations	91 (16%)	48
Research, educational and health institutions	118 (20%)	76
	580	368 (63%)

Table 4. The distribution of investigations between 4 categories up to December 1981

1. Possible non-uniform exposure in which the relationship between the dose to the physical dosemeter and that to the body is uncertain.	395
2. Suggested over-exposure to persons not wearing a dosemeter.	118
3. Over-exposure where satisfactory estimates of the whole-body dose can be made from physical reconstructions	5
4. Chronic internal and external exposures.	62
	580

Four main categories of persons are described for whom cytogenetic investigations are undertaken. These are listed in Table 4 together with the number of cases in each category. The first group comprises the majority because most possible exposures are first appreciated when a dosemeter is returned for routine processing and found to have been exposed. Doubt is then raised as to whether the dose on the badge truly reflects the dose to the wearer or indeed whether the badge was even being worn at the time it was irradiated. Usually chromosome analysis can resolve these doubts.

The second largest category is where there may have been an overexposure to persons not wearing a dosemeter. Such instances may pass unnoticed unless there are reasons to suspect a radiation accident and persons at risk can be identified, or where effects such as skin burns become apparent. In these cases chromosome dosimetry can be particularly valuable as it may be the only method of dose assessment available.

The third category is a small group where overexposures have occurred which were well-defined and serious enough to warrant a detailed reconstruction of the incident. Such situations are rare but they are particularly valuable because they provide an opportunity for direct comparison of the biological and physical methods for evaluating the exposure. The general experience of our laboratory and others who have reported such cases is that there is a good measure of agreement between the two methods.

In the final category, samples are sometimes obtained from persons who have been chronically exposed to radiation. In many of these cases, especially where radionuclides have been incorporated, it is not possible to make an estimate of the equivalent whole body dose. This is because of the localised distribution of the nuclide, and the protracted exposure to dose rates which are reducing because of either radioactive decay or metabolism and excretion. For exposures protracted over months or years one needs to consider the turnover of PHA-responsive lymphocytes which must vary between individuals depending on factors such as infection. Accidental overexposure to tritium seems to be an exception as its incorporation results in a more or less uniform exposure of the whole body with a biological half-life of 10 days. Several incidents involving tritium gas escapes in factories have been investigated and remarkably close agreement has been found between cytogenetic dose estimates and committed dose equivalents as calculated from tritium excreted in the urine.

Introduction of Automated Metaphase Finding

The labour intensive and tedious part of chromosome dosimetry is clearly the time spent at the microscope and it has long been recognised that this is an area which is at least partly amenable to automated techniques. Although precise data are not available, questioning of several experienced scorers has produced the information that on average about 50% of their time at the microscope is spent in searching for the metaphases and perhaps 5% in moving objectives, refocussing etc. when changing back and forth from low magnification searching to higher magnification analysis. They also agreed that the searching phase is the greater contributor to fatigue. This leads to a probably conservative estimate that the introduction of a metaphase finder built around one microscope would approximately double the number of cells that a person can score in a given time. This approximation will be used in speculating on the likely consequences of the introduction of a metaphase finder into the laboratory.

In vitro Dose Effect Curves

The principal advantage in the production of dose effect curves will be that the time saved would allow attempts to score responses to low doses to be made more readily. This is an important area where improved precision in the yield coefficients is desirable. This is not only because most exposures are to less than 0.5 Gy but also there is a more fundamental interest in the radiobiological effects of low doses. It is appreciated that chromosome damage is one of the most sensitive end-points yet demonstrated. Regarding the yield coefficients given in Table 2, it has been calculated that doubling the number of cells that were scored to produce these data would have reduced the standard errors on the coefficients by about 1.4. For the acute gamma curve in Table 2 the smallest dose investigated was 0.25 Gy because of the daunting prospect of the effort required to extend the curve to lower doses. The alpha coefficient has a fairly large standard error of around 20%. Metaphase finding could probably enable this to be reduced to 10%. The low dose RBE difference between X and gamma rays of about 3.0 based on the acute alpha coefficients in Table 2 carries a standard error of ± 1.0 i.e. about 30%. Most of this is due to the large standard error on the linear coefficient of the gamma relationship and with improved statistics could be reduced from 30% to about 15%.

This would be a worthwhile increase in the confidence with which it might be asserted that there is a real difference in low dose effects of these two qualities of low LET radiation.

Individual Dose Estimates

In practice the decrease in the uncertainties on the *in vitro* yield coefficients described above are of no consequence in comparison with the uncertainties on dose estimates due to only 500 cells being scored for an individual suspected of having received an overexposure. The effect of the extra scoring capacity here might improve the lower limit of sensitivity of the technique and may also narrow the confidence limits on estimates of high doses. These points are perhaps best illustrated by a few worked examples.

One dicentric in 500 cells corresponds to about 0.1 Gy on the acute gamma dose effect curve. This estimate carries a large upper 95% confidence limit of 0.34 Gy (Table 1). If one were to double the scoring and found one dicentric in 1000 cells the dose estimate now falls to 0.07 Gy with an upper confidence limit of 0.24 Gy. In practice there is very little to be gained from the extra information. Certainly a physician to whom this may be referred would not view the patient any differently. The improvement is negligible by comparison with the millirems level of uncertainty of a film badge, which is the kind of accuracy required for the precise definition of overexposure needed by those responsible for enforcing radiological protection legislation. Moreover 1 in 1000 cells is approaching the spontaneous level for dicentrics of about 1 in 2000, so that the confidence with which one might conclude that a person has even been exposed above his normal background is weakened.

In the event of a large exposure to gamma rays for which one might observe say 20 dicentrics in 500 cells, the estimated equivalent whole body dose would be 0.75 Gy with 95% confidence limits of 0.55 and 0.96 Gy. The importance and interest of such a case would justify increasing the scoring to 1000 cells with present conventional methods. If we were now to find 40 dicentrics in 1000 cells the confidence limits would close in to 0.62 and 0.90 Gy. With a metaphase finder using the same scoring time one could perhaps find 80 dicentrics in 2000 cells and the confidence limits would now come down to 0.66 and 0.84 Gy. Thus scoring extra cells reduces the statistical uncertainties but as with the low dose example given above the improvement is of marginal value. The smaller confidence limits would not influence the clinical management of the patient. It would seem therefore that the main advantage of a metaphase finder which permits the scoring rate to be doubled is a halving of the labour cost. In our experience scoring 500 cells by conventional methods occupies two man days which would therefore reduce to one.

Nevertheless there are situations where extra scoring may produce a worthwhile improvement in dose estimation. The examples given above assume a Poisson distribution which results from a whole body homogeneous exposure. However, most exposures are to only part of the body and this produces a non-Poisson distribution of damage amongst the cells analysed. Dolphin [13] showed how this will result in over dispersion and suggested that in principle the degree of departure from a Poisson distribution might indicate the fraction of the body irradiated and its dose. Such data would be very approximate although they could be improved by chromosome examinations of fixed cells in the skin [14] or hair sheaths. [15] Provided sufficient sites on the body were sampled, such studies could delineate the actual areas of the body exposed and recent research [16] has shown how this information together with lymphocyte aberration data might be used to interpret partial body exposures, taking into account the uneven distribution of lymphocytes in the body. A further complication is that highly irradiated lymphocytes are at a selective disadvantage due to interphase death and mitotic delay but there are experimental data which would permit allowance to be made for these effects. [17]

A simplified example (Table 5) using only the dicentric aberration distribution data shows how one can determine what proportion of the body might have been irradiated. In an actual overdose case [18] in which the exposure was clearly uneven a non-Poisson distribution of aberrations was observed in 500 cells which were scored initially (Table 5, line 1). Making naive assumptions. one could consider the patient's body as comprising two compartments, irradiated and unirradiated and that within the irradiated section the dose was uniform so that the distribution of aberrations in the exposed cells was Poisson. Using the method of maximum likelihood the fraction of cells irradiated (22%) and their mean dicentric yield (0.378 per cell) may be estimated (Table 5, line 2). This yield corresponds to a dose to the irradiated fraction of 2.6 Gy gamma radiation. The distribution of dicentrics in the two compartments are shown in lines 2 and 3. By scoring more cells these calculations may be improved and in this particular case a further 500 cells were examined. These produced a similar dicentric yield (line 4) but indicated that a lower fraction (14%) of the body was exposed and therefore to a higher average dose (3.35 Gy). The difference between the first and second sets of 500 cells was mainly due to two heavily damaged cells with four and five dicentrics respectively being found in the second group. Combining the two sets of scoring (line 7) produced a final estimate (line 8) of the irradiated fraction as 18% of the body with an average dose of 3.0 Gy. In order to attempt properly this type of calculation it is really advisable to analyse several thousands of cells and this is only practical with the assistance of automated metaphase finding.

Survey of Populations

There have been a number of published surveys of chromosome aberrations in groups of irradiated people such as radiographers and nuclear power industry workers, compared with controls. In many instances too few cells and too few subjects were sampled for any far reaching conclusions. However, in two recent studies [19, 20] sufficient effort was expended to demonstrate that workers exposed over many years to radiation doses within the ICRP recommended limits do have an enhanced aberration level. Also a dose effect relationship could be discerned when comparing the aberration frequency against dose as determined from film badge records. This relationship was comparable with published **in** *vitro* dose effect curves.

As a result of these and similar studies one detects a strong pressure on some laboratories to continue looking at 'at risk' populations such as radiation or chemical industry workers. At low levels of exposure to mutagens the health consequences which concern

Table 5. The calculation of the irradiated fraction of the body and its average dose using the non-Poisson distribution of dicentrics observed in a case of high partial-body exposure to γ radiation

	No. of dicentrics	Dicentrics per cell	Dose (Gy)	Distribution of dicentrics					
				0	1	2	3	4	5
1st 500 cells scored	42	.084	1.15	465	29	5	1	0	0
irradiated .22	42	.378	2.60	76	29	5	1	0	0
Postulated Fractions									
Unirradiated .78	0	0	0	389	0	0	0	0	0
2nd 500 cells scored	44	.088	1.20	467	27	4	0	1	1
irradiated .14	44	.611	3.35	39	27	4	0	1	1
Postulated Fractions									
unirradiated .86	0	0	0	428	0	0	0	0	0
Combined 1000 cells scored	86	.086	1.17	932	56	9	1	1	1
irradiated .18	86	.489	3.00	108	56	9	1	1	1
Postulated fractions									
unirradiated .82	0	0	0	824	0	0	0	0	0

us are the induction of malignant disease and genetic abnormalities. At present we have no understanding of the relationship between the development of these diseases from mutated cells and the incidence of dicentric or other aberrations in lymphocytes. Until it can be established whether such a relationship even exists and, if it does, to quantify it, there must be considerable misgivings about the scientific merit of such surveys. Nevertheless, some researchers will wish, or be unable to withstand pressure, to sample such populations. To be worthwhile a survey must look at a sufficient number of subjects and controls and clearly the availability of a metaphase finder would be invaluable.

Conclusions

Chromosome aberration analysis is used in a number of centres around the world and experience over the past 20 years has shown that in practice it works. Credible estimates of dose are obtained when compared with other available sources of information on the circumstances of an overexposure. Some recent advances in cytogenetic technique, such as FPG, have led to improvements for dosimetry and it is highly probable that an end to such developments has not yet been seen. The introduction of automation is likely to provide another advance. It should relieve a considerable proportion of the laborious and fatiguing aspects of microscopy. A metaphase finder which permits a doubling of the scoring rate would not lead to a worthwhile improvement in the statistical uncertainties associated with individual dose estimates. The main advantage here lies in a reduction of labour costs. However, the ability to score more cells may assist in resolving the long recognised problem of estimating doses in cases of partial body exposure. Any experimental studies such as examination of the dose effect relationship at low doses or population surveys must also benefit from the ease with which larger numbers of cells will be scored.

Acknowledgements. Statistical advice from Mr. A. Edwards and Mr. D. Papworth is gratefully acknowledged.

References

1. Bender MA, Gooch PC (1966) Somatic chromosome aberrations induced by human whole-body irradiation: the "Recuplex" criticality accident. Radiat Res 29:568–582
2. Buckton KE, Evans HJ (1973) Methods for the analysis of human chromosome aberrations. WHO Geneva
3. Crossen PE, Morgan WF (1977) Analysis of human lymphocyte cell cycle time in culture measured by sister chromatid differential staining. Exp Cell Res 104:453–457
4. Andrews GA (1980) Medical management of accidental total-body irradiation. Hübner KF, Fry SA (eds) The medical basis of radiation accident preparedness. Elsevier, Amsterdam, pp 297–310
5. Buckton KE, Langlands AO, Smith PG, Woodcock GE, Looby PC, McLelland J (1971) Further studies on chromosome aberration production after whole body irradiation in man. Int J Radiat Biol 19:369–378
6. Lloyd DC, Purrott RJ, Dolphin GW, Bolton D, Edwards AA, Corp MJ (1975) The relationship between chromosome aberrations and low LET radiation dose to human lymphocytes. Int J Radiat Biol 28:75–90
7. Lloyd DC, Purrott RJ, Dolphin GW, Edwards AA (1976) Chromosome aberrations induced in human lymphocytes by neutron irradiation. Int J Radiat Biol 29:169–182
8. Catcheside DG, Lea DE, Thoday JM (1946) The production of chromosome structural changes in *Tradescantia* microspores in relation to dosage, intensity and temperature. J Genet 47:137–149
9. Purrott RJ, Vulpis N, Lloyd DC (1981) Chromosome dosimetry: the influence of culture media on the proliferation of irradiated and unirradiated human lymphocytes. Radiation Protection Dosimetry 1:203–208
10. Lloyd DC, Edwards AA (in press) Chromosome aberrations in human lymphocytes: effect of radiation quality, dose and dose rate. In: Ishihara T, Sasaki MS (eds) Radiation-induced chromosome damage in man. A.R. Liss Inc New York
11. Littlefield LG, Joiner EE, DuFrain RJ, Hübner KF, Beck WF (1980) Cytogenetic dose estimates from *in vivo* samples from persons involved in real or suspected radiation exposures. In: Hübner KF, Fry SA (eds) The medical basis for radiation accident preparedness. Elsevier Amsterdam pp 375–390
12. Lloyd DC, Prosser JS, Lelliott DJ, Moquet JE (1982) Doses in radiation accidents investigated by chromosome aberration analysis. XII A review of cases investigated: 1981. UK National Radiological Protection Board Report R-128

13. Dolphin GW (1969) Biological dosimetry with particular reference to chromosome aberration analysis. In: Handling of radiation accidents IAEA Vienna pp 215–224
14. Savage JRK, Bigger TRL (1978) Aberration distribution and chromosomally marked clones in X-irradiated skin. In: Evans HJ, Lloyd DC (eds) Mutagen-induced chromosome damage in man. Edinburgh University Press, Edinburgh pp 155–169
15. Wells J, Charles MW (1982) Biological dosimetry of non-uniform radiation exposure. In: Radiation protection advances in theory and practice. Society for Radiological Protection pp 352–357
16. Ekstrand KE, Dixon RL (1982) Lymphocyte chromosome aberrations in partial-body fractionated radiation therapy. Phys Med Biol 27:407–411
17. Lloyd DC (1978) The problems of interpreting aberration yields induced by in vivo irradiation of lymphocytes. In: Evans HJ, Lloyd DC (eds) Mutagen-induced chromosome damage in man. Edinburgh University Press, Edinburgh pp 77–88
18. Basson JK, Hanekom AP, Coetzee FC, Lloyd DC (1981) Health physics evaluation of an accident involving acute overexposure to a radiography source. PEL-279 Atomic Energy Board of South Africa
19. Evans HJ, Buckton KE, Hamilton GE, Carothers A (1979) Radiation-induced chromosome aberration in nuclear dockyard workers. Nature 277:531–534
20. Lloyd DC, Purrott RJ, Reeder EJ (1980) The incidence of unstable chromosome aberrations in peripheral blood lymphocytes from unirradiated and occupationally exposed people. Mutation Res 72:523–532

Cytogenetic Effects in Human Lymphocytes as a Dosimetry System

Manfred Bauchinger

Gesellschaft für Strahlen- und Umweltforschung, D-8042 Neuherberg, F. R. G.

Introduction

To enable personnel monitoring of occupational radiation exposure, dosimetric data are usually obtained from individual dosimeters such as film badges and thermoluminescent dosimeters. Over a number of years, biological dose estimation has been used to support the physical dosimetry in order to receive adequate information on uncertain or unavailable physical dosimetry in cases of actual or suspected over-exposure. Vastly differing biological indicator systems of radiation exposure have been investigated, such as hematological and biochemical parameters, the sperm abnormality assay, the micronucleus analysis, and last but not least, the analysis of structural chromosome changes in peripheral human lymphocytes.

Since 1962, when Bender and Gooch [7] suggested that measurement of aberration yield in peripheral lymphocytes could be used to estimate the dose to a radiation exposed individual, it has been established as the so called 'chromosome' or 'cytogenetic dosimetry' [3, 5, 8, 11, 12, 16]. This system is now senstive enough to detect an exposure of about 0.1 Sv of sparsely ionizing radiation as equivalent whole body dose, provided that representative calibration curves are available.

Method and in vitro Calibration

T-lymphocytes from whole blood are stimulated to proliferate in standard microcultures. To provide a discrimination between first and later post-irradiation mitoses, cultures may be treated with BrdU. After about 44 h, cultures are stopped and chromosome preparations are carried out.

The basis for a cytogenetic dose estimation are dose-effect curves for radiation-induced chromosome aberrations. For this reason, whole blood is irradiated *in vitro* with different radiation qualities at different doses. After culture of lymphocytes and preparation of chromosomes, the chromosome analysis is done by light microscopy.

The resulting calibration curves comprise about 10–12000 cells and are mainly for dicentric chromosomes, although similar curves can be obtained for acentric elements, such as terminal or interstitial deletions. From 24000 first division cells of 47 unexposed individuals analysed in our laboratory, we have a background frequency of 3.8 dicentrics and 30 acentric per 10^4 cells.

Biological Dosimetry. Edited by W. G. Eisert and M. L. Mendelsohn
© Springer-Verlag Berlin Heidelberg 1984

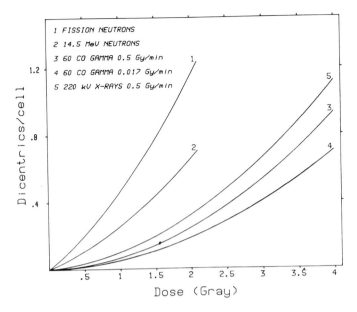

1 FISSION NEUTRONS
2 14.5 MeV NEUTRONS
3 60 CO GAMMA 0.5 Gy/min
4 60 CO GAMMA 0.017 Gy/min
5 220 kV X-RAYS 0.5 Gy/min

Fig. 1. Linear-quadratic dose-effect curves for dicentrics induced by different radiation qualities.

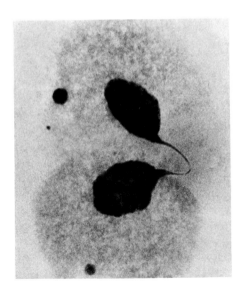

Fig. 2. Persisting anaphase bridge connecting two daughter cells after irregular mitosis. The two daughter cells contain one and 2 micronuclei, respectively

Figure 1 gives an example of dose-response curves for dicentrics induced in vitro by different radiation qualities. For the mathematical description of the curves, theoretical models are used which take into account mechanisms of aberration induction, as well as biophysical principles. Depending on radiation quality and dose, one frequently finds a linear or linear-quadratic dependence between the yield of chromosome aberrations and dose. Table 1 shows the linear-quadratic relations for the corresponding curves. The coefficients a and β have to be determined by least squares approximation applying appropriate statistical weights for the different data points [14]. From the ratio a/β of the two coefficients, the relative size of the linear and quadratic component can be deduced. It is evident that the relative contribution of the linear component increases with increasing LET.

Table 1. Estimates of parameters, a, and β, $\pm$ s.d. of linear-quadratic dose-relations ($Y=c+aD+\beta D^2$) for dicentrics in human lymphocytes exposed to different radiation qualities.

Radiation quality	$a\pm$s.e. $[10^{-1}\ Gy^{-1}]$	$\beta\pm$s.e. $[10^{-2}\ Gy^{-2}]$	a/β $[Gy]$	Ref.
^{60}Co γ, (0.5 Gy min^{-1})	0.107 ± 0.041	5.55 ± 0.28	0.19	6
^{60}Co γ, (0.017 Gy min^{-1})	0.090 ± 0.040	4.17 ± 0.28	0.22	6
220 kV, X-rays, (0.5 Gy min^{-1})	0.404 ± 0.030	5.98 ± 0.17	0.68	
14.5 MeV (d+T) neutrons	1.79 ± 0.15	7.40 ± 0.14	2.43	4
Fission neutrons, $\bar{E} \sim 2$ MeV	3.43 ± 0.20	11.68 ± 1.48	2.93	

Aberration Yield and Cell Proliferation

Radiation induced aberration yields *in vitro* and *in vivo* depend on several physical and biological factors. In order to accomplish a high sensitivity of biological dose estimates, these parameters have to be investigated and taken into account. Among them, cell proliferation is one of the most important.

During successive cell division in lymphocyte cultures, dicentrics can be lost owing to irregular mitoses and acentrics may be included into micronuclei (Fig. 2). Additionally, the formation of 'derived chromosome-type aberrations' or duplicated aberrations during DNA-synthesis can lead to mis-interpretations when conventionally stained cells are scored. Therefore, dose estimations can be misleading in any quantitative analysis which does not use exclusively first division cells.

Conventional staining methods do not allow a discrimination between cells which have replicated for 1, 2 or 3 cycles. This can be attained, however, by the FPG-staining tech-

nique after BrdU treatment of lymphocyte cultures [1, 19]. The chromosome analysis
can then be carried out exclusively in first-division cells as suggested by Crossen and
Morgan [10], Scott and Lyons [22]. This is quite essential, since in a fast culture system, despite a culture time of 48 h, we found that in blood samples of 80 persons the
frequency of cells in second or later divisions ranged between 5.0 and 65.0%. Similar
fluctuations can be observed in repeated samples of the same individual.

Figure 3 gives an impression of such differences in the frequency of cells beyond
first division in samples of 7 donors [15]. It further shows that with increasing concentrations of BrdU an exponential decline of the number of harlequin-stained metaphases occurs, but with different slopes for the various donors. At concentrations lower
than about 10^{-5}M, there is no inhibitory effect of BrdU on cell proliferation. Hence,
the concentration of 3.3×10^{-5}M which is widely used in cytogenetic experiments,
and which is indicated by an arrow in Fig. 3, can be reduced for optimum culture conditions.

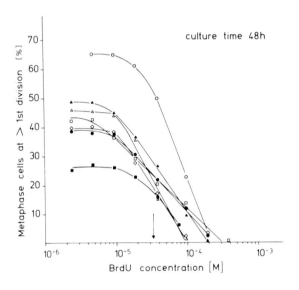

Fig. 3. Percentage of metaphases beyond first division in blood samples of 7 donors (identified
by harlequin-staining) at different concentrations of BrdU and a culture time of 48 h

The influence of X-irradiation on cell proliferation is shown in Fig. 4. A dose-dependent
decrease of the frequency of cells beyond first division is observed in blood samples
from 2 donors with either a high or a low number of second division cells. This means
that with increasing dose the mitotic delay increases as well. At low levels of dose,
delay is far lower or even unimportant, cell proliferation is undisturbed, and the loss
of aberrations should be more pronounced. Consequently, a source of error in dose-
response data can be excluded by scoring of exclusively first division cells.

The results of irradiation experiments with fast neutrons (Fig. 5) and with ^{60}Co-γ—
rays (Fig. 6) of different dose rates reveal significant differences for the yields of di-

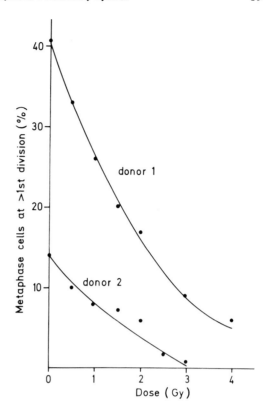

Fig. 4. Percentage of lymphocytes beyond first division in blood samples of 2 donors after X-irradition (220 keV) *in vitro* (culture time 48 h)

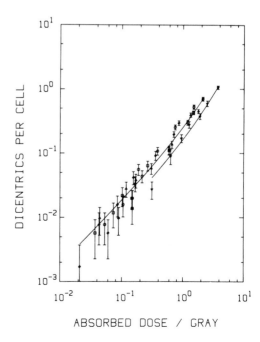

Fig. 5. Linear-quadratic dose-effect curves for dicentrics induced by fast neutrons at different positions of a man phantom. The *upper curve* for 14.5 MeV neutrons is established by FPG-staining, the *curve below* for 15.0 MeV neutrons is established by conventional staining

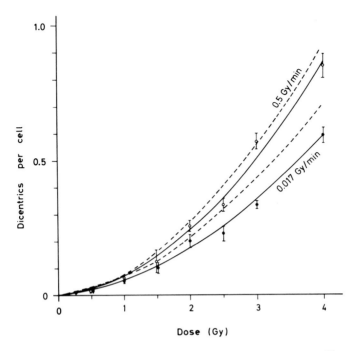

Fig. 6. Linear-quadratic dose-effect curves for dicentrics induced by ^{60}Co-γ-rays at different dose-rates. *Full line=* conventional staining; *broken line=* FPG-staining

centrics as well as for the calibration curves, depending on the application of either conventional or FPG-staining. At comparable doses for fast neutrons, the overall difference in the dicentric yield amounts to about 46%.

Data from experiments at low levels of dose (0.05–0.5 Gy) established with exclusively first division cells are demonstrated in Figs. 7 and 8. For 220 kV X-rays [24, 25], the assumption of a linear dose-response for both dicentrics and acentrics is supported within this dose range (Fig. 7). There is no evidence for a plateau between 0.1 and 0.3 Gy, as was described in studies with conventionally stained cells and a far lower total number of cells [17, 23]. From preliminary data at a low dose range, distinct dose-RBE relations for the induction of dicentrics can be demonstrated for various radiation qualities (Fig. 8). It is clearly indicated that the RBE of neutrons increases with decreasing dose.

Analysis of Micronuclei

It is obvious from chromosome aberration data that 'cytogenetic dosimetry' provides a sensitive means of estimating an individual's exposure to radiation even at levels of dose low enough to be relevant in occupational exposure. Certainly high numbers of cells have to be analysed for statistical significance. This is a time-consuming procedure and a disadvantage of this method.

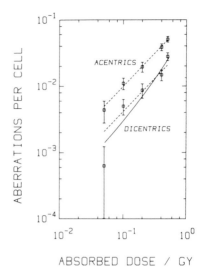

Fig. 7. Linear dose-response for acentrics and dicentrics *(broken line)* and linear-quadratic dose-response for dicentrics *(full line;* quadratic component significant at 68.3% confidence limits only) induced by 220 keV X-rays. (FPG-staining) [26]

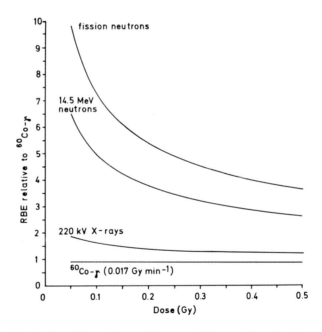

Fig. 8. Dose-RBE relations for the induction of dicentrics by different radiation qualities. The data are based on a total of 80 000 first-division metaphases

Therefore, an effort was made to introduce another biological or cytogenetic response to radiation for dosimetric purposes: the analysis of micronuclei in peripheral lympho-cytes [9]. Micronuclei can be regarded mainly as equivalents of acentric chromatin ele-ments (Fig. 2). By means of careful preparation techniques they can be easily detected in the cytoplasm and can be directly associated to the main nucleus [13].

Similarly, as with structural chromosome aberrations, their intercellular distribution can be analysed and their dose-response can be calculated.

Figure 9 demonstrates dose-effect curves for micronuclei at different culture times after X-irradiation of human lymphocytes in vitro. The curves are based on 2000 analysed cells per point at 0 and 0.25 Gy and on 1000 cells per point at the higher doses. Although micronuclei can be scored very rapidly, the method is unsuitable for dose estimation for two main reasons. First, the maximum response occurs at different culture times for various subjects, and second, at low levels of dose, (i.e. below 0.3 Gy), the method is not sensitive enough to detect a significant increase in the incidence of micronuclei as compared to controls.

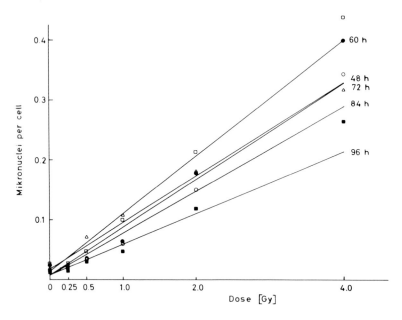

Fig. 9. Linear dose-effect curves for micronuclei induced by 220 keV X-rays in human lymphocytes

However, a variant of the method might be useful in detecting potential radiosensitive radiation workers. In this approach, the worker's cells are exposed to radiation at high levels of dose *in vitro* and the yield of micronuclei is checked for excessively high responses. Additionally, the micronucleus analysis might provide a means of rapid prescreening of acute exposure to high doses.

Conclusions

The micronucleus analysis cannot replace chromosome analysis, at least not at low levels of dose. 'Cytogenetic dosimetry' is at present the method of choice. It has been successfully applied in some hundreds of cases of occupational exposure by the NRPB-cytogenetics laboratory, (see contribution by D. Lloyd to this volume), as well as in atomic bomb survivors [2, 18, 20]. Nevertheless, the method can be further improved. The complex influences of physical and biological factors on the aberration yield in human lymphocytes are by far not yet fully elucidated. For a meaningful dose estimation, we still need further information, e.g. on how the aberration yields *in vivo* depend on partial-body or inhomogeneous exposure and on radiation-induced interphase death of lymphocytes. Studies on the analysis of a potential interindividual variability of radiosensitivity and of the time factor should be carried out by means of differential staining techniques. Finally, a further development of automated systems for rapid pre-selection, registration, and analysis of high cell numbers could contribute to a further improvement of this system.

List of Abbreviations

FPG – fluorescence plus Giemsa;
BrdU – 5-bromodeoxyuridine;
LET – linear enrgy transfer;
RBE – relative biological effectiveness.

References

1. Apelt F, Kolin-Gerresheim I, Bauchinger M (1981) Azathioprine, a clastogen in human somatic cells? Analysis of chromosome damage and SCE in lymphocytes after exposure in vivo and in vitro. Mutation Res 88:61–72
2. Awa AA, Sofuni T, Honda T, Itoh M, Neriishi S, Otake M (1978) Relationship between the radiation dose and chromosome aberrations in atomic bomb survivors of Hiroshima and Nagasaki. J Radiat Res 19:126–140
3. Bauchinger M, Schmid E, Hug O (1971) The relevance of chromosome aberration yields for biological dosimetry after low-level occupational irradiation: Proceedings of Symposium-Advances in physical and biological radiation detectors. Vienna 1971; IAEA-SM-143/13
4. Bauchinger M, Kühn H, Dresp J, Schmid E, Streng S (1983) Dose effect relationship for 14.5 MeV (d+T) neutron-induced chromosome aberrations in human lymphocytes irradiated in a man phantom. Int J Rad Biol 43:571–578
5. Bauchinger M (1978) Chromosome aberrations in human lymphocytes as a quantitative indicator of radiation exposure. In: Evans HJ, Lloyd DC (eds) Mutagen-induced chromosome damage in man. University Press, Edinburgh pp 9–13
6. Bauchinger M, Schmid E, Streng S, Dresp J (1983) Quantitative analysis of the chromosome damage at first division of human lymphocytes after $^{60}Co\gamma$-irradiation. Rad an Environm Biophys 22:225–229
7. Bender MA, Gooch PC (1962) Types and rates of X-ray-induced chromosome aberrations in human blood irradiated in vitro. Proc Natl Acad Sci (U.S.) 48:522–532
8. Bender MA (1969) Human radiation Cytogenetics. Adv. Radiat Biol 3:215–275
9. Countryman P, Heddle J (1976) The production of micronuclei from chromosome aberrations in irradiated cultures of human lymphocytes. Mutation Res 41:321–331

10. Crossen PE, Morgan WF (1977) Analysis of human lymphocyte cell cycle time in culture measured by sister chromatid differential staining. Exp Cell Res 104:453–457
11. Dolphin GW (1969) Biological dosimetry with particular reference to chromosome aberration analysis: Proceedings of Symposium – Handling of radiation accidents. Vienna 1969; IAEA-SM-119/4
12. Dolphin GW, Lloyd DC, Purrot RJ (1973) Chromosome aberration analysis as a dosimetric technique in radiological protection. Hlth Phys 25:7–15
13. Huber R, Streng S, Bauchinger M (1983) The suitability of the human lymphocyte micronucleus assay system for biological dosimetry. Mutation Res 111:185–193
14. Kellerer AM, Brenot J (1974) On the statistical evaluation of dose-response function. Rad and Environm Biophys 11:1–13
15. Kolin-Gerresheim I, Bauchinger M (1981) Dependence of the frequency of harlequin-stained cells on BrdU concentration in human lymphocyte cultures. Mutation Res 91:251–254
16. Lloyd DC, Purrott RJ (1981) Chromosome aberration analysis in radiological protection dosimetry. Radiation Protection Dosimetry 1:19–28
17. Luchnik NV, Sevankaev AV (1976) Radiation-induced chromosomal aberrations in human lymphocytes. I. Dependence on the dose of γ-rays and an anomaly at low doses. Mutation Res 36:363–378
18. Otake M (1979) The nonlinear relationship of radiation dose to chromosome aberrations among atomic bomb survivors, Hiroshima and Nagasaki. Report on Radiation Effects Research Foundation, Research Project 2–66, RERF TR 19–78 pp 1–19
19. Perry P, Wolff S (1974) New Giemsa method for the differential staining of sister chromatids. Nature (London) 251:156–158
20. Sasaki MS, Miyata H (1968) Biological dosimetry in atomic bomb survivors. Nature (London) 220:1189–1193
21. Schmid E, Bauchinger M, Mergenthaler W (1976) Analysis of the time relationship of the interaction of X-ray-induced primary breaks in the formation of dicentric chromosomes. Int J Rad Biol 30:339–346
22. Scott D, Lyons CY (1979) Homogeneous sensitivity of human peripheral blood lymphocytes to radiation-induced chromosome damage. Nature (London) 278:756–758
23. Takahashi E, Hirai M, Tobari I, Utsugi T, Nakai S (1982) Radiation-induced chromosome aberrations in lymphocytes from man and crab-eating monkey. The dose-response relationships at low doses. Mutation Res 94:115–123
24. Wagner R (1982) Comparative analysis of the dose-response at low doses of radiation-induced chromosome aberrations in human lymphocytes applying Feulgen-Orcein-Acetic Acid and FPG-staining. Thesis, Faculty for Biology, Ludwig Maximilians-Universität, München

Flow Cytometric Detection of Aberrant Chromosomes*

J.W. Gray, J. Lucas, L.C. Yu, and R. Langlois

Lawrence Livermore National Laboratory, Biomedical Sciences Division, University of California, P.O. Box 5507 L-452, Livermore, CA 94550, U.S.A

Summary

This report describes the quantification of chromosomal aberrations by flow cytometry. Both homogeneously and heterogeneously occurring chromosome aberrations were studied. Homogeneously occurring aberrations were noted in chromosomes isolated from human colon carcinoma (LoVo) cells, stained with Hoechst 33258 and chromomycin A3 and analyzed using dual beam flow cytometry. The resulting bivariate flow karyotype showed a homogeneously occurring marker chromosome of intermediate size. Heterogeneously occurring aberrations were quantified by slit-scan flow cytometry in chromosomes isolated from control and irradiated Chinese hamster cells and stained with propidium iodide. Heterogeneously occurring dicentric chromosomes were detected by their shapes (two centromeres). The frequencies of such chromosomes estimated by slit-scan flow cytometry correlated well with the frequencies determined by visual microscopy.

Introduction

Flow cytometry has been shown to be a powerful tool for classification of chromosomes from several mammalian species including humans. In this approach, chromosomes are isolated from mitotic cells, stained with one or two DNA specific fluorescent dyes and processed through a flow cytometer which excites and measures the fluorescence from the dye(s) in each chromosome. We, and others, have shown previously that Chinese hamster (Gray et al., 1975a, b; Stubblefield et al., 1975), mouse (Disteche et al., 1981), muntjac (Carrano et al., 1976), kangaroo rat (Stöhr et al., 1980) and human (Gray et al., 1979a; Carrano et al., 1979; Yu et al., 1981; Young et al.,1981) chromosomes can be classified according to their relative fluorescence intensities. Successful classification of normal chromosomes, of course, suggests the possibility of detecting chromosome aberrations.

Two classes of chromosomes may be quantified by flow: 1) Homogeneous chromosome aberrations that occur in most or all of the cells from which the chromosomes

*Work performed under the auspices of the U.S. Department of Energy by the Lawrence Livermore National Laboratory under contract number W-7405-ENG-48 with support from USPHS grants HD 17665 and GH 25076.

were isolated and 2) Heterogeneous aberrations that occur randomly and at low frequency in the cells from which the chromosomes were isolated. Detection of homogeneous aberrations, such as translocations or trisomies, is useful for identification of inherited chromosome aberrations that may signal genetic disease in the affected individual or fetus or that may cause genetic disease in future offspring. Homogeneous aberrations have also been identified in human malignancies (e.g., chronic myelogeneous leukemia, retinoblastoma, etc.) and may be diagnostically important. Quantification of heterogeneous aberrations, like dicentric chromosomes, acentric fragments and chromosome breaks, is important for assessment of the genetic damage accumulated by an individual or organism exposed to clastogenic agents (Lloyd, 1983).

We report here two studies that illustrate the utility of flow cytometry for quantifying homogeneous and heterogeneous chromosome aberrations. Specifically, we describe the identification of a homogeneously occurring marker chromosome in human colon carcinoma (LoVo) cells and we describe the measurement of the frequencies of dicentric chromosomes in Chinese hamster cells exposed to doses of X-irradiation up to 4 Gy.

Homogeneous Aberrations

Dual beam flow cytometry has proved particularly useful for classification of human chromosomes (Gray et al., 1979a; Langlois et al., 1982). In this approach, chromosomes are isolated and stained with the adenine-thymine specific dye Hoechst 33258 (HO) and the guanine-cytosine specific dye chromomycin A3 (CA3; Langlois et al., 1980). The dual stained chromosomes are processed through a dual beam flow cytometer as illustrated in Fig. 1. During dual beam cytometry, the two DNA specific dyes are independently and sequentially excited; HO by illumiunation at about 350nm and

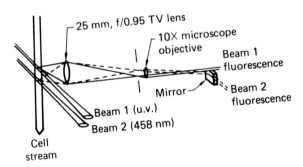

Fig. 1. Dual Beam Flow Cytometry. The chromosomes flow sequentially through two laser beams. One beam operates in the ultraviolet (∿350 nm) to excite Hoechst 33258 and the other operates at 458 nm to excite Chromomycin A3. The fluorescence intensities from the two laser beam crossing are measured for each chromosome and added to an accumulating bivariate distribution (bivariate flow karyotype). About 10^5 chromosomes per minute can be analyzed by this technique

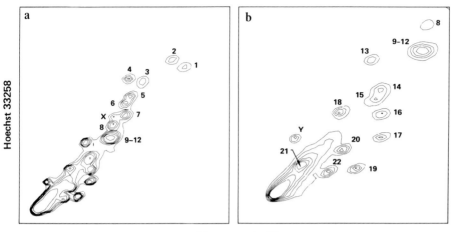

Fig. 2 a,b. Bivariate Flow Karyotype. **a** is a contour plot showing the distribution of HO and CA3 fluorescence intensities (actually the fluorescence intensities measured as chromosomes stained with both HO and CA3 passed through the UV and 458 nm laser beams) for chromosomes isolated from normal human lymphocytes. **b** shows the bivariate flow karyotype for the smaller human chromosomes. The *numbers* indicate the chromosomes that produced each peak. From Langlois et al., 1982

CA3 by illumination at 458nm. Two fluorescence intensities are measured for each chromosome. The results of measuring several hundred thousand chromosomes are accumulated to form bivariate HO/CA3 distributions like those shown in Fig. 2. Each homogeneously occurring chromosome type produces a peak in the bivariate distribution. Distinct and separate peaks are produced by all of the human chromosomes except numbers 9 through 12 and sometimes 14 and 15. The peak means are highly reproducible for analyses made on different days or for different individuals (Langlois et al., 1982). For example, Fig. 3 summarizes the results of analyzing chromosomes for 10 individuals; the ellipses represent 95% confidence intervals for the bivariate peak means for the 10 individuals. Importantly, the ellipses rarely overlap. Thus, a normal chromosome producing a peak in a bivariate flow HO/CA3 distribution (hereafter called a bivariate flow karyotype) can be identified by its relative location. The relative frequency of occurrence of the chromosome type producing each peak is proportional to the volume of the peak.

Bivariate flow karyotypes have several properties that make them particularly useful for detection of marker chromosomes in human malignancies: 1) Marker chromosomes with base composition or DNA content changes larger than the cross in Fig. 3 (i.e., larger than about one band) may produce abnormally located peaks in bivariate flow karyotypes, 2) bivariate flow karyotypes are insensitive to karyotypic instability so that chromosomes occurring at high frequency in the population produce peaks and randomly occurring aberrant chromosomes like translocations, trisomies, monosomies, acentric fragments, etc., have little effect (this concept is illustrated schematically in Fig. 4), and 3) the chromosomes need not be bandable.

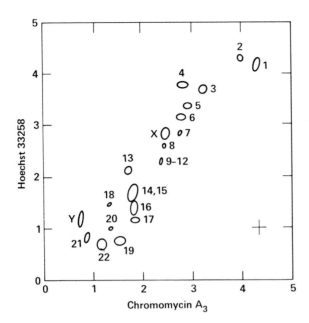

Fig. 3. Bivariate Flow Karyotype Variability. This figure shows 95% confidence ellipses determined from measurements of bivariate flow karyotypes measured for chromosomes from 10 normal individuals. The size of the ellipses indicates the variability observed for various chromosome peak means among the 10 individuals. The *cross* indicates the amount of change expected in a chromosome peak mean caused by a gain or loss of 1/600th of the genomic DNA (approximately equivalent to a $\pm$ one band change). From Langlois et al., 1982

We have applied bivariate flow karyotyping to human colon carcinoma (LoVo) cells to demonstrate its potential in marker chromosome detection. These cells, grown in *vitro*, were characterized by conventional cytogenetic analysis to be karyotypically unstable with chromosome numbers per cell varying from 37 to 55 per cell. The average number of chromosomes per cell was 49. Drewinko et al., (1982) previously reported 48 to 51 chromosomes per cell with a mode of 49 for this cell line. Interestingly, production of high quality banded karyotypes for these cells has proved very difficult so far. Chromosomes were prepared for bivariate flow karyotyping as previously described (Gray et al., 1979a). Figure 5 shows the bivariate flow karyotype measured for the LoVo cells. Panel a shows the entire bivariate flow karyotype and Panel b shows the bivariate flow karyotype for chromosomes smaller the number 8. Peaks corresponding to all of the normal chromosomes have been identified and marked in the figure. In addition, a peak corresponding to a homogeneously occurring marker chromosome has also been identified and labeled M in the figure. Work is now underway in our laboratory to definitively identify the chromosome producing the atypical peak. We conjecture that this is the altered B group chromosome previously reported for this cell line by Drewinko et al. (1982).

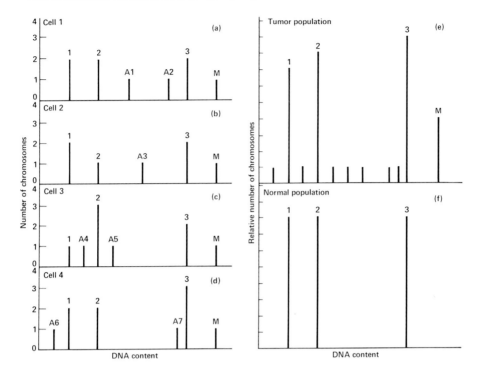

Fig. 4 a–f. Flow Karyotypic Detection of Marker Chromosomes. a–d illustrate hypothetical univariate flow karyotypes that might be measured for 4 cells from a karyotypically unstable population. In this hypothetical situation, each cell is presumed to contain one or more of each of the normal chromosomes designated 1, 2, and 3 as well as some random chromosome aberrations. In addition, each cell contains at least one marker chromosome designated M. f shows the flow karyotype that would be measured for a completely normal cell population while Panel e shows the flow karyotype that would be measured for chromosomes isolated from cells 1 through 4. Note that the homogeneously occurring normal and marker chromosomes form peaks while the randomly occurring aberrant chromosomes form a continuum underlying the peaks. Aberrant marker chromosomes can be located by comparing peak locations in flow karyotypes measured for normal (f) and test (e) populations

Heterogeneous Aberrations

The insensitivity of flow karyotypes to karyotypic instability, as described above, makes the method inherently insensitive to heterogeneously occurring aberrations. Slit-scan flow cytometry (Gray et al., 1979b, 1980; Cram et al., 1979), on the other hand, seems ideally suited to detection of certain classes of heterogeneous aberrant chromosomes. In this approach; illustrated in Fig. 6, the isolated, stained chromosomes are forced to flow lengthwise across a thin (typically 2 μm) laser beam. A sequence of fluorescence intensities is recorded for each chromosome as it passes through the beam. The resulting profile shows the distribution of the fluorescent DNA-specific dye along

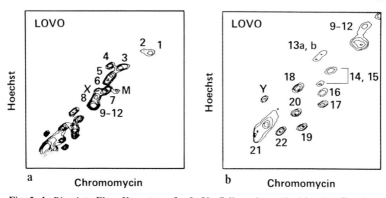

Fig. 5a,b. Bivariate Flow Karyotype for LoVo Cells. **a** shows the bivariate flow karyotype measured for chromosomes isolated from LoVo cells. **b** shows the bivariate flow karyotype measured for the smaller LoVo chromosomes. The obvious difference between these data and the normal bivariate flow karyotypes shown in Fig. 2 is the peak designated M

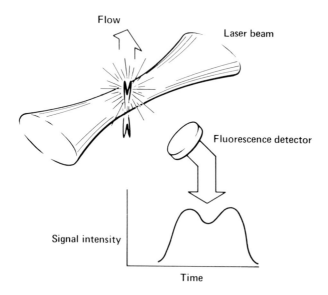

Fig. 6. Slit-Scan Flow Cytometry. During slit-scan flow cytometry, chromosomes stained with a DNA specific dye are forced to flow lengthwise through a thin (∼ 2 μm thick) laser beam. The resulting time varying fluorescence intensity is recorded as a measure of the distribution of fluorescent dye along the length of the chromosome. Recorded profiles typically dip when the centromere with its reduced DNA content crosses the laser beam. From Gray et al. (1980)

the chromosome. A typical profile measured by this technique for a normal propidium iodide stained Chinese hamster chromosome is shown in Fig. 7a. Of special interest, is the centrally located dip produced when the chromosome centromere with its reduced DNA content crosses the beam. Normal chromosomes have only one such dip while aberrant dicentric chromosomes have two. Fig. 7b shows a profile measured for a

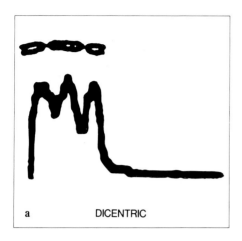
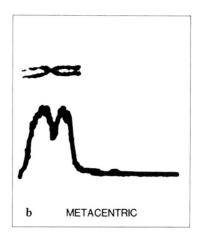

Fig. 7a,b. Slit-Scan Profiles. **a** shows a slit-scan profile measured for a normal metacentric Chinese hamster chromosome. **b** shows a slit-scan profile measured for a dicentric Chinese hamster chromosome

putative dicentric chromosome. It is a straightforward matter to detect and count profiles with two dips. The frequency of dicentric chromosomes in the total chromosomal population is estimated as the ratio of the number of profiles with two dips to the number having only one.

We have used slit-scan flow cytometry to quantify the frequency of occurrence of dicentric chromosomes in chromosomes from Chinese hamster M3-1 cells exposed to 0, 0.5, 1.0, 2.0 and 4.0 Gy of X-irradiation. In this experiment, the cells were growing exponentially in monolayer culture at the time of irradiation. The cells were returned to culture for 20 h after irradiation and then treated for 4 h with colcemid. The chromosomes were isolated as previously described (Aten et al., 1980) and stained with the DNA specific dye propidium iodide. At least 4000 profiles were recorded for chromosomes from cells taken at each dose point using the Livermore slit-scan flow cytometer. During these analyses, the slit-scan flow cytometer was adjusted so that profiles were recorded only for chromosomes larger than the number two chromosome. Metaphase cells were also collected for each dose point and prepared for conventional chromosome analysis.

The profiles recorded for each sample were analyzed by computer to determine the number of dicentric and normal No. 1 chromosome profiles. Dicentric profiles were recognized as having two distinct centromeric dips of approximately equal depth. Number one chromosome profiles were recognized as having a single centrally located centromeric dip. The frequency of dicentric chromosomes in the population was calculated as the ratio of the number of dicentric chromosomes profiles to the number of No. 1 chromosomes profiles. The frequency of dicentric chromosomes per cell was taken to be half of this ratio since there are two No. 1 chromosomes per Chinese hamster M3-1 cell.

The frequency of dicentric chromosomes per cell also was determined by visual analysis of metaphase spreads prepared for the irradiated samples. Only chromosomes

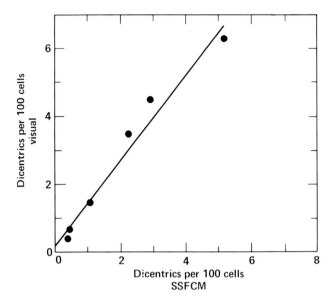

Fig. 8. Visual and Slit-Scan Flow Cytometric Estimates of Dicentric Chromosome Frequencies. These data show the correlation between dicentric chromosome frequency estimates made visually and by slit-scan flow cytometry for X-irradiated Chinese hamster M3-1 cells. Only those chromosomes as large or larger than the number 2 chromosome were scored. The data are shown as *solid points.* The line is a linear regression to the data (r = 0.97)

larger than the number 2 Chinese hamster chromosomes were scored. This is relatively easy in the Chinese hamster M3-1 cells used in this experiment because chromosomes 1 and 2 are obviously larger than the other chromosomes and are morphologically distinct.

Figure 8 shows the correlation between the frequency of dicentric chromosomes determined by visual scoring with the frequency of dicentric chromosomes determined by slit scan flow cytometry. The correlation between the two measurement techniques is quite high (r = 0.97). Further, the slope of the regression between the two sets of data is nearly unity (actually 1.25) suggesting that the two procedures were both measuring dicentric chromosomes larger than the number 2 chromosome. Figure 9 shows the dose response curve measured by slit scan flow cytometry for irradiated Chinese hamster chromosomes from three separate experiments. The frequency of occurrence of dicentric chromosomes increases continuously. The close agreement between the frequencies measured in the three experiments shows excellent reproducibility for the slit-scan flow cytometric approach to dicentric chromosome analysis. These data also suggest that the response to doses at least as low as 0.5 Gy can be quantified by this approach.

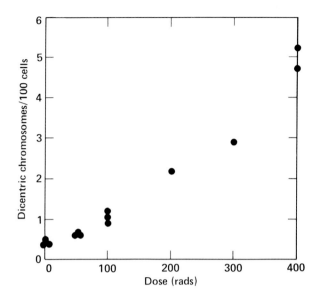

Fig. 9. Dicentric Frequency Dose Response Curve. X-ray dose response curves were measured for Chinese hamster M3-1 cells on three separate occasions. Doses ranged from 0 to 4 Gy (1 Gy = 100 rads). The individual frequency estimates for each experiment are shown separately as solid dots

Future Directions

These two examples illustrate our initial efforts in the application of flow cytometry in the study of chromosome aberrations. Bivariate flow karyotyping is a highly sensitive procedure for detection of homogeneously occurring structural chromosome aberrations. However work in several areas is still required before this procedure can be routinely applied to population screening, prenatal diagnosis or tumor classification. The most formidable challenge remaining is to reduce the number of mitotic cells required for analysis. Our procedures currently require at least 1×10^6 mitotic cells. However straightforward miniaturization procedures should reduce this by at least one order of magnitude thus substantially reducing the time and labor required for sample preparation. It also remains to improve the accuracy of the data analysis procedures used to estimate the relative frequency of occurrence of each chromosome type so that monosomies and trisomies can be detected with confidence. Frequency estimates now made by calculating the volume of each chromosome peak, are complicated by the continuum in the distributions produced by fluorescent debris. More robust data analysis procedures now under development in our laboratory promise to minimize this source of inaccuracy.

Continued development is also necessary for dicentric chromosome analysis by slit-scan flow cytometry. The extension of the procedure to analysis of human cells is formidable because of the low frequencies at which dicentric chromosomes can be expected. In the Chinese hamster example described above, the background dicentric frequency was about 0.5/100 cells. The background dicentric chromosomes frequency in human cells is about 0.25/1000 cells (Lloyd, 1983); almost 20 times lower. Thus almost 2×10^6 chromosomes must be processed to allow a statistically significant detection of a doubling in the background frequency. The analysis speed of the present slit scan flow cytometer must be increased substantially from its current 10 profiles/s before this becomes possible. The necessary speed increase seems feasible, however, using special purpose microprocessors and this is currently under development in our laboratory.

References

Aten J, Kipp JBA, Barendsen GW (1980) Flow-cytofluorometric determination of damage to chromosomes from x-irradiated Chinese hamster cells. In: Laerum O, Lindmo T, Thorud E (eds) Flow Cytometry IV, Universitetsforlaget, p 287–292

Carrano AV, Gray JW, Langlois RG, Burkhart-Schultz K, Van Dilla MA (1979) Measurement and purification of human chromosomes by flow cytometry. Proc Natl Acad Sci USA 76:1382–1384

Carrano AV, Gray JW, Moore II DH, Minkler J, Mayall BH, Van Dilla MA, Mendelsohn ML (1976) Purification of the chromosomes of the Indian muntjac by flow sorting. J Histochem Cytochem 24:348–354

Cram LS, Arndt-Jovin DJ, Grimwade BG, Jovin TM (1979) Fluorescence polarization and pulse width analysis of chromosomes by a flow system. J Histochem Cytochem 27:445–453

Dean PN, Pinkel D (1978) High resolution dual laser flow cytometry. J Histochem Cytochem 26:622–627

Disteche CM, Carrano AV, Ashworth LH, Burkhart-Schultz K, Latt SA (1981) Flow sorting of the cattanach x-chromosome in an active or inactive state. Cytogenet Cell Genets 29:189–197

Drewinko B, Romsdahl M, Yang L, Ahern M, Trujillo J (1982) Establishment of a human carcino-embryonic antigen-producing colon adenocarcinoma cell line. Cancer Res 36:467–475

Gray JW, Carrano AV, Moore II DH, Steinmetz LL, Minkler J, Mayall BH, Mendelsohn ML, Van Dilla MA (1975a) High-speed quantitative karyotyping by flow microfluorometry. Clin Chem 21:1258–1262

Gray JW, Carrano AV, Steinmetz LL, Van Dilla MA, Moore II DH, Mayall BH, Mendelsohn ML (1975b) Chromosome measurement and sorting by flow systems. Proc Natl Acad Sci USA 72:1231–1234

Gray JW, Langlois RG, Carrano AV, Burkhart-Schultz K, Van Dilla MA (1979a) High resolution chromosome analysis: One and two parameter flow cytometry. Chromosoma 73:9–27

Gray JW, Lucas J, Pinkel D, Peters D, Ashworth L, Van Dilla MA (1980) Slit scan flow cytometry: Analysis of Chinese hamster M3-1 chromosomes. In: Laerum OD, Lindmo T, Thorud E (eds) Flow Cytometry IV, Universitetsforlaget, Bergen Norway, p 485–491

Gray JW, Peters D, Merrill JT, Martin R, Van Dilla MA (1979b) Slit scan flow cytometry of mammalian chromosomes. J Histochem Cytochem 27:441–444

Langlois R, Yu L-C, Gray JW, Carrano AV (1982) Quantitative karyotyping of human chromosomes by flow cytometry. Proc Natl Acad Sci USA 79:7876–7880

Langlois RG, Carrano VA, Gray JW, Van Dilla MA (1980) Cytochemical studies of metaphase chromosomes by flow cytometry. Chromosoma 77:229–251

Lloyd DC (1983) An overview of radiation dosimetry by conventional cytogenetic methods. This volume.

Stöhr M, Hutter K-J, Frank M, Futterman G, Goerttler K (1980) A flow cytometric study of chromosomes from rat kangaroo and Chinese hamster cells. Histochemistry 67:179–190

Stubblefield E, Deaven LL, Cram LS (1975) Flow microfluorometric analysis of isolated Chinese hamster chromosomes. Exp Cell Res 94:464–468

Young BD, Ferguson-Smith MA, Sillar R, Boyd E (1981) High-resolution analysis of human peripheral lymphocyte chromosomes by flow cytometry. Proc Natl Acad Sci USA 78:7727–7731

Yu L-C, Aten J, Gray JW, Carrano AV (1981) Human chromosome isolation from short-term lymphocyte culture for flow cytometry. Nature 293:154–155

Flow Cytometric Measurement of Cellular DNA Content Dispersion Induced by Mutagenic Treatment

F.J. Otto,[1] H. Oldiges,[1] and V.K. Jain[2]

[1]Fraunhofer-Institut für Toxikologie und Aerosolforschung, D-5948 Schmallenberg,
 Federal Republic of Germany
[2]National Institute of Mental Health and Neuro Sciences, Bangalore 560029, India

Recent improvements in flow-cytometric instrumentation and techniques enable us to make very precise measurements of distributions of cellular DNA content in populations of suspended cells. Flow-cytometric measurements of cellular DNA content are being used in tumour research for the detection and monitoring of aneuploid cell lines [1, 47, 49] and for studying the cell kinetics [15, 32, 33]. Similar techniques could also be used for the assessment of cytogenetic damage induced by environmental agents like radiations or chemicals; for example, by measuring dispersion in the DNA content of chromosomes or whole cells in a cell population [34–36, 39]. Such measurements are based on the fact that the chromosomal aberrations and dysfunction of the mitotic apparatus due to the action by mutagenic agents lead to an unequal distribution of the DNA amount in the daughter cells and consequently to an increased variation of the cellular DNA content in the whole cell population. This effect can be observed in the flow-cytometric DNA distributions as broadening of the G_0/G_1- and G_2+M-phase peaks and can be quantitated. Work done along these lines will be briefly reviewed in the following.

1 Measurement of the Cellular DNA Content Dispersion by Flow-Cytometry

1.1 Basic Principles

Dispersion in the cellular DNA content distributions of cell suspensions measured by flow-cytometry could arise due to a number of factors. In general, these could be classified into 3 categories as follows

$$\text{Observed Dispersion} = \text{Intrinsic Dispersion} + \text{Measurement Dispersion} + \text{Induced Dispersion}$$

Taking variance (σ^2) as a quantitative measure of dispersion and assuming that the deviations from a central value are small and follow a normal distribution, we can write

$$\sigma^2_{ob} = \sigma^2 + \sigma^2_m + \sigma^2_i \tag{1}$$

Biological Dosimetry. Edited by W.G. Eisert and M.L. Mendelsohn
© Springer-Verlag Berlin Heidelberg 1984

The intrinsic dispersion, σ_o^2, reflects the biological variability of the sample and may depend on its physiological state. Measurement dispersion, σ_m^2, is introduced during various steps involved in the process of measurement; for example, random variations in the staining procedures, the flow rate, light intensity and electronic processing. Finally, induced dispersion σ_i^2, is due to genotoxic treatments to which the cell populations might be exposed. Under optimal working conditions, the intrinsic dispersion and dispersion due to measurement could be minimized and held constant within certain limits in untreated control populations, so that

$$\sigma_m^2 \;+\; \sigma_o^2 \;=\; \sigma_c^2 \tag{2}$$

The dispersion induced by treatments could then be given by

$$\sigma_i^2 \;=\; \sigma_{ob}^2 \;-\; \sigma_c^2 \tag{3}$$

In general, the magnitude of the induced dispersion will depend upon the type and the dose of treatment as well as on the interval of time that has elapsed between the treatment and observation (see sect. 2.1 and 2.2).

For comparisons in the different cell populations, it is convenient to use relative standard deviation or the coefficient of variation (V) as an index of dispersion:

$$V = \frac{\sigma}{m} \tag{4}$$

whereas σ is the standard deviation around the mean value m. The increase in dispersion due to treatment is then given by the expression

$$V_i = \sqrt{(V_{ob}^2 - V_c^2)} \tag{5}$$

The coefficients of variation V_{ob} and V_c can be calculated from the corresponding flow-cytograms.

One of the advantages of the flow-cytometric measurement of the cellular DNA content dispersion is that this assay can also be performed using cells from animals exposed *in vivo*. As an illustration, the flow-cytograms of bone marrow cells from mice irradiated with various doses of X-rays are shown in Fig. 1. The cells were isolated and prepared for flow measurement 24 h after exposure. Figure 1 shows that increasing doses of X-rays produce a broadening of the G_1-peak and an increase in V is apparent.

Before using this assay in mutagenicity testing and biological dosimetry, its reliability and reproducibility have to be investigated.

1.2 Reliability and Reproducibility

The reliability of the V measurement from the statistical point of view can be specified by giving the confidence interval which can be calculated by the following formulae [43]:

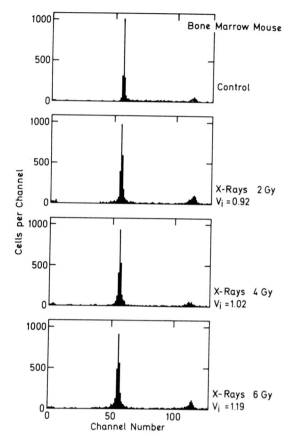

Fig. 1. Flow-cytograms of bone marrow cells from mice irradiated with various doses of x-rays, showing a broadening of the G_1-peak and an increase in V_i

$$V_{\text{lower limit}} = \frac{1}{1 + k\sqrt{(1+2V^2)}} \cdot V \tag{6}$$

and

$$V_{\text{upper limit}} = \frac{1}{1 - k\sqrt{(1+2V^2)}} \cdot V \tag{7}$$

k is calculated from

$$k = \frac{z}{\sqrt{2(n-1)}} \tag{8}$$

where n is the number of events in the distribution and z is a factor which depends on the desired level of confidence. Using these formulae, the dependence of confidence interval of V on the number of cells measured can be studied; a numerical example for

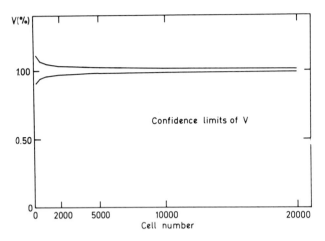

Fig. 2. Dependence of the confidence interval of V on the number of cells measured

the 95% level of confidence is shown in Fig. 2. When the number of cells measured is small, the width of the confidence interval is large and the reliability of the measured V is poor. However, when a few thousand cells are measured the confidence interval can be reduced and does not exceed $\pm$ 2% when 5000 cells are measured. It may be pointed out that these considerations are valid for V only and have nothing to do with the peak position or the relative DNA content. These are the statistical considerations to provide an estimate of cell numbers required for one measurement.

The reproducibility of the assay should be established by examining the experimental data. Figure 3 shows the values of V measured from the DNA distributions of bone marrow cells of 8 control mice and 8 mice treated with cyclophosphamide (30 mg/kg), a strong carcinogen. There are some individual variations in each group but the values are consistently higher in the group of mice treated with cyclophosphamide. The mean values of V obtained from each group are statistically significant as shown by the U-test [56]. Results of 8 measurements made from the same sample of one control mouse are also shown in Fig. 3; the standard deviation of the mean is in the range of V indicating that the dispersion due to the measurement is quite small. In order to obtain this degree of reproducibility, a few methodological precautions should be observed:
a) The cell samples should be fixed and stored until the whole series is complete
b) For preparation and staining an appropriate procedure should be used
c) All samples belonging to one series should be stained and measured in one run.

A new staining procedure developed by us [34, 36] based on a method described by Pinaev et al. [37] has been found satisfactory for work with mouse and rat bone marrow cells, mouse thymocytes, mouse and human lymphocytes, chicken and frog erythrocytes and various cultured cell lines. The procedure involves treatment of ethanol-fixed cells in one volume of 0.2 molar citric acid with 0.5% Tween 20 at pH 1.8 for 20 min. Subsequently, 9 volumes of a 0.4 molar solution of disodium hydrogen phosphate containing 5 μmol of 4,6 – diamidino-2-phenylindole (DAPI) are added to raise the pH to 7.4 and to stain the DNA. Available evidence shows that using this procedure the fluorescence intensity really reflects nuclear DNA content and is not influenced by confor-

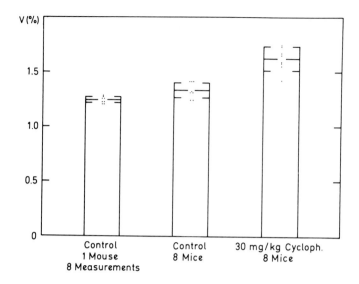

Fig. 3. Individual values, means, and standard deviations of V measured from the DNA content distributions of the bone marrow cells of 1 control mouse in 8 subsequent measurements (*left*), of 8 control mice (*middle*), and 8 mice treated with 30 mg/kg of cyclophosphamide (*right*)

mational changes of the chromatin. This is illustrated in Fig. 4 which shows flow-cytograms of two parallel subcultures of mouse L-cells, one of which was treated with colcemid for 6 h to accumulate cells in mitosis. It can be seen that G_2 + M-peak in both the samples is located exactly at the same position, indicating that the mitotic cells containing condensed chromosomes exhibit the same fluorescence intensity as G_2-cells and double the intensity compared with G_1-cells. The only known exception are the mammalian spermatozoa which have chromatin of extremely high density so that quantitative dye binding cannot be achieved using the staining procedure described above.

2 DNA Content Dispersion Induced by Mutagens

2.1 Time Dependence

The unequal distribution of DNA content per cell resulting from clastogenic events and dysfunction of the mitotic apparatus occurs during the process of cell division. Thus, a variable DNA content becomes manifest at first in the daughter cells and the effect becomes measurable in the whole cell population when a sufficiently large number of cells have divided. Consequently, the effect is time dependent and if agents with cytostatic properties are used, a dose related delay of the effect should be expected. This is true for ionizing radiations as well as for cytostatic agents.

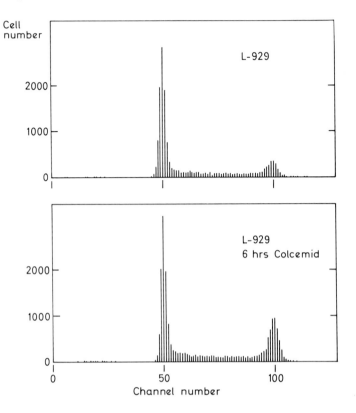

Fig. 4. Flow cytograms of two parallel subcultures of mouse L-cells with increased M-peak after treatment with colcemid

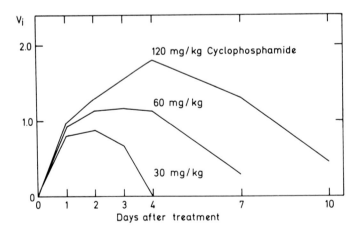

Fig. 5. Time dependence of V_i in the bone marrow cells of mice after treatment with cyclophosphamide

As an example, the time dependence of V_i in the bone marrow cells of mice after treatment with cyclophosphamide is shown in Fig. 5. After administration of a low dose of the drug (30 mg/kg), V_i increases during the first two days and comes back to normal after the 4th day. At higher doses (60 and 120 mg/kg), the maximum increases are found at day 3 and 4 and it takes 7 and 10 days respectively to come back to normal. Similar effects have been observed in cell cultures of human leukocytes and human cancer cells.

From Fig. 5 it can also be seen that the increased DNA dispersion can be observed only for a limited time after treatment. Applied to mutagenicity testing, the assay has to be considered as a short-term test and in biological dosimetry it reflects the acute response to exposure to environmental hazards. The assay, in its present form, seems to be incapable of detecting long lasting damage.

2.2 Dose Dependence

Dose response curves for the induced DNA content dispersion due to treatment with various chemical mutagens are shown in Fig. 6. Measurements were made on the bone marrow cells of mice, two days after *in vivo* treatment with the chemicals. V_i was calculated according to Eq. 5.

The dose range for the chemicals was chosen taking into consideration the available data on the toxicity. The doses of various chemicals used are comparable with regard to toxicity, the highest doses being below the LD_{50} value for mice.

Figure 6 shows that there may be large differences in the pattern of dose-response curves for various mutagens. These could be, however, grouped into 3 main types:

a) At low doses, the effect increases rather rapidly with dose but the rate of increase becomes less at higher dose, for example, dose-effect curves obtained with cyclophosphamide and trenimon. The saturation effects could be due to the cytostatic effects of the chemicals used. Because of the increased division delay at higher doses, smaller number of cells would be able to complete the division 2 days after treatment. Therefore, the effect does not achieve its maximum value (see Fig. 5) 2 days after treatment.

b) The dose-effect curves show a maximum; at higher doses, the effects decrease. Poly-cyclic hydrocarbons, benzo(a)pyrene and methylcholanthrene, for example, show this kind of behaviour. The observed decrease at higher doses could be due to en-hanced cytotoxic effects. At high doses, most of the cells could be damaged to such an extent that they would be unable to divide and therefore should not contribute to the increase in DNA dispersion.

c) The dose-effect curve shows a threshold below which no effects are observable (example colchicine).

2.3 Comparison with Other Indices of Cytogenetic Damage

Increase in DNA content dispersion after treatments with mutagens could be observed in different laboratories employing various *in vivo* and *in vitro* test systems. Some

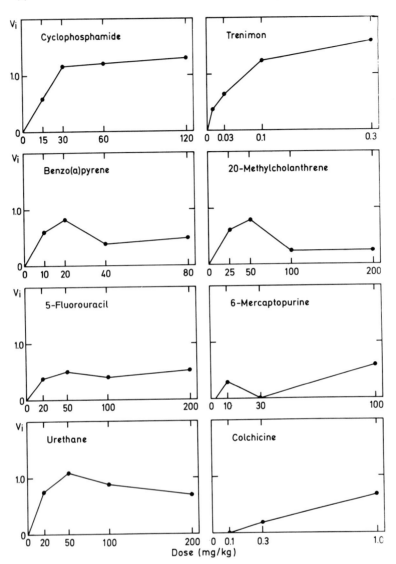

Fig. 6. Dose response curves for the induced DNA content dispersion due to the treatment with various chemical mutagens

examples are listed in Table 1, indicating that this assay could be useful in mutagen screening. It would be desirable therefore, to compare it with the conventional and well accepted indices for cytogenetic damage, for example, chromosomal aberrations and micronuclei formation. Such a comparison is shown in Table 2 for measurements on the bone marrow cells of mice treated *in vivo* with various mutagenic and non-mutagenic agents. In general, there appears to be a good agreement between the results obtained by flow-cytometric measurements of DNA content dispersion and microscopic

Table 1. List of references showing increased DNA content dispersion in various test systems after treatment with physical and chemical mutagens

Agent	Test System	References
X-Rays	Chinese hamster cells in vitro	34
	Ehrlich ascites cells in vitro	32, 33
	HeLa Cells in vitro	Kalia: Personal communication
	Human leukocytes in vitro	20
	Mouse bone marrow	34
	Mouse spermatocytes	16
	Mouse spermatozoa	38
UV	Human Leukocytes in vitro	20
Adriamycin	Chinese hamster cells in vitro	15
Isoniazid	Mouse bone marrow	Bhanumathi: Personal communication

observations of the frequencies of chromosomal aberrations and micronuclei. It should be pointed out, however, that different fractions of the cell population are being examined by these methods. For chromosomal analysis, cells are observed at the metaphase whereas micronuclei are scored from cells in G_1, S and G_2 phases of the cell cycle and increase in the DNA content dispersion can be measured either from the G_1- or the G_2- phase population.

The chromosomal analysis is laborious, and usually based on the examination of a small number of cells. The micronucleus test is less time-consuming but also less sensitive. However, measurement of DNA content dispersion proved to be much faster and may have a wider and more general applicability since formation of micronuclei as well as chromosomal aberrations could contribute to DNA dispersion and therefore the probability of getting false negative results is expected to be smaller. An example is furnished by UV light which does not induce micronuclei but still leads to an increase in the DNA content dispersion [20]. On the basis of the micronucleus test alone, one could wrongly conclude that UV is non-mutagenic.

In conclusion it may be stated that high resolution flow-cytometric measurement of DNA content dispersion could provide a sensitive and quick method for screening mutagens. Further studies directed towards improvement and application of this technique should be therefore, undertaken.

Table 2. Results of the DNA Content Dispersion Assay in Mouse Bone Marrow Compared with Literature Data on the Cytogenetic Effectiveness of Various Chemical Mutagens and Non-Mutagens

Agent	DNA Content Dispersion	Literature Data on Cytogenetic Effectiveness	
		Chromosomal Aberrations	Micronucleus Formation
Cyclophosphamide	++	++ (4, 8, 11, 13, 14, 27, 40, 42, 46)	++ (3, 5, 6, 8, 14, 21, 24, 26, 40, 44, 55, 58)
Trenimon	+++	++ (11, 39)	+++ (21, 25, 26)
5-Fluorouracil	+	+ (11, 12, 28)	+ (24, 58)
6-Mercaptopurine	+	+ (10, 11, 17, 30)	+ (24, 58)
4-Nitroquinoline-1-oxide	–	+ (11, 18, 23, 29, 31)	+– (5, 22, 45, 53, 55)
Benzo(a)pyrene	+	+ (2, 11, 31, 41, 48, 54)	+– (5, 22, 45, 53)
Pyrene	–	– (7, 31)	– (22, 45, 53)
20-Methylcholanthrene	+	+– (7, 11, 48, 54)	+– (5, 9)
Urethane	++	+ (18, 23)	++ (5, 45, 53, 55)
Isopropyl-N(3-chlorophenyl)-carbamate	–		– (45, 53)
Colchicine	+	+– (11, 26, 50, 51, 52)	+– (5, 19, 50, 51, 57, 58)

+++ = High effectiveness; ++ = Moderate effectiveness; + = Low effectiveness; – = No effectiveness; +– = Contradictory data

References

1. Barlogie B, Göhde W, Johnston DA, Smallwood L, Schumann J, Drewinko B, Freieich EJ (1978) Determination of ploidy and proliferative characteristics of human solid tumors by pulse cytophotometry. Cancer Res 38:3333–3339
2. Basler A, Röhrborn G (1976) Chromosome aberrations in oocytes of NMRI mice and bone marrow cells of Chinese hamster induced with 3,4-benzpyrene. Mutat Res 38:327–332
3. Bauknecht T, Vogel W, Bayer U, Wild D (1977) Comparative in vivo mutagenicity testing by SCE and micronucleus induction in mouse bone marrow. Hum Genet 35:299–307
4. Benedict WF, Banerjee A, Venkatesan N (1978) Cyclophosphamide-induced oncogenic transformation, chromosomal breakage, and sister chromatid exchange following microsomal activation. Cancer Res 38:2922–2924
5. Bruce WR, Heddle JA (1979) The mutagenic activity of 61 agents as determined by the micronucleus, Salmonella, and sperm abnormality assay. Can J Genet Cytol 21:319–334
6. Chrisman CL, Baumgartner AP (1979) Cytogenetic effects of diethyl-stilbestrol-diphosphate (DES-dp) on mouse bone marrow monitored by the micronucleus test. Mutat Res 67:157–160

7. Dean BJ, Hodson-Walker G (1979) An in vitro chromosome assay using cultured rat-liver cells. Mutat Res 64:329–337
8. Frank DW, Trzos RJ, Good PI (1978) A comparison of two methods for evaluating drug-induced chromosome alterations. Mutat Res 56:311–317
9. Friedman MA, Staub J (1977) Induction of micronuclei in mouse and hamster bone-marrow by chemical carcinogens. Mutat Res 43:255–262
10. Frohberg H, Schulze-Schenking M (1975) In vivo cytogenetic investigations in bone marrow cells of rats, Chinese hamsters and mice treated with 6-mercaptopurine. Arch. Toxicol 33:209–224
11. Gebhart E (1977) Chemische Mutagenese. Gustav Fischer Verlag, Stuttgart
12. Gebhart E, Lösing J, Wopfner F (1980) Chromosome studies on lymphocytes of patients under cytostatic therapy. I. Conventional chromosome studies in cytostatic interval therapy. Hum Genet 55:53–63
13. Gebhart E (1981) Sister chromatid exchange (SCE) and structural chromosome aberration in mutagenicity testing. Hum Genet 58:235–254
14. Goetz P, Sram RJ, Dohnalova J (1975) Relationship between experimental results in mammals and man. I. Cytogenetic analysis of bone marrow injury induced by a single dose of cyclophosphamide. Mutat Res 31:247–254
15. Göhde W, Meistrich M, Meyn R, Schumann J, Johnston D, Barlogie B (1979) Cell-cycle phase-dependence of drug-induced cycle progression delay. J Histochem Cytochem 27:470–473
16. Hacker U, Schumann J, Göhde W, Müller K (1981) Mammalian spermatogenesis as a biologic dosimeter for radiation. Acta Radiol Oncol 20:279–282
17. Holden HE, Ray VA, Wahrenburg MG, Zelenski JD (1973) Mutagenicity studies with 6-mercaptopurine: I. Cytogenetic activity in vivo. Mutat Res. 20:257–263
18. Ishidate M, Odashima S (1977) Chromosome tests with 134 compounds on Chinese hamster cells in vitro – a screening for chemical carcinogens. Mutat Res. 48:337–354
19. Jenssen D, Ramel C (1980) The micronucleus test as part of a short-term mutagenicity test program for the prediction of carcinogenicity evaluated by 143 agents tested. Mutat Res 75:191–202
20. Kalia VK, Jain VK, Otto FJ (1982) Optimization of cancer therapy. Part IV – Effects of 2-deoxy-D-glucose on radiation induced chromosomal damage in PHA-stimulated peripheral human leukocytes. Indian J Exp Biol 20:884–888
21. King MT, Wild D (1979) Transplacental mutagenesis: the micronucleus test on fetal mouse blood. Hum Genet 51:183–194
22. Kirkhart B (1981) Micronucleus test on 21 compounds. In: DeSerres FJ, Ashby J (eds) Evaluation of short-term test for carcinogens. Elsevier/North-Holland, New York Amsterdam Oxford, pp 698–704
23. Kurita Y, Shisa H, Matsuyama M, Nishizuka Y, Tsuruta R, Yosida TH (1969) Carcinogen-induced chromosome aberrations in hematopoietic cells of mice. Gann 60:91–95
24. Maier P, Schmid W (1976) Ten model mutagens evaluated by the micronucleus test. Mutat Res 40:325–338
25. Matter B, Schmid W (1971) Trenimon-induced chromosomal damage in bone-marrow cells of six mammalian species, evaluated by the micronucleus test. Mutat Res 12:417–425
26. Matter B, Grauwiler J (1974) Micronuclei in mouse bone-marrow cells. A simple in vivo model for the evaluation of drug-induced chromosomal aberrations. Mutat Res 23:239–249
27. Miltenburger HG, Engelhardt G, Röhrborn G (1981) Differential chromosomal damage in Chinese hamster bone-marrow cells and in spermatogonia after mutagenic treatment. Mutat Res 81:117–122
28. Musilova J, Michalova K, Urban J (1979) Sister chromatid exchanges and chromosomal breakage in patients treated with cytostatics. Mutat Res 67:289–294
29. Nakanishi Y, Schneider EL (1979) In vivo siter chromatid exchange as a sensitive measure of DNA damage. Mutat Res 60:329–337
30. Nasjleti CE, Spencer HH (1966) Chromosome damage and polyploidization induced in human peripheral leukocytes in vivo and in vitro with nitrogen mustard, 6-mercaptopurine, and A-649. Cancer Res 26:2437–2443

31. Natarajan AT, van Kesteren-van Leeuwen AC (1981) Mutagenic activity of 20 coded compounds in chromosome aberrations/sister chromatid exchanges assay using Chinese hamster ovary (CHO) cells. In: DeSerres FJ, Ashby J (eds) Evaluation of short-term test for carcinogens. Elsevier/North-Holland, New York Amsterdam Oxford, pp 551–559

32. Nüsse M (1981) Cell cycle kinetics of irradiated synchronous and asynchronous tumor cells with DNA distribution analysis and BrdUrd-Hoechst 33258-technique. Cytometry 2:70–79

33. Nüsse M (1984) Cell cycle kineticy of synchronized EAT-cells after irradiation in various phases of the cell cycle studied by DNA distribution analysis in combination with two parametric flow cytometry (mitotic index) using acridine orange. In: Eisert W, Mendelsohn ML (eds) Proceedings of the international symposium on biological dosimetry: cytometric approaches to mammalian systems. Springer, Berlin

34. Otto FJ, Oldiges H (1980) Flow cytogenetic studies in chromosomes and whole cells for the detection of clastogenic effects. Cytometry 1:13–17

35. Otto F, Oldiges H, Göhde W, Dertinger H (1981) Flow cytometric analysis of mutagen induced chromosomal damage. Acta Pathol Microbiol Scand Section A Suppl 274:284–286

36. Otto FJ, Oldiges H, Göhde W, Jain VK (1981) Flow cytometric measurement of nuclear DNA content variations as a potential in vivo mutagenicity test. Cytometry 2:189–191

37. Pinaev G, Banyopadhyay D, Glebov O, Shanbhag V, Johansson G, Albertsson PA (1979) Fractionation of chromosomes. Exp Cell Res 124:191–203

38. Pinkel D, Gledhill BL, VanDilla MA, Lake S, Wyrobek AJ (1982) Radiation-induced DNA content variability in mouse sperm. Radiat Res, in press

39. Renner HW (1979) Zur Bewertung des in-vivo-SCE (sister chromatid exchange)-Test, dargestellt an verschiedenen Zytostatika. Arzneim Forsch 29:1871–1875

40. Röhrborn G, Basler A (1977) Cytogenetic investigations of mammals. Comparison of the genetic activity of cytostatics in mammals. Arch Toxicol 38:35–43

41. Röhrborn G (1980) Assessment of mutagenicity tests. Arch Toxicol Suppl 4:3–9

42. Roszinsky-Köcher G, Röhrborn G (1979) Effects of various cyclophosphamide concentrations in vivo on sister chromatid exchanges (SCE) and chromosome aberrations of Chinese hamster bone-marrow cells. Hum Genet 46:51–55

43. Sachs L (1971) Statistische Auswertungsmethoden. Springer, Berlin

44. Salamone M, Heddle J, Stuart E, Katz M (1980) Towards an improved micronucleus test. Studies on 3 model agents, mitomycin C, cyclophosphamide and dimethylbenzanthracene. Mutat Res 74:347–356

45. Salamone MF, Heddle JA, Katz M (1981) Mutagenic activity of 41 compounds in the in vivo micronucleus assay. In: DeSerres JF, Ashby J (eds) Evaluation of short-term tests for carcinogens. Elsevier/North-Holland, New York Amsterdam Oxford, pp 686–697

46. Schmid W, Arakaki DT, Breslau NA, Culbertson JC (1971) Chemical mutagenesis: the Chinese hamster bone marrow as an in vivo test system. I. Cytogenetic results on basic aspects of the methodology, obtained with alkylating agents. Humangenetik 11:103–118

47. Schumann J, Tilkorn H, Göhde W, Ehring F, Straub C (1981) Zytogenetik maligner Melanome. Hautarzt Suppl V:62–66

48. Sugiyama T (1973) Chromosomal aberrations and carcinogenesis by various benz(a) anthracene derivatives. Gann 64:637–639

49. Tribukait B, Gustafson H, Esposti P (1979) Ploidy and proliferation in human bladder tumors as measured by pulse-cytofluorometric DNA-analysis and its relationship to histopathology and cytology. Cancer 43:1742–1751

50. Tsuchᵒto T, Matter BE (1977) Comparison of micronucleus test and chromosome examination in detecting potential chromosome mutagens. Mutat Res 46:240

51. Tsuchimoto T, Matter BE (1979) In vivo cytogenetic screening methods for mutagens, with special reference to the micronucleus assay. Arch Toxicol 42:239–248

52. Tsuchimoto T (1980) Micronucleus test, chromosome analysis and SCE: evaluation of the in vivo cytogenetic methods for screening chemical mutagens. Mutat Res 74:236

53. Tsuchimoto T, Matter BE (1981) Activity of coded compounds in the micronucleus test. In: DeSerres FJ, Ashby J (eds) Evaluation of short-term tests for carcinogens. Elsevier/North-Holland, New York Amsterdam Oxford, pp 705–711

54. Weinstein D, Katz ML, Kazmer S (1977) Chromosomal effects of carcinogens and non-carcino-
    gens on WI-38 after short term exposures with and without metabolic activation. Mutat Res
    46:297–304
55. Wild D (1978) Cytogenetic effects in the mouse of 17 chemical mutagens and carcinogens
    evaluated by the micronucleus test. Mutat Res 56:319–327
56. Winne D (1964) Zur Auswertung von Versuchsergebnissen: Die verteilungsfreien Rangteste
    von Wilcoxon. Arzneim Forsch 14:119–121
57. Yamamoto KI, Kikuchi Y (1980) A comparison of diameters of micronuclei induced by clasto-
    gens and by spindle poisons. Mutat Res 71:127–131
58. Yamamoto KI, Kikuchi Y (1981) Studies on micronuclei time response and on the effects of
    multiple treatments of mutagen on induction of micronuclei. Mutat Res 90:163–173

Flow Cytometric Analysis of Chromosome Damage After Irradiation: Relation to Chromosome Aberrations and Cell Survival

J.A. Aten, M.W. Kooi, J.Th. Bijman, J.B.A. Kipp, and G.W. Barendsen

Laboratory for Radiobiology, University of Amsterdam, Plesmanlaan 121, 1066 CX Amsterdam, The Netherlands

Summary

Reproductive death of cultured cells is commonly assessed by measurement of clonogenic capacity which requires a culture period equivalent to about ten cell doubling times. Chromosome structural changes can be observed microscopically in stained preparations of mitotic cells but this requires tedious counting For a rapid determination of cellular sensitivity which might provide predictions of responses of tumors to various treatments, a new technique would be valuable if the dependence of responses on dose and radiation quality would correlate well with other cellular responses.

Flow cytometry has provided a technique for the rapid determination of DNA content of individual chromosomes of mammalian cells and of changes induced by various treatments. This technique involves selection of mitotic cells, the preparation of monodisperse chromosome suspensions, measurement of DNA content histograms and the analysis of these histograms by a computer program. Using this method we have determined changes in the distributions of chromosome frequency versus DNA content as a function of X-ray dose. The results are compared with dose effect relations for chromosome structural changes assessed by conventional cytogenetic techniques and for impairment of cellular clonogenic capacity.

1 Introduction

Changes taking place in living cells or organisms after exposure to ionizing radiation are complex and cannot be analysed merely in terms of quantitiy and quality of the absorbed radiation. Physical dosimetry of ionizing radiation alone, therefore, is of limited value in the prediction of radiation damage in biological material. Factors such as cell type, cell cycle phase and oxygen supply, strongly influence the reaction of cells to irradiation. Hence a direct correlation between radiation dose and biological effects must be established with respect to their dependence on these factors, before questions relevant to biological research and to medical applications can be adressed. Loss of proliferative capacity and cell transformation are endpoints that can be quantified as a function of radiation dose.

Biological Dosimetry. Edited by W. G. Eisert and M. L. Mendelsohn
© Springer-Verlag Berlin Heidelberg 1984

To obtain the parameters which describe radiosensitivity, cells must be cultured to form clones but the technique for the culturing of cells from a variety of mammalian tissue and tumors is still in development. Moreover, the time necessary to assess these effects is in the order of several weeks to several months. The information that has been obtained with this method has until now been limited to test-systems that were developed specifically, for this type of application, e.g. experimental transplantable tumors.

The primary lesions induced by ionizing radiation are assumed to be DNA strand breaks. When DNA strand breaks are not well repaired by the cell, they may be expressed through loss of proliferative capacity or through mutation. At the sub-cellular level DNA-double strand breaks may lead to chromosome aberrations which can be detected by microscopic analysis. A large volume of data has been published on the relation between radiation dose and chromosome aberrations and it is now generally accepted that the frequencies of cell death and of chromosome aberrations are correlated. This correlation could be applied to monitor the radiation treatment of cancer.

A fundamental parameter governing tumor growth is the proliferative capacity of the individual tumor cells. Reduction, therefore, in the number of proliferating cells is a measure of the succes of the therapy. Analysis of chromosome aberrations by the scanning of metaphase spreads of tumor cells can be performed within a few days of the beginning of the therapy, which is an advantage compared to the assay of clonogenic capacity. The counting of chromosome aberrations, on the other hand, is a very time consuming procedure. A method for the analysis of chromosome damage that does not involve extensive human effort might therefore be of interest as a technique for the assessment of radiobiological sensitivity.

The observation that chromosome damage can be detected by flow cytometry has led several groups to analyse the effect of clastogens on mammalian cells through changes in chromosome flow histograms [1—4]. When cell cultures are used as a test-system, other biological endpoints can be studied to determine the validity of this parameter as a biological indicator of damage. In this paper we present a technique for the analysis of radiation damage by the flow cytometry of chromosome suspensions. We compare the results with data on chromosome aberrations and on cloning capacity. In addition a computer generated histogram is presented in which chromosome damage has been induced artifically through a model calculation.

It was found that the analysis of chromosome damage by flow cytometry is similar in sensitivity and in reliability to conventional cytogenetic techniques.

2 Materials and Methods

The experiments were carried out with cultured V-79 Chinese hamster cells. Cultures were grown to a confluent plateau phase in order to obtain a population of $G_1 + G_0$ cells only. After irradiation cells in G_0 and G_1 exhibit chromosomal aberrations when induced to proceed to mitosis. After irradiation with different doses of up to 8 Gy of 200 kV X-rays, cultures were incubated for 30 min at 37°C, subsequently trypsinised and replated in appropriate dilutions to allow progression through the cell cycle to mitosis or to produce clones after 8 days.

Cultures from dilutions to be analysed by flow cytometry were incubated for 16 h and subsequently vinblastine was added to attain a concentration of .3 $\mu g/ml$ medium in order to accumulate cells in mitosis for 5 h. The mitotic cells were collected by shake off, spun down and resuspended at room temperature in 1.0 ml of hypotonic KCl (0.075 M) + Propidium Iodide (5×10^{-5} g.ml^{-1}). After 10 min of swelling, .5 ml of KCl (0.075 M) + Propidium Iodide (5×10^{-5} g.ml^{-1}) + Triton-X 100 (10^{-2} g.ml^{-1}) was added. Three min later the cells were disrupted by shearing the suspension twice through a 21 gauge needle [2, 5]. As shown by tests with vital stain, débris from dead cells was kept at a minimum by the selection and preparation procedure. Measurements of chromosomal DNA histograms were carried out by the ORTHO FC 4800A-FC 200 system operated in the slow mode, at a laser power reading of 30 mW at 488 nm wavelength.

Cultures used for studies of chromosome aberrations were incubated for 16 h after irradiation. Subsequently vinblastine was added for 5 h to collect metaphase cells. After hypotonic treatment with 0.075 M KCl for 10 min, the cells were fixed in methanol-acetic acid, 3:1, pipetted onto slides, air dried and stained with Giemsa. Well spread metaphases were scored for dicentrics and centric rings.

Cells from dilutions employed for measurement of clonogenic capacity were cultured for 8 days, fixed, stained with Giemsa and colonies containing 50 or more cells were counted as survivors. Details of experimental procedures have been described elsewhere [6].

3 Analysis of Chromosome Histograms Using the Fourier Transform

In the flow cytometric analysis of chromosome suspensions the DNA content of a large number of chromosomes is determined individually as they pass through the laser beam. The resulting distribution is a flow karyogram in which the number of chromosomes is represented as a function of their relative DNA content. Chromosomes with identical DNA content are grouped as peaks, as shown in Fig. 1a. In the flow karyograms of chromosome suspensions prepared from irradiated cells, the shapes of the peaks are altered. To quantify the changes observed in the histograms the distribution of peaks can be analysed by assuming that they represent oscillations along the horizontal axis.

An efficient way to analyse oscillating functions is the application of fourier analysis. In this technique, graphs or histograms are not analysed as a composition of separate entities, but as the superposition of oscillations with different frequencies, thus reducing the influence of statistical fluctuations present in each of the peaks separately. To explain the procedure a schematic histogram is presented in Fig. 2a. It is composed of a rapid oscillation representing the chromosome peaks super-imposed on a slowly oscillating background. By applying fourier analysis all oscillations are extracted from the histogram and sorted in order of increasing frequency. The result is given in the form of another histogram, Fig. 2b, the fourier power spectrum, representing the power of

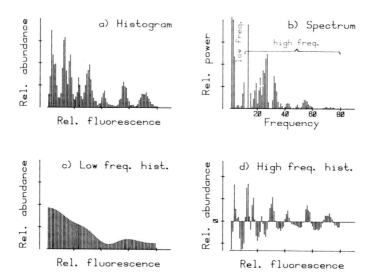

Fig. 1a–d. Application of fourier analysis to DNA-histograms of V79 chromosomes. **a)** Chromosome histogram of V79 cells. **b)** Corresponding fourier spectrum. **c)** Back transform of the low frequency part of the fourier spectrum. **d)** Back transform of the high frequency part of the spectrum

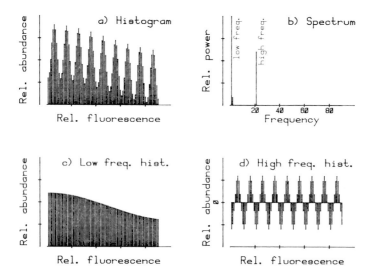

Fig. 2a–d. The application of fourier analysis to DNA-histograms of chromosomes. **a)** Schematic chromosome histogram. **b)** The corresponding fourier spectrum. **c)** Result of the back transform of the low frequency part of the fourier spectrum. **d)** back transform of the high frequency part of the spectrum

the extracted oscillations. The low and high frequencies are located on the left and right side of the spectrum respectively. Because Fig. 2a consists of not more than two frequencies, the spectrum shows only two narrow distributions. The chromosome peaks are represented by the high frequency part and the slow variation in the histogram by the low frequency part, as can be seen from Figs. 2c and 2d where the low and high frequency parts of the fourier spectrum have been transformed back into a "slow variation" and into a "chromosome peaks" histogram. Together they reconstitute the original histogram.

In Fig. 3 the analysis is applied to an actual chromosome histogram. In the spectrum the low frequency "slow variation" and the high frequency "chromosome peaks" distribution are wider, but they can still be separated from each other as is demonstrated in the two bottom histograms. In order to obtain a measure for the size of the chromosome peaks, the surface of the high frequency contribution of the fourier power spectrum, Fig. 2b, can be compared with the total surface of the histogram represented by the zero frequency component in the spectrum.

The fourier analysis was programmed in HP-basic and was performed on a HP-85 personal computer.

4 Results and Discussion

We have analysed radiation damage induced in mammalian cells by three endpoints: by changes in chromosome DNA histograms, by the frequency of chromosome aberrations and by loss of reproductive capacity.

In Figs. 3a, c, e and g chromosome histograms are shown for cells irradiated with 0, 2, 4 and 8 Gy of X-rays. The shapes of the peaks are altered as a result of the irradiation. With increasing dose the heights of the peaks are reduced and their widths are increased. In addition the areas between the peaks are filled in, presumably with fragments of damaged chromosomes. To quantify these changes we applied the technique described in the previous section. The Figs. 3b, d, f and h show the changes in the fourier spectra of the histograms caused by increasing doses of X-rays. The surface of the high frequency part of the spectra is used as a measure for the amount of intact chromosomes. With increasing dose the high frequency components in the spectra are progressively suppressed. This corresponds with the lowering of the chromosome peaks which by the fourier analysis, is interpreted as a damping of the high frequency oscillations in the DNA histograms.

The changes in the fourier spectra are summarized in Fig. 4a. The stars represent the results of 5 experiments; the standard deviations are indicated with bars. The data on chromosome aberrations and on cell survival are shown in the Figs. 4b and 4c. The three methods for the detection of radiation effects can now be compared with respect to their sensitivity. It appears that the flow cytometry of chromosome suspensions is as good a biological indicator of damage in V79 cells as the reduction of cell survival and that it may be more sensitive than the detection of chromosome aberrations by microscope.

Gray et al. [7] have reported data on the relative number of chromosome aberrations in chromosome suspensions prepared from irradiated cells. They analysed these suspensions by slit-scan flow cytometry. This technique offers the possibility to determine the number of centromeres per chromosome, thus functioning as a dicentrics detector.

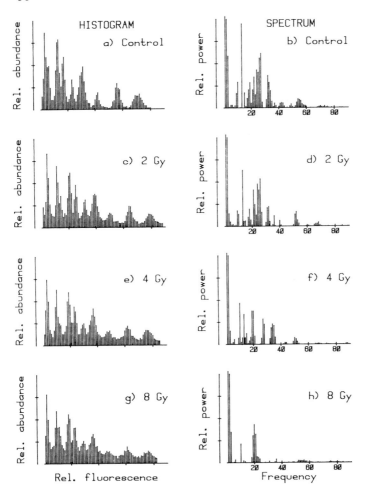

Fig. 3a–h. DNA histograms of V79 Chinese hamster chromosomes, irradiated with 0, 2, 4 and 8 Gy of X-rays. In each of the histograms $\simeq 2 \times 10^4$ chromosomes were analysed (**a, c, e, g**). The effect of radiation on the fourier power spectra of the chromosome histograms (**b, d, f, h**)

The data which they obtained, indeed showed an increase in the number of dicentrics in the valleys of the chromosome DNA histograms, correlated with the radiation dose. In principle it should be possible to use this kind of data for a mathematical model describing changes in the histograms in terms of damaged chromosomes. It would be interesting if in that way a quantitative comparison could be made of the relative efficiencies of these two criteria for the detection of chromosome damage.

In Fig. 5 we present a preliminary result of model calculations simulating the effect of chromosome damage on DNA histograms. In Figs. 5a and b chromosome histograms are shown from unirradiated cells and from cells irradiated with 6 Gy of X-rays. It appears that the peaks on the right side of the histogram in Fig. 5b have been affected by the irradiation more strongly than the peaks on the left side. This

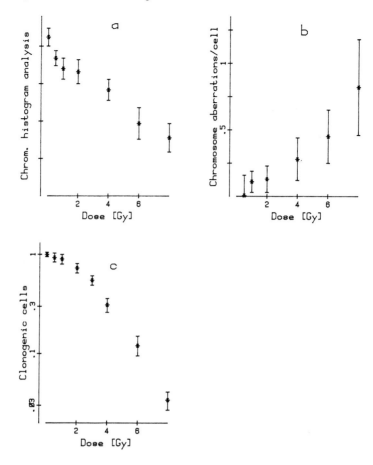

Fig. 4a–c. The effect of X-irradiation on V79 cells. **a)** The sum of the high frequency components of the fourier power spectra as a function of radiation dose. **b)** The number of dicentrics + rings observed in V79 metaphases as a function of dose. **c)** The dose-effect curve for clonogenicity of V79 cells

would be in agreement with the assumption that the risk of breakage and of malformation is proprotional to the amount of DNA in the chromosomes. With this assumption we calculated the change that would be observed in the histogram of Fig. 5a if 20% of the chromosomes would have been damaged. The result of this calculation is shown in Fig. 5c. The changes in the histograms 5b and 5c due to real and to artificial chromosome damage are of the same magnitude. The number of dicentrics + rings per cell irradiated with 6 Gy of X-rays was determined independently and can be read from Fig. 4b. The value of .45 aberrations per cell is equivalent to 4.0% of damaged chromosomes; the V79 cells having $\pm$ 23 chromosomes and taking into account that in the formation of each dicentric or ring two chromosomes are involved. Comparing the number of microscopically observed chromosomes aberrations, 4.0%, and the amount of chromosomes damage, 20%, necessary to simulate the observed changes, it appears

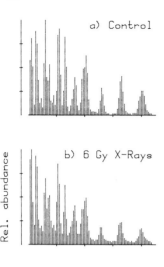

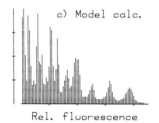

Fig. 5a–c. Model calculation simulating radiation damage in DNA histograms. a) Chromosome histogram from control cells. b) Histogram from cells irradiated with 6 Gy of X-rays. c) The histogram from Fig. 5a, changed by simulating damage in 20% of the chromosomes

that by flow cytometric analysis of chromosome suspensions, a large amount of chromosome damage is detected that, in microscopic analysis, is not recognized as dicentrics or rings.

References

1. Carrano AV, Gray JW, Van Dilla MA (1977) Flow cytogenetics: Progress toward chromosomal aberration detection. Proceedings of the Symposium on Actions of Physical and Chemical Mutagens on the Somatic Chromosomes of Man. Edinburgh United Kingdom 7-8 July 1977
2. Aten JA, Kipp JBA, Barendsen GW (1980) Flow cytofluorometric determination of damage to chromosomes from X-irradiated Chinese Hamster cells. Acta Microbiol Scand Suppl 1980: 287–292
3. Otto FJ, Oldiges H, Göhde W, Dertinger H (1980) Flow cytometric analysis of mutagen induced chromosomal damage. Acta Pathol Microbiol Scand Suppl 1980:284–286
4. Otto FJ, Oldiges H (1980) Flow cytogenetic studies in chromosomes and whole cells for the detection of clastogenic effects. Cytometry 1:13
5. Buys CHCM, Koerts T, Aten JA (1982) Well-identifiable human chromosomes isolated from mitotic fibroblasts by a new method. Hum Genet 61:157–159

6. Barendsen GW (1980) Analysis of tumour responses by excision and in vitro assay of cellular clonogenic capacity. The Brit J Cancer 41: Suppl IV 209–216
7. Gray JW, Peters D, Lucas J, Aten JA, Van Dilla M (1980) Slit scan flow cytometry of mammalian chromosomes. Basic and Applied Histochem 24:345

Chromosome Analysis of Leukemia and Preleukemia: Preliminary Data on the Possible Role of Environmental Agents

Herman van den Berghe

Centre for Human Genetics, University of Leuven, Capucienenvoer, 35, B-3000 Leuven, Belgium

Introduction

There is substantial evidence indicating that non-random chromosome changes are specifically associated with a number of human malignancies, and the bulk of information in this respect comes from the study of chromosome patterns in hematologic disorders, particularly chronic and acute leukemia [1]. Before the advent of banding techniques in 1970, the Philadelphia chromosome (Phl) had been described by Nowell and Hungerford [2] in chronic myelocytic leukemia (CML) and in about 50% of the acute leukemias chromosome abnormalities had been found, but the resolution of the techniques did not allow for a more precise identification of these anomalies.

Introduction of banding techniques has changed the picture and the amount of information thus gathered has been considerable.

For the purpose of this meeting I propose:
1. to present an overview of the characteristic chromosome changes in overt acute non-lymphoblastic leukemia
2. to compare those results with the data obtained in socalled preleukemia
3. To evaluate these data with regard to the possible relationship between some of these characteristic chromosome anomalies on the one hand, and environmental agents on the other hand.

1 Characteristic Chromosome Changes in Acute Non-Lymphoblastic Leukemias (ANLL)

Before the advent of banding techniques it was recognized that about 50% of acute non lymphocytic leukemias showed more or less gross chromosomal abnormalities, and banding techniques did not substantially change this proportion. On the other hand, the use of these techniques has markedly increased our understanding of the type of chromosome abnormalities, while missing or extra chromosomes as well as structural rearrangements have been identified in terms of the particular chromosomes or chromosome band involved [3, 4].

The most frequent numerical chromosome aberrations are trisomy 8 and monosomy 7, occurring either alone, or in combination with other anomalies. Monosomy 7, in some cases, is preceded by a deletion of part of the long arm of this chromosome suggesting that the deleted 7 may be unstable. A deleted 7q on the other hand can also be found as the sole anomaly or in combination with other abnormalities, without loss of

Biological Dosimetry. Edited by W. G. Eisert and M. L. Mendelsohn
© Springer-Verlag Berlin Heidelberg 1984

the deleted chromosome. Other cases of ANLL are characterized by a reciprocal translocation between the long arms of chromosome 8 and 21, involving band q22 of each of these chromosomes. In about 50% of the cases there is also loss of a sex chromosome, in addition to the t (8; 21), males losing the Y, females losing the inactive X. Moreover, a subtype of this anomaly exists, in which, in addition to the t(8; 21) ± sex chromosome loss, an interstitial deletion is present in the long arm of chromosome 9. A t (15; 17) (q26; q22) is characteristic of the vast majority of ANLL of the promyelocytic type. Variants involving the same chromosomes may exist [5]. This particular chromosome anomaly was not found sofar in hematologic disorders other than APL. Some APL seem to have a normal karyotype although less than was initially thought, and a few may present with other chromosome anomalies, including trisomy 8. The occurrence of a Phl chromosome in typical ANLL, t(9; 22) as well as variant translocations is well established.

A substantial number of ANLL cases is characterized by an intersititial deletion of the long arm of chromosome 5, with a proximal break occurring either in band q12 or q24, and a distal break in q32 [6]. In contrast to most other deletions, there never is any translocation of this interstitial 5q material upon another chromosome. Primary 5q-ANLL may be but is usually not preceded by documented 5q- refractory anemia and as a rule, additional chromosome anomalies are present. Other non-randomly occurring anomalies include deletions of the 11q and of the 20q and a partial trisomy of the 1q in which cases the bands 1q25 to 1q32 are always included. Very recently a t (6;9) was found to be characteristic of some ANLL [7].

With regard to the association of these anomalies with specific types of ANLL, trisomy 8, occurring as the sole anomaly, has been found in different types of ANLL, including APL, and cannot be associated with any specific sub-type. The 7q deletion/7 monosomy also shows no predilection for any specific sub-type of ANLL. On the other hand, a considerable number of secundary leukemias show a deleted 7q/7 monosomy as the sole anomaly or associated with other abnormalities. This is also the case for the 5q- interstitial deletion, usually accompanied by other anomalies.

As emphasized above, t(15; 17) is very specific for APL, including the socalled hypogranular morphological variant. The t(8; 21) apparently characterizes a type of ANLL, the typical bone marrow morphology of which is that of M2, i.e. myeloblastic with some maturation. Overall, the t(8; 21) ANLL is associated with a good prognosis. A considerable proportion of acute Monoblastic leukemias (M5) present a deletion 11q23 [8]. Further clinical correlations show that patients with a completely normal karyotype at diagnosis may have a better prognosis than those with only abnormal cells, and that the presence of some normal cells may also be a better prognostic sign.

2 Chromosome Changes in Preleukemia

In socalled preleukemia or smoldering leukemia chromosome studies have shown that the frequency and type of chromosomal abnormalities observed are similar to those found in overt ANLL. If preleukemia is considered to be merely an early preclinical phase of ANLL, this result would not be unexpected. However, there are a number of indications that not all leukemias go through such a preleukemic, preclinical phase. Anomalies such as t(8; 21), t(15; 17) and t(9; 22) are notoriously absent in the material, suggesting that leukemias with those particular chromosome anomalies do not have a similar if any preleukemic phase. On the other hand, evidence is accumulating that clonal proliferation characterized by an interstitial 5q- deletion, and involving at least the myelocytic, erythrocytic and megakaryocytic precursurs may manifest itself as refractory anemia with or without excess of myeloblasts, without manifestations, even after 10 years, of ANLL. Very remarkably this 5q- refractory anemia without excess of blasts at presentation, occurs predominantly in females beyond the age of menopause. The male/female ratio is 1:3. Even when the 5q- is found in RAEB or smoldering leukemia at presentation, an excess of females remains. This finding is in contrast with the overall sex ratio in preleukemia, in which males are more frequently affected than females. In 5q- RAEB and smoldering leukemia, transformation into acute leukemia is more common, and male sex plus the presence of additional chromosome anomalies may be unfavorable prognostic parameters in this respect. The general picture of 5q- shows more than superficial similarity with that of the Ph1 positive disorder. Both the 5q- and the Ph1 apparently characterize a family of clinically different conditions: (1) a chronic myeloproliferative disorder which may or may not terminate in an acute phase, usually though not necessarily accompanied at that time, by additional chromosome anomalies; (2) acute myelogenous leukemia: (3) acute lymphoblastic leukemia.

The possible involvement of lymphoid cells in 5q- refractory anemia and in 5q- acute myeloid disorders remains to be demonstrated. On the other hand, there is one striking difference between the Ph1 and the 5q- family: never has a Ph1 chromosome been convincingly demonstrated in secundary leukemia.

3 Is There Any Relationship Between Chromosome Changes Observed and Possible Environmental Agents

The only way a relationship between the chromosome changes and environmental agents or factors can be established, is through well designed and specifically directed epidemiological studies. Thusfar, no studies of that kind have been carried out, but some preliminary data indicate that they may come up with very interesting and meaningful correlations.

A first body of information is gradually emerging from the cytogenetic study of secundary leukemias, i.e. leukemias occurring in individuals who have been treated with radiation and/or alkylating agents for a previous malignancy, and the following preliminary conclusions are gradually emerging.

In de novo occurring primary leukemias, not more than 60% show chromosome changes in the leukemic cells with the usual banding techniques. In contrast more than 90% of secundary leukemias present with chromosome abnormalities. In complicating leukemias however, occurring in patients with previous malignancy not treated by radiation or alkylating agents, chromosome changes are much more rare [9].

Chromosome 3 and 17, but especially 5 and 7 must be stressed as being non-randomly involved in complicating leukemia [10]. The 7q- deletion and 7 monosomy and the 5q- interstitial deletion together may probably make up for 50% or more of all secundary acute leukemias, of which some show a combined 5 and 7 monosomy. In contrast t(8; 21), t(15; 17) and t(9; 22) are virtually not found in secundary leukemia.

It is possible that there is an interaction between the type of drugs and the whole of the genetic make-up of the patient, and the whole process may very well be multifactorial and/or under the influence of various other factors. The drugs may play a major role in leading to non-random changes in chromosomes 3, 5, 7 and 17. It may well be that for any spontaneous leukemia with involvement of the chromosomes 5 and 7, 3 and 17, particularly if they occur together, the suspicion should be raised that such a leukemia is not "spontaneous" in nature but related to toxic agents of one type or another.

The data obtained from the study of secundary leukemias are in support of the possible relationship between a causative agent (petroleumproducts, insecticides etc.) and the karyotypic picture in ANLL as reported by Mitelman et al. [11, 12] and studied in a prospective study at the 4th workshop on chromosomes in leukemia in Chicago. Patients that were classified as exposed to those agents had significantly more chromosome anomalies in the leukemic cells than the non-exposed.

In view of the preceding, it is of particular interest to note that the "spontaneous" 5q- anomaly occurs predominantly in females-housewives, above the age of 60. Possible exposure to household products should therefore be thoroughly investigated. It is quite likely that such an approach may ultimately lead to subclassification of various leukemias according to possible causative agents.

References

1. Sandberg AA (1980) The chromosomes in human cancer and leukemia. Elsevier North Holland
2. Nowell PC and Hungerford AA (1960) Chromosome studies on normal and leukemic human leucocytes. JNCI 25:85
3. Chromosomes in Acute Non-Lymphocytic Leukemia (1978) First International Workshop on Chromosomes in Leukemia. Brit J Haemat 39:311
4. Second International Workshop on Chromosomes in Leukemia (1979) Cancer Genet Cytogenet 2:89
5. Van den Berghe H, Louwagie A, Broeckaert-van Orshoven A, David G, Verwilghen R, Michaux JL, Ferrant A and Sokal G (1979) Chromosome Abnormalities in Acute Promyelocytic Leukemia (APL). Cancer 43:558
6. Van den Berghe H, David G, Michaux JL, Sokal G and Verwilghen R (1976) 5q- Acute Myelogenous Leukemia. Blood 48:4, 624
7. Vermaelen K, Michaux JL, Louwagie A and Van den Berghe H (1983) Reciprocal translocation t(6;9)(p21;q33): New characteristic chromosome anomaly in myeloid leukemias (Cancer Genet Cytogenet, in press)
8. Vermaelen K, Michaux JL, Tricot G, Casteels-van Daele M, Noens L, Van Howe W, Drochmans A, Louwagie A and Van den Berghe H: Anomalies of the Long Arm of Chromosome 11 in Human Myelo- and Lymphoproliferative disorders. I. Acute Non-Lymphocytic Leukemia (Cancer Genet Cytogenet, in press)

9. Fourth Workshop on Chromosomes in Leukemia, Chicago 1982. Cancer Genet Cytogenet, in press

10. Sandberg AA, Abe S, Kowalczyk JR, Zedgenidze A, Takeuchi J, Kakati S (1982) Chromosomes and Causation of Human Cancer and Leukemia. L. Cytogenetics of Leukemias complicating other Diseases.Cancer Genet Cytogenet 7:95

11. Mitelman T, Brandt L, Nilsson PO (1978) Relation among occupational exposure to mutagenic/carcinogenic agents; Clinical findings and bone marrow chromosomes in acute nonlymphocytic leukemia. Blood 52:1229

12. Mitelman F, Nilsson FG, Brandt E, Alimena G, Gastaldi R, Dallapiccola B (1981) Chromosome pattern, occupation and clinical features in patients with acute non-lymphocytic leukemia. Cancer Genet Cytogenet 4:197

Radiation Dosimetry Using the Methods of Flow Cytogenetics

Daryll K. Green, Judith A. Fantes, and George Spowart

MRC Clinical and Population Cytogenetics Unit, Western General Hospital, Crewe Road, Edinburgh EH4 2XU, Scotland

Summary

The measurement of whole body radiation dose by manually counting the proportion of damaged chromosomes in a sample of cultured peripheral blood lymphocytes requires many hours of skilled technician time. A low dose of approximately 15 rads could only be measured with certainty after 500 metaphase cells had been scored and one technician would spread this task over a period of 2-3 days. Flow cytogenetics offers an alternative scoring technique which is potentially more than an order of magnitude faster and which does not involve technician fatigue. We have studied the effects of in vitro radiation doses ranging from 25–400 rads on human peripheral blood lymphocytes by measuring the distortion of the Hoechst 33258 fluorescence distribution of human chromosomes in flow. The level of low intensity background signals, which increases with increasing radiation dose, correlates numerically with the radiation dose administered and the amount of chromosome damage scored by manual means. A human chromosome centromeric staining technique has been investigated which could lead to a method of recognising dicentric chromosomes in flow and a further enhancement to the radiation dose sensitivity of flow cytogenetics.

Introduction

The aim of the work reported here is to develop a routine, reliable method for measuring the effect of mutagenic agents on human peripheral lymphocytes (Evans et al. 1979; Evans and Lloyd 1979; Lloyd et al. 1980) by employing the techniques of flow cytometry. Most flow machines, whether they are commercially manufactured or specially built claim chromosome analysis rates ranging from 500 to 5000 chromosomes per second, which exceeds by several orders of magnitude the best manual scoring rate of chromosomes on a microscope slide of 5000 per hour. Speed of analysis in the case of flow is balanced against information content in the case of slide analysis.

We shall show that the fluorescence distribution of 10^5 chromosomes measured with a flow cytofluorimeter contains sufficient information to indicate a radiation dose of 100 rads and that the total time taken to achieve this result from the moment a fluorescently stained suspension of human chromosomes was presented to the flow machine was 30 min. This can be compared with the information regarding dicentric chromosomes, acentric fragments, etc., gathered from the manual analysis of 5×10^3 chromo-

Biological Dosimetry. Edited by W. G. Eisert and M. L. Mendelsohn
© Springer-Verlag Berlin Heidelberg 1984

somes and which took a total time of 75 min from the moment a slide was mounted on to a microscope. This example does not show a dramatic improvement of radiation dosimetry performance for flow analysis compared with manual analysis. The potential advantage over manual methods which flow analysis can offer will be at the low dose level, where in order to achieve a statistically significant result more chromosomes must be scored. Scoring 5000 more chromosomes by slide analysis required a further 75 min (neglecting scorer fatigue); analysing 10^5 more chromosomes by a flow machine requires a further 1 to 5 min, depending on the actual flow rate.

The results presented and discussed in this paper show that the distortion of the fluorescence distribution of Hoechst 33258 stained human chromosomes is related to the radiation dose administered in vitro to the peripheral blood lymphocytes from which they were prepared. Separate experiments using the peripheral blood of a single donor but carried out on different occasions using slightly different preparation techniques show that preparation variability can mask the effect of radiation dose. Similar experiments using the fluorescence of ethidium bromide stained chromosomes are reported elsewhere (Fantes et al. 1982). The reduced tendency of the dye Hoechst 33258 to stain cell debris and the generally increased amount of detail in the Hoechst fluorescence distribution (Gray et al. 1979) as appose to that of ethidium bromide for human chromosomes indicate that Hoechst stained chromosomes offer the greatest potential for measuring low radiation dose by flow cytofluorimetry.

The results of both the ethidium bromide and the Hoechst experiments show that a further measurement parameter is needed in flow analysis of chromosomes for reliable low dose measurements. Recording the position or simply the existence of a chromosome centromere would lead to the counting of dicentric chromosomes and this is the ultimate goal of workers striving to slit-scan chromosomes in flow (Gray and Peters et al. 1979; Wheeliss and Potter 1973). Here we briefly discuss the results of some experiments which attempt to attach a fluorescent dye specifically to the centromere region of human chromosomes (Moroi et al. 1980). A successful conclusion to these experiments would ultimately lead to the counting in flow of acentric fragments, normal chromosomes and dicentric chromosomes corresponding to the measurement of zero units, 1 unit and 2 units of centromere stain respectively.

Materials and Methods

Irradiation and Chromosome Isolation

Lymphocytes were separated from human blood on a ficoll hypaque gradient and were irradiated at 20ºC by a Siemans Stabilipan Xray machine (250kV; Th2 filter; 67 rads/minute). Doses ranged from 25 to 400 rads. Irradiated cells and controls were cultured at 5×10^5 lymphocytes/ml in RPMl 1640 medium supplemented with 15% foetal calf serum, 100U/ml penicillin, 100µg/ml streptomycin and 1% phytohaemagglutinin for 52 h at 37ºC. Colcemid was added to the cultures for the final 6h at a concentration of 0.1µg/ml.

Chromosomes were isolated from irradiated and control cultures using the method of Sillars and Young (Sillars and Young 1981; Young et al. 1981). Cells were swollen in hypotonic 75mM KCl, and resuspended after centrifugation in ice-cold buffer containing 15mM Tris-HCl, 0.2mM spermine, 0.5mM spermidine, 2mM EDTA, 0.5mM EGTA, 80mM KCl, 20mM NaCl and 14mMβ-mercaptoethanol, final ph 7.2. After centrifugation out of buffer the cells were disrupted by mild detergent lysis in buffer + 0.1% digitonin followed by vortex mixing and this mixture of isolated chromosomes, interphase nuclei and debris was stored overnight on ice.

Flow Cytometry

Each sample contained 10^7 chromosomes/ml and had been prepared from 10 mls of blood. Coded and randomised samples were stained with 0.5μg/ml of Hoechst 33258 and injected into a flow cytofluorimeter (Green and Fantes 1982) with a motor driven syringe at 0.3ml/h. Prior to injecting each chromosome sample a sample consisting of buffer and Hoechst dye was flushed through the sample delivery syringe and tubing for approximately 10 min. Following this procedure a given chromosome sample was allowed a 10 minute period of stabilisation before a fluorescence intensity distribution (512 channels) of approximately 10^5 chromosomes was recorded. The Hoechst dye was excited with a 350nm UV Spectra Physics 171-18 argon laser and the fluorescence emission was detected in the 475nm to 550nm wavelength band.

In experiment EXP1, purified PHA rather than reagent grade PHA was used to transform the lymphocytes. It was used at 1.5X the recommended concentration and this would account for the low mitotic indices observed. In experiment EXP2 dividing cells were accumulated with colcemid between 45 and 52 h to ensure that more of the dividing cells were blocked at metaphase than in previous experiments and the suspensions were centrifuged to remove undividing nuclei.

Microscope Scoring

Aliquots were taken of each sample of cells in 75mM KCl. These were fixed in 3:1 methanol:acetic acid and air dried slides were prepared and stained in orcein. Slides were coded and randomised before 100 cells from each sample were scored for dicentric chromosomes and acentric fragments.

Kinetochore Staining

Chromosomes, isolated by the procedure described above, were spun onto a microscope slide in a slide centrifuge. A drop of dilute solution of "active" serum from a scleroderma patient was placed on the chromosomes, covered with a coverslip and incubated at room temperature for 30 min. The slide was then rinsed with saline. A drop of dilute Fluorescein-conjugated rabbit immunoglobulins to human (IgG) was placed on the chromosomes, a coverslip added, and a further incubation at room tem-

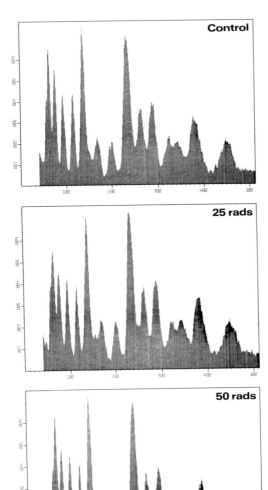

Fig. 1. Normalised fluorescence intensity distributions for Hoechst 33258 stained chromosomes prepared from peripheral blood lymphocyte cultures which had received prior radiation doses of zero, 25, 50, 100, and 200 rads. Frequency is plotted vertically and fluorescence intensity horizontally

perature was carried out for 45 min. After rinsing in saline the chromosomes were counterstained in Ethidium bromide (10µg/ml) for 5 min, given a final saline rinse and mounted in a 1:1 mixture of glycerol:Tris-HCl buffer pH8.5 for examination by fluorescence microscopy.

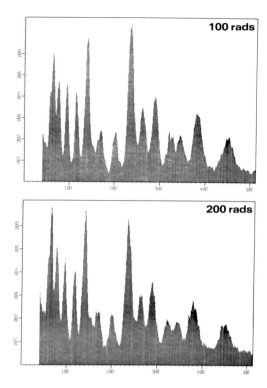

Fig. 1 (Continuation)

Results

In order to make both visual and numerical comparison between fluorescence distributions of different chromosome suspensions a computerised normalisation process was operated on each distribution such that the total area between channels 40 and 511 amounted to 10^5 units and the peaks corresponding to chromosome groups (9–12) and groups (1, 2) were centred on channels 235 and 450 respectively. The normalised series of fluorescence distributions for experiment R2 are shown in Fig. 1. A visual inspection of Fig. 1 is only helpful in distinguishing between a low radiation dose and a high radiation dose and suggests to the observer that the amount of background, (Otto and Oldiges 1980) particularly at low fluorescence intensity increases with increasing radiation dose. The results of experiment R1 are very similar except for a lower level of background signals over the whole range of intensities and for every sample in the series.

A numerical description of the level of background signals has been confined to a simple measurement of the frequencies of signals at the valley between the peaks for chromosomes 17 and 20 (V_1) and the valley between the peaks for chromosomes 13 and 14 (V_2). The frequencies V_1 and V_2, which are illustrated in Fig. 2 depend not only on the actual level of background signals but also on the extent to which the surrounding peaks overlap. In order to partially compensate for the latter effect, the size of which depends on the uniformity of chromosome staining, constitutional chromosome polymorphisms and machine performance, the following "background factor" is proposed:

background factor = $(V_1+V_2) / (C.V.)_{20}$

where $(C.V.)_{20}$ is the coefficient of variation of the chromosome group 20 peak, calculated from a least squares fit of a Gaussian distribution. The step by step procedure for calculating a given background factor were therefore:

(a) normalisation to 10^5 area units and centering of chromosome group (9–12) on channel 235 and chromosomes (1, 2) on channel 450
(b) drawing of a background line by smoothly joining several valleys of a given distribution. (this is neccessary for the gaussian fitting algorithm)
(c) measurement of the valleys V_1 and V_2
(d) fitting a gaussian distribution to chromosome group 20 peak and calculation of $(C.V.)_{20}$

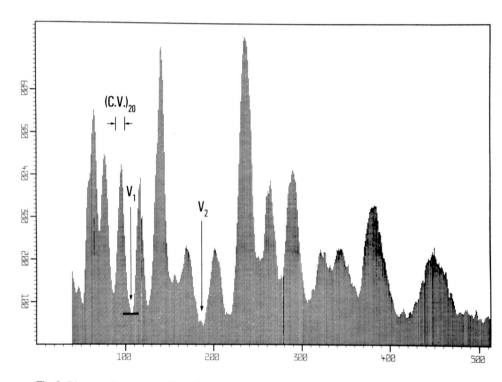

Fig. 2. Diagramatic representation of the measurements V_1, V_2 and $(C.V.)_{20}$. Each valley was computed as the minumum value of a smoothed fit to the points surrounding the lowest point between user specified right and left limits. $(C.V.)_{20}$ was computed from a least squares fit of a gaussian distribution to the peak corresponding to group 20 chromosomes

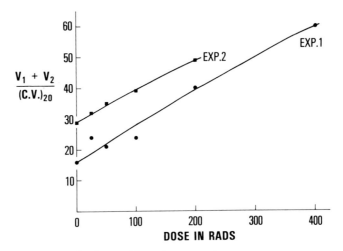

Fig. 3. Graph showing the change in the 'background factor' $(V_1+V_2) / (C.V.)_{20}$ with radiation dose for experiments R1 and R2. Each set of results show a general trend of increased background with increased radiation dose but are offset by a major amount due to different preparation methods

Figure 2 shows the results of steps (a) to (d) diagramatically. The background factor plotted against dose for experiments EXP1 and EXP2 are shown in Fig. 3.

In parallel with the flow cytometry approach manual scoring of the chromosome damage sustained in all of the samples was scored from microscope slides. Table 1 shows for each sample the number of cells scored, the counts of dicentric chromosomes and acentric fragments and for comparison the corresponding 'background factor' derived from the flow results.

Fluorescently stained chromosomes prepared according to the kinetochore staining method and photographed through a microscope are shown in Fig. 4. Here the chromatids were counterstained with ethidium bromide. The brighter kinetochores clearly contrast with the chromatids though it is not certain whether the integrated ethidium bromide fluorescence for a single chromosome will exceed the FITC fluorescence at the centromere. Two stains with similar excitation wavelengths but differing emission bands would be more ideal for detection of chromosome centromeres in flow, for example, ethidium bromide and rhodamine conjugated immunoglobulin both of which can be excited with the strong 488nm wavelength emission of an argon-ion laser but which emit into sufficiently different bands to enable the separate detection of their fluorescence. Difficulties have arisen when the centomere staining technique was applied to suspended chromosomes. The most severe of these is a strong tendency for chromosomes to aggregate after treatment. The effect of centromere staining has therefore not been observed yet in flow.

Table 1. Chromosome damage sustained by the lymphocyte nuclei in experiments R1 and R2 scored by manual inspection of microscope slide preparations. Alongside each score is shown the corresponding 'background factor' measured from the flow cytometry results

Manual scoring results				Flow background factor $(V_1+V_2) / (C.V.)_{20}$
Exp. R1	No. of cells scored	Fragments + dicentrics etc.	Abnormal items / cell	
control	100	1	0.01	16
25 rads	200	15	0.075	24
50 rads	200	29	0.145	21
100 rads	100	49	0.49	24
200 rads	100	106	1.06	40
400 rads	100	354	3.54	60
Exp. R2				
control	100	1	0.01	29
25 rads	100	6	0.06	32
50 rads	100	23	0.23	35
100 rads	100	39	0.39	39
200 rads	100	148	1.48	49

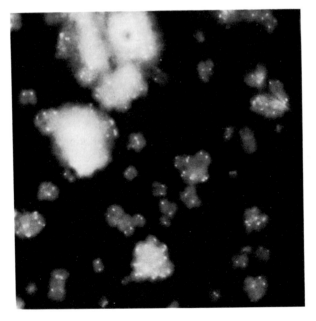

Fig. 4. Photomicrograph of isolated human chromosomes which have been stained at the kineto-chores with antibody bound FITC. The chromatids were stained with ethidium bromide

Discussion

At a radiation dose of 50rads an average of one item of chromosomal origin, whether it be an acentric fragment or a dicentric chromosome, occurs for every 250 normal chromosomes. In a batch of 10^5 chromosomes therefore 400 abnormal objects will arise and if their fluorescence emission is spread over the whole range of chromosome fluorescence intensity, which for the experiments described here is linearly divided into a useful range of 400 channels, then a background level of one event per channel would occur. In practice it would appear that a higher level of background occurs, perhaps due to other disturbances to the whole cell behaviour or a non-linear size distribution of abnormal events, following exposure to radiation. It is nevertheless not surprising that measuring a low radiation dose by simply measuring the frequency of background events in a fluorescence intensity distribution of chromosomes in flow is beset with uncertainty. The uncertainty could be reduced by carefully standardising the preparation of chromosome suspensions from peripheral blood and by keeping a library of control fluorescence distributions for individuals who might at some later time suffer radiation exposure. The evidence of the results presented here and those of other workers (Gray and Peters et al. 1979) points to the need for a method of isolating the abnormal objects in flow. It is our aim therefore to focus attention on detecting chromosome centromeres either by specifically binding a second fluorochrome to the chromosome kinetochores or by a slit scanning arrangement.

Acknowledgements. The authors wish to thank Patricia Malloy for her assistance in preparing the chromosome suspensions and John Elder for writing computer programs to analyse the fluorescence distribution data.

References

Evans HJ, Buckton KE, Hamilton GE and Carothers A (1979) Radiation Induced Chromosome Aberrations in Nuclear Dockyard Workers. Nature. 277:531–534

Evans HJ and Lloyd DC (1978) Mutation Induced Chromosome Damage in Man. (Eds. Evans HJ and Lloyd DC) University Press, Edinburgh

Fantes Judith A, Green DK, Elder JK, Mallow Patricia and Evans HJ (1983) Detecting Radiation Damage to Human Chromosomes by Flow Cytometry Mutation Res. Letters: 119:161–168

Gray JW, Langlois RG, Carrano AV, Burkhart-Schulte K and Van-Dilla MA (1979) High Resolution Chromosome Analysis: One and Two Parameter Flow Cytometry. Chromosoma 73:9–27

Gray JW, Peters D, Merrill JT, Martin R and Van Dilla MA (1979) Slit-scan Flow Cytometry of Mammalian Chromosomes. J. Histochem. Cytochem 27:441–444

Green DK and Fantes Judith A (1983) Improved Accuracy of In-Flow Chromosome Fluorescence Measurements by Digital Processing of Multi-parameter Flow Data. Signal Proc: 5:175–186

Lloyd DC, Purrott RJ and Reeder EJ (1980) The Incidence of Unstable Chromosome Abberrations in Peripheral Blood Lymphocytes from Unirradiated and Occupationally Exposed People. Mutation Res 72:523–532

Moroi Y, Peebles C, Fritzler MJ, Steigerwald J and Tan EM (1980) Autoantibody to Centromere (Kinetochore) in Scleraderma Sera. Proc. Natl. Acad. Sci. 77:1627–1631

Otto FI and Oldiges H (1980) Flow Cytogenetic Studies in Chromosomes and whole Cells for the Detection of Clastogenic Effects. Cytometry 1:13–17

Sillars R and Young BD (1981) A New Method for the Preparation of Metaphase Chromosomes for Flow Analysis. J Histochem. Cytochem 29:74–78

Wheelis LL, Hardy JA and Balasubramanian N (1975) Slit-scan Flow System for Automated Cytopathology. Acta. Cytol 19:45–52

Young BD, Ferguson-Smith MA, Sillars R and Boyd E (1981) High Resolution Analysis of Human Peripheral Lymphocyte Chromosomes by Flow Cytometry. Proc. Nat. Acad. Sci 78:7727–7731

Preliminary Reanalysis of Radiation-Induced Chromosome Aberrations in Relation to Past and Newly Revised Dose Estimates for Hiroshima and Nagasaki A-Bomb Survivors

Akio A. Awa[1], Toshio Sofuni[1], Takeo Honda[1], Howard B. Hamilton[1], and Shoichiro Fujita[2]

[1]Department of Clinical Laboratories, [2]Department of Epidemiology and Statistics, Radiation Effects Research Foundation (RERF), 5-2 Hijiyama Park, Minami-ku, Hiroshima 730, Japan

Summary

A comparison of the chromosome data from 408 atomic bomb survivors, 229 in Hiroshima and 179 in Nagasaki, was performed in terms of the frequency of cells with radiation-induced chromosome aberrations plotted against the existing A-bomb dosimetry system (T65D), and the new dosimetry systems from ORNL (Oak Ridge National Laboratory) and LLNL (Lawrence Livermore National Laboratory). Preliminary analysis shows that the inter-city difference in aberration frequencies is still apparent at every T65D dose level, while the aberration frequencies based on both ORNL and LLNL systems do not differ strikingly between the two cities, particularly in the dose range below 200 rad.

In our previous cytogenetic study of atomic bomb survivors in Hiroshima and Nagasaki (Awa et al. 1978), we found a close relationship between radiation dose and frequency of cells with residual chromosome aberrations. Furthermore, we noted a difference in aberration frequencies between Hiroshima and Nagasaki: the frequency of aberrant cells was consistently higher in every dose range in Hiroshima than in Nagasaki. One possible explanation for this difference was thought to be the contribution of the neutron component released from the bomb at Hiroshima. Estimated radiation doses for individual survivors used in the cytogenetic study were derived from the T65D (tentative 1965 dose) system, currently in use at RERF, calculated by Auxier et al. (1966), and Milton and Shohoji (1968).

Recently, re-evaluation of A-bomb radiation air doses was made by Loewe and Mendelsohn (1981) of the Lawrence Livermore National Laboratory (LLNL), and independently by Kerr (1981) and his associates of the Oak Ridge National Laboratory (ORNL). The major difference between the existing T65D and the two new dosimetry systems is that for Hiroshima, the estimates for the relative magnitude of the neutron component are much lower for the ORNL and LLNL than for the T65D system.

The purpose of this study is to compare the somatic chromosome aberration frequencies from Hiroshima and Nagasaki A-bomb survivors, based on the T65D, ORNL and LLNL dose estimates.

Biological Dosimetry. Edited by W. G. Eisert and M. L. Mendelsohn
© Springer-Verlag Berlin Heidelberg 1984

Material and Methods

Samples

Subjects of the present study were selected from participants in the RERF Adult Health Study sample, who visit the RERF clinic biennially to receive periodical physical examinations as part of regular procedure based on the A-bomb Survivors Medical Treatment Law. Cytogenetic samples consisted primarily of exposed and control groups; in the former were survivors exposed proximally, or within 2,000 m from the hypocenter at the time of bombings, with an estimated T65D of 1 rad or more, while the latter were distally exposed survivors at a distance of more than 2,500 m from the hypocenter, with an estimated dose of less than 1 rad.

Exclusions from this study were (1) those who had received radiotherapy or radioisotope exposure at any time prior to the cytogenetic examination, (2) those whose doses were estimated to have exceeded 1,000 rad by the T65D system, and (3) those whose scorable metaphases were found to be less than 30.

For comparison of the T65D, ORNL and LLNL dosimetry systems, the samples were further limited to survivors who were exposed either in the open or in Japanese houses. Marcum's shielding factors were employed as dose attenuation factors for survivors exposed in Japanese houses (Marcum, cited from Kerr, 1981).

Table 1 shows the number of cases finally eligible for this analysis, 408 in the two cities, 229 in Hiroshima and 179 in Nagasaki, together with breakdown by city, and 100-rad-dose intervals for each of the T65D, ORNL and LLNL systems. It is evident, although the number of cases is fairly reasonably distributed in each dose category by the T65D system, there is a large deficit in the number of survivors in the high dose ranges for both the ORNL and LLNL systems, particularly those over 400 rad in Hiroshima and 300 rad in Nagasaki. Further data collection is obviously needed in the future on survivors with high radiation dose for detailed statistical analysis.

Blood samples from eligible cases were collected between 1970 and 1971, when all chromosome slides from both cities were examined by the same observers in Hiroshima without knowledge of exposure status. Thus, any inter-city difference due to errors in identification and scoring of induced chromosome aberrations would be minimized.

Cytogenetic Methods

Blood samples drawn from each donor were cultured for 52 hours, and chromosome slides prepared according to the flame-dry method using conventional Giemsa stain, the details of which have been described elsewhere (Awa et al. 1978). Cultures were judged successful when a minimum of 30 metaphases were analyzable cytogenetically, but no more than 100 metaphases were examined for each case. No banded preparations were available for the present study.

Table 1. Frequency of cells with chromosome aberrations by dose for each A-bomb dosimetry system, Hiroshima and Nagasaki

Dose group (rad)	T65D			ORNL			LLNL		
	No. of cases	Aberrant cells (%)	Mean total dose	No. of cases	Aberrant cells (%)	Mean total dose	No. of cases	Aberrant cells (%)	Mean total dose
Hiroshima									
0	82	1.6	0.0	82	1.6	0.0	82	1.6	0.0
1– 99	17	3.3	35.9	36	5.1	60.8	23	4.1	53.0
100–199	44	7.2	153.9	69	11.5	144.4	62	9.3	152.3
200–299	35	13.6	255.1	27	19.8	231.0	39	16.4	249.9
300–399	21	14.6	347.7	10	27.0	329.7	13	25.0	346.4
400–499	19	23.3	434.7	4	11.0	449.8	5	28.8	439.6
500+	11	22.8	744.7	1	28.0	513.0	5	14.4	578.8
Nagasaki									
0	77	1.6	0.0	77	1.6	0.0	76	1.6	0.0
1– 99	9	1.6	47.2	36	2.8	55.1	37	2.7	57.5
100–199	27	3.5	144.9	39	9.5	147.2	40	9.7	152.6
200–299	24	6.5	264.7	21	11.5	227.7	20	11.2	231.5
300–399	17	8.9	351.3	5	13.0	342.8	4	14.3	327.3
400–499	17	14.4	436.5	0	–	–	1	8.0	404.0
500+	8	18.1	586.5	1	28.0	516.0	1	28.0	502.0

Results and Discussion

As reported previously (Awa et al. 1978), exchange-type chromosome aberrations demonstrable in the cultured lymphocytes of A-bomb survivors consisted almost exclusively of the symmetric type, commonly referred to as "stable" type aberrations, such as reciprocal translocations, inversions, insertions, and so on, all of which are identifiable as abnormal monocentrics. In striking contrast to the predominance of stable type aberrations, asymmetric exchanges, or "unstable" type aberrations such as dicentrics and rings with accompanying acentrics, were far less frequent. The ratio of unstable to stable aberrations was 1:10–20.

A preliminary test of our culture system showed that, when the culture was terminated after 52 h of incubation, approximately two-thirds of the observed metaphases were found to be in their first in vitro cell division, while the remaining cells were already in the second cell cycle. The culture conditions used would thus yield a decreased frequency of metaphases that carry unstable type aberrations, since it is known that the cells with such aberrations are likely to be eliminated during mitosis due to structural disadvantages. However, the majority of aberrant cells scored in the present study were of the stable type, occurring with a frequency which appeared to remain constant regardless of the number of in *vitro* cell divisions. Thus, the culture conditions used in this study did not seem to affect the frequencies of aberrant cells.

The number of aberrations (or exchange events) per aberrant cell increased proportionally with increasing radiation dose to the survivors. The dose-dependent increase in the frequency of cells with complex chromosome aberrations among survivors is also apparent in the high dose range. This phenomenon has been confirmed by an analysis of G-banded preparations (Sofuni et al. 1978, Ohtaki et al. 1981).

There remain a number of unresolved problems in the use of the stable type aberration as a biological dosimeter for irradiated persons, since the type and frequency of these aberrations cannot be determined with precision in non-banded preparations. In addition, there are no standardized criteria for the identification of stable type aberrations. It thus seems difficult at present to make any inter-laboratory comparisons for the yield of stable type aberrations at least when non-banded chromosome preparations are used. These technical difficulties can be eliminated by the use of various banding methods, but a great deal of time and experience are required to obtain banded preparations of high quality for detailed karyotype analysis.

Because of these underlying difficulties, we thus examined the metaphases in conventionally stained preparations to determine only whether or not structurally rearranged chromosomes were present. The dose-response analysis in the present study was consequently based on the frequency of aberrant cells among the total metaphases examined, and not on the frequency of aberrations per se.

Table 1 shows the frequencies of aberrant cells by city, and by dose groups for T65D, ORNL and LLNL systems. Inter-city comparisons by each dose estimate, aberration frequencies are plotted graphically, shown in Fig. 1. Dashed lines indicate that the number of samples is too scanty to represent the corresponding dose categories.

As shown in these data, for T65D dosimetry an inter-city difference in the frequency of aberrant cells is apparent: those for Hiroshima are consistently higher than for Nagasaki in every dose range. This again confirms our previous findings, reported elsewhere (Awa et al. 1978).

When the frequencies are plotted against the recalculated dose estimates of the ORNL and LLNL systems, the dose-response curves for the two cities appear to approach one another more closely than for the T65D system, particularly so in the dose range below 200 rad for ORNL and LLNL (Fig. 1). Although the number of cases among high dose ranges for ORNL and LLNL systems is not sufficient to warrant strict statistical analysis, aberration frequencies are generally higher in Hiroshima than in Nagasaki in the high dose ranges even by the ORNL and LLNL systems. The reason for this inter-city difference remains unclear.

This preliminary analysis seems to indicate that on the basis of the revised dose estimates by ORNL and LLNL, the dose-response curves for Hiroshima and Nagasaki resemble each other more closely than do those derived from the T65 dosimetry.

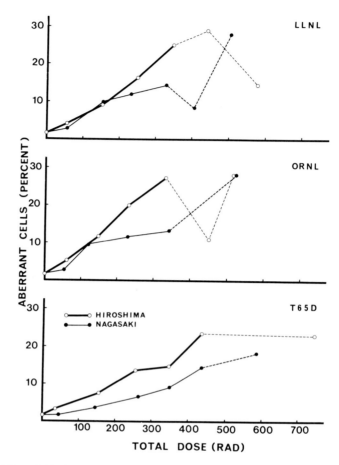

Fig. 1. Graphs, composited from Table 1, showing the frequency of aberrant cells in percent plotted against the total radiation dose in rad, for each of LLNL (*top*), ORNL (*middle*), and T65D (*bottom*) systems

Remarks

In our previous study, a difference in aberration frequencies was observed between Hiroshima and Nagasaki, and the shape of the dose-response curve appeared to be linear for Hiroshima and exponential for Nagasaki. One possible interpretation for this inter-city difference was thought to be the different radiation spectra between the two cities based on T65 dosimetry. The increased level of the neutron component in Hiroshima was thus considered in part responsible for the elevated aberration frequency at every dose range.

Recent re-evaluation of the T65 dosimetry system together with shielding factors for individual survivors suggested that the previously reported dose-response curve for Nagasaki may not be exponential, but linear. This supposition is reinforced if the cytogenetic data derived from a subgroup of survivors who were irradiated in a factory at Nagasaki are excluded from consideration, because of large ambiguities in the shielding factors.

Given the uncertainties (shielding and attenuation among others) inherent in the current (and earlier) dose estimates for the Hiroshima and Nagasaki survivors, any conclusion concerning the "correctness" of any one in favor of the others is premature. Further collection of cytogenetic data, now in progress, constitutes a basic part of the biological dose reappraisal program currently underway at RERF.

Acknowledgments. We are grateful to the cytogenetic laboratory staff in Hiroshima and Nagasaki RERF, without whose technical assistance the present study would have been impossible.

References

Auxier JA, Cheka JS, Haywood FF, Jones TD, Thorngate JH (1966) Free-field radiation-dose distributions from the Hiroshima and Nagasaki bombings. Health Phys 12:425–429

Awa AA, Sofuni T, Honda T, Itoh M, Neriishi S, Otake M (1978) Relationship between the radiation dose and chromosome aberrations in atomic bomb survivors of Hiroshima and Nagasaki. J Radiat Res 19:126–140

Kerr GD (1981) Review of dosimetry for the atomic bomb survivors. Paper presented at Fourth Symposium on Neutron Dosimetry, Gesellschaft für Strahlen- und Umweltforschung, Munich-Neuherberg, June 1–5, 1981

Loewe WE, Mendelsohn E (1981) Revised dose estimates at Hiroshima and Nagasaki. Health Phys 41:663–666

Milton RC, Shohoji T (1968) Tentative 1965 radiation dose (T65D) estimation for atomic bomb survivors, Hiroshima-Nagasaki. ABCC Tech Rep 1–68

Ohtaki K, Shimba H, Sofuni T, Awa AA (1981) Comparison in the types and frequencies of chromosome aberrations by conventional and G-staining methods in Hiroshima A-bomb survivors. RERF Tech Rep 24–81

Sofuni T, Shimba H, Ohtaki K, Awa AA (1978) A cytogenetic study of Hiroshima atomic bomb survivors. In: Evans HJ, Lloyd DC (eds.) "Mutagen-induced Chromosome Damage in Man" Edinburgh Univ Press, 108–114

Reproductive Effects

Biological Dosimetry of Sperm Analyzed by Conventional Methods: Cautions and Opportunities*

Barton L. Gledhill

Lawrence Livermore National Laboratory, Biomedical Sciences Division, University of California, P.O. Box 5507 L-452, Livermore, California 94550, USA

1 Introduction

Increasing concern for heritable consequences of exposure to mutagens, carcinogens, and teratogens has encouraged development of semen bioassays. Today, these bioassays are based largely on conventional andrological methods. But analytical cytology offers many new approaches with potential to increase the speed, accuracy, sensitivity, and resolution of measuring biologically relevant reproductive effects of exposure to noxious agents. The objective of this article is to review the dynamic processes leading to the formation of mammalian sperm, to illustrate how some reproductive disorders affect semen quality, to summarize some of the more pertinent dosimetric applications of conventional (not automated cytologic) semen assays and to point out significant differences in species characteristics that can influence hazard evaluation based on changes in semen quality.

2 Spermatogenesis

2.1 General Considerations

The testis influences through its exocrine and endocrine activities, practically every organ and function in the male body. Insofar as production of semen is concerned, the exocrine activity of the testis produces sperm (spermatogenesis) whereas the endocrine activity, as well as influencing spermatogenesis, stimulates secretory function of the male accessory organs thereby determining the output of seminal plasma. Moreover, testicular influence on overall "maleness" is governed by an intricate control and feedback system, centered in the brain and regulated by the hypophysis and the hypothalamus.

Intricacies of gametogenesis in the testis and quantitative aspects of the dynamic alterations in cell population during active spermatiogenesis are available for a number of species (Amann 1981; Mann and Lutwak-Mann 1981; Steinberger and Steinberger 1975).

---

*Work performed under the auspices of the U.S. Department of Energy by the Lawrence Livermore National Laboratory under contract number W-7405-ENG-48

2.2 Seminiferous Tubules

Mammalian spermatogenesis is a complex sequence of cellular proliferations and morphologic transformations ending with the production of spermatozoa (sperm). It takes place in the epithelial lining of the seminiferous tubules, which are long folded loops approximately 0.1 to 0.4 mm in diameter opening at both ends into the rête testis. Two main types of cells are found in the epithelium: the fixed population of Sertoli cells extending from the basal membrane to the lumen of the tubule and the proliferating population of germ cells, whose anatomic relation to the Sertoli cells continuously changes during differentiation and migration toward the lumen.

Mitotic division of the undifferentiated spermatogonia either gives rise to more spermatogonia, thereby ensuring the continuous renewal of stem cells, or to committed primary spermatocytes. Spermatogonia destined to become spermatocytes divide mitotically several times, progressing through successive generations of type A spermatogonia, intermediate-type spermatogonia and type-B spermatogonia. One more mitotic division yields two primary spermatocytes. The final round of DNA synthesis in preparation for meiosis occurs during the preleptotene stage. Meiosis follows as two successive divisions of the primary spermatocyte with one duplication of chromosomes. The product of meiosis is four spermatids, each containing half the somatic number of chromosomes. In most mammals, as a result of segregration of the sex chromosomes, half of the spermatids contain the female-determining X chromosome, the others the male Y chromosome.

Thereafter, during spermiogenesis the four spermatids differentiate into sperm. It is during this metamorphosis that the acrosome forms and overlaps the anterior portion of the nucleus. Early in spermiogenesis the two centrioles begin to move; the proximal centriole takes up a position at the posterior pole of the nucleus and the distal centriole sprouts a flagellum. Soon the nucleus takes on the species-specific size and shape; the nuclear chromatin begins to condense into a dense, granular mass. The bulk of cytoplasm is moved distad to surround the proximal part of the flagellum and the mitochondria move from the periphery of the cytoplasm toward the flagellum to begin formation of the mid-piece mitochondrial sheath. During the final phase of spermiogenesis the flagellum completes its 9+9+2 fiber-filament pattern and the mitochondria arrange themselves in a spiral around the mid-piece. When sperm are released into the lumen of the seminiferous tubule (spermiation) most of their cytoplasm separates in the form of a residual body and is absorbed by Sertoli cells but a small portion, the cytoplasmic droplet, remains attached to each flagellum.

2.3 Kinetics of the Spermatogenic Process

Spermatogenesis has several unusual kinetic features. First, the time required for spermatogenesis is constant for any given species but varies between species. In man it takes 64 days (Heller and Clermont 1963), 34 days in the mouse (Oakberg 1957), 48 days in the Sherman strain of rat (Clermont et al. 1959) and the Long-Evans strain of rat (Steinberger 1962), 35 days in the hamster (Clermont and Trott 1969), 43 days in the rabbit (Swierstra and Foote 1963) and 49 days in the bull (Ortavant 1956). Each step of spermatogenesis also is of precisely fixed duration. As an example, in the mouse 8 days are used for the many mitotic divisions of spermatogonia, 13 days for meiosis and 13.5 days for spermiogenesis (Monesi 1974).

Another characteristic of spermatogenesis is the synchronous evolution of germ cell generations. In any cross section of a tubule containing spermatids the remaining less differentiated cells of the seminiferous epithelium form well-defined associations of specific germinal-cell types. When combined with an orderly, cyclical succession of the various cell types, a "cycle of the seminiferous epithelium" is defined as the complete series of cellular changes in a particular cross-sectional area of seminiferous tubule between two appearances of the same developmental stages (Leblond and Clermont 1952). The formation of specific cellular associations and the sequence of their appearance in a given area of the seminiferous tubule are highly synchronized. Moreover, the numerical relationships between the various cell types within a cellular association and their absolute numbers within a cross section of tubule are highly consistent (Clermont 1962), and have led to exceedingly precise techniques for quantitative evaluation of cellular changes in the tubule after damage to the germinal epithelium (Amann 1981; Oakberg 1956; Steinberger 1962).

2.4 Duration of Spermatogenesis

The cellular associations characteristically seen in the seminiferous tubule correspond to successive stages of development and usually occur in an orderly fashion along the length of the tubule. The exact number of associations differs among species and ranges from 6 to 14, depending on the criteria used for their description. A complete series, i.e., from one association to the next identical one, is referred to as a "wave of the seminiferous epithelium". In most mammals, each association occupies considerable area along the basement membrane of the tubule and the adjacent cohorts of germ cells develop synchronously. In humans, however, many germ cells die and each death breaks the chain of differentiating germ cells interconnected by intercellular bridges (Amann 1981). Presumably as a result, each association takes up only a small area of the tubule epithelium. A cross section through a human seminiferous tubule likely will contain several associations; usually only one is seen in most other mammalian species.

The interval required, if a fixed point within the tubule were to be observed over time, for one complete series of cellular associations to appear at that point is termed the "duration of the cycle of the seminiferous epithelium". This duration is constant for a species, cannot be altered by external means, and varies berween species. It requires about 15 days in man, 13 in bull and rat, 12 in stallion and 10 in ram and rabbit. From evidence for several species, spermatogenesis typically requires between 4.3 and 4.7 cycles of the seminiferous epithelium. Since cohorts of differentiating germ cells within a testis complete spermatogenesis asynchronously sperm are released continuously from the mammalian germinal epithelium.

3 Abnormal Spermatogenesis

3.1 Degenerative Changes

Defects arise in spermatogenesis from a multitude of causes. Important causes include
irradiation, trauma, and chemical insult to the testis. Radiation is a well recognized
health hazard to man and animals. The testis is particularly radiosensitive. Its exposure
to a relatively small dose of x-rays results in extensive damage to the seminiferous epithelium and breakdown of spermatogenesis, owing principally to distruction of type A
spermatogonia (Huckins 1978; Oakberg 1975). Minor mechanical trauma to the testis
or blockage of its excurrent duct system can be detrimental to spermatogenesis. In man
and farm animals, epididymal lesions and varicocele often coincide with degenerative
changes in the seminiferous epithelium, oligospermia (reduced sperm numbers) and
subfertility (Amelar 1966; MacLeod 1971; Schirren 1971). Elevating the interior temperature of the testis to body temperature for not more than 15 min or ligating the cor-
pus epididymis will disrupt spermatogenesis; many weeks are needed for restoration of
normal function of the germinal epithelium. When considering the adverse effects of
radiation, heat or other types of injury to the germ cells, the possibility must be considered that the damage may have been initiated by changes elewhere in the gonad, or
for that matter, in the body. Changes in blood flow, hypoxia, alterations in enzymes,
coenzymes and general tissue metabolism, disturbances in pituitary-hormone releasing
mechanisms, testicular steroidogenesis, and Sertoli cell function as well as other even
unrecognized factors must be taken into account when confronted with a spermiogram
indicative of abnormal spermatogenesis.

3.2 Aneuploidy

Chromosomal abnormalities can adversely influence spermatogenesis. Meiosis generally
suffers most acutely although spermatogenesis can fail at a variety of points owing to
chromosomal irregularities. Sex chromosome aberrations are numerically the most frequent type of chromosomal irregularity associated with subfertility or sterility in man;
in particular the presence of an extra X chromosome causes spermatogenic arrest (Mann
and Lutwak-Mann 1981). Klinefelter's syndrome (Klinefelter et al. 1942) is characterized by two or more X chromosomes in the karyogram of a phenotypically male individual with varying degrees of eunuchoidal or female habitus. The most frequent karyotype is of the 47/XXY but other types, such as 48/XXXY, 48/XXYY, and 49/XXXXY
have been seen. Small, non-functional testis with sclerosing tubular degeneration, under-
developed secondary sex characteristics, poor muscular development and frequently
gynaecomastia are seen as part of Klinefelter's syndrome.
   Turner's syndrome, associated with a defect or absence of the second sex chromosome (45/X0, 45/Y0) was formerly thought to be limited to chromosomal females but
is now known to affect chromosomal males (Herting and Mansell 1957). The internal
and external genitalia are of infantile female type except that the gonad is composed
of a minute fibrous mass or ridge lacking either female or male elements. Such severe

lesions emphasize the hormonal control of embryonic genital ducts by their corresponding gonads. In the absence of an embryonic gonad, the Müllerian system (female) pre-
dominates.

It is clear that all misshaped sperm are not chromosomally defective, (Wyrobek et al.
1975) but it is unsettled if chromosomally defective sperm must be malformed. Some
reports have noted a correlation between high proportions of sperm with abnormal
shapes and habitual abortions (Jöel 1966) or spontaneous abortions (Furuhjelm et al.
1962) but the correlations are not seen in all studies (Homonnai et al. 1980). Until quite
recently virtually all information on the chromosome constitution of both male and
female germ cells was gotten by inference from a consideration of the chromosome constitution of the conceptus. Although very little direct information about the chromosome constitution of sperm exists, a hamster egg-human sperm fusion technique (Rudak
et al. 1978) now permits the direct observation of human sperm chromosomes (Brandriff et al. 1982, 1983; Martin et al. 1982).

Should a chromosomally abnormal sperm fertilize an ovum, it would seem to have a
higher likelihood of being an impediment to normal embryogenesis than if fertilization
were with a normal sperm. At least 15% of recognized human pregnancies end as spontaneous abortions, a high proportion of which are associated with detectable chromosomal aberrations (Carr 1967). In about 7% of all human pregnancies a chromosomal
abnormality is present (Jacobs 1972): 1.2% with an abnormal number of sex chromosomes; 3.7% with autosomal aneuploidy; 0.3% with a structural rearrangement of the
autosomes, and 1.5% with either triploidy or tetraploidy. The majority of those with
sex-chromosome abnormalities had the 45/XO karyotype of Turner's syndrome and in
this particular situation the maximum contribution from sperm with abnormal chromosome constitutions was calculated to be 75% of the total.

3.3 Immune Orchitis

Immunological agents are also capable of seriously interfering with spermatogenesis.
Auto-or isoimmune orchitis is a good example of such interference. It arises when a
male is immunized with testicular homogenates prepared from tissue of his own or
another member of his species. Damage is limited to the seminiferous epithelium and
destroys spermatocytes and spermatids; testicular sperm are phagocytized by intratub-
ular macrophages. Three potent antigens (a protein and two glycoproteins) capable
of inducing allergic aspermatogenic orchitis have been isolated from guinea pig testes
(Hagopian et al. 1976).

4 Conventional Assay of Semen Quality

4.1 Semen Evaluation

Semen is most often collected from mammals by masturbation, coitus interruptus, or
by employing an artificial vagina, electroejaculation or various massage techniques.

Sperm are collected from the vas deferens and cauda epididymidis of mice and other laboratory rodents after they are killed. The method employed depends upon the clinical or experimental circumstances and, above all else, the species and the individual being investigated. Characteristics of more than one ejaculate must be evaluated to obtain a reasonable understanding of semen quality.

Experience from medical and veterinary practice indicates that fertilization is generally conditional upon semen attaining certain widely accepted qualitative and quantitative standards. Each component of the ejaculate originates in a different part of the male reproductive tract. Dysfunction of any one will have profound effects on the quality of the ejaculate. The general appearance of any semen sample is usually the first criterion examined. Freshly ejaculated semen normally is a creamy fluid and white, yellowish or grayish in color. Semen from men, monkeys, stallions, boars, and rodents undergoes coagulation or gelification and sperm are entraped for varying lengths of time. Both semen volume and sperm concentration vary markedly between and within species. For example, daily sperm production per gram of testicle in the human is lower by a factor of 2.5 to 6 than that of other mammals (Amann 1981) and the average volume of an ejaculate in the ram is less than 1 ml while the boar will ejaculate more than 500 ml of semen. Frequency of ejaculation, length of abstinence period, method of collection, season, temperature, light, nutrition and other factors, some hormonally conditioned and partly dependent on diurnal rhythms potentially affect semen volume and sperm diversity. The relative proportion of motile sperm is routinely evaluated, and sometimes the intensity of motility is examined. The percentage and type of abnormal forms are prime criteria used in semen evaluation. Less commonly determined are degree of seminal contamination by microorganisims; presence of non-germinal cells; deviation of chemical composition; and sperm penetrability of cervical mucus (Mann and Lutwak-Man 1981). None of these, used singly or in combination, offers total certainty in predicting the fertilizing potential of a semen sample. Likewise, none offer a wholly satisfactory basis for measuring normalcy of testicular function. At best, the laboratory tests provide guidelines or limits outside which a semen sample no longer rates as being of good quality. Because of the inherent variability in semen, multiple evaluations over a period of days or weeks must be done to maximize predictive/diagnostic powers of the tests. Two broad dosimetric applications are made of sperm assays: testing agents for their ability to induce spermatogenic damage and monitoring the germinal effects of occupational, accidental or therapeutic exposure of men to noxious agents.

4.2 Motility

Sperm motility is often claimed to be one of the best correlates with fertility, and therefore sperm functionality. However, its measurement requires a fresh semen sample, careful control of temperature and standardization of time between collection and assay. Sperm motility, in practice, is difficult to determine and is frequently unreliable because of varying external influences. Large scale measurements are especially difficult.

Most commonly the proportion of motile sperm is determined by direct visual microscopic estimation. However, there have been many attempts to speed up the visual measurement of sperm motility and make it objective. Spectrophotometry (Timourian and Watchmaker 1970), time-lapse photometry and cinematographic recording (Amann and Hammerstedt 1980; Overstreet et al. 1979; Rikmenspoel 1978), scattered laser light (Finsy et al. 1979), scanning by television camera (Suarez and Katz 1982), scanning with an image analyzing computer-linked microscope (Katz and Dott 1975), have been reported to reduce subjectivity in evaluation of sperm motility. Visual assessment, however, continues to be the dominant method used to judge sperm movement and scores are assigned to samples on a purely arbitrary scale developed through practice and experience. We can expect more interesting research in this field, and the contribution by Evenson et al. (this volume) is one intriguing example.

4.3 Sperm Count (density)

Today, hemocytometers, absorptiometers in various configurations and Coulter Counters are used most commonly to determine sperm count, reported as the number of sperm per milliliter of ejaculate or total number of sperm ejaculated. Flow cytometers may find important application on account of high precision, sensitivity and speed of measurement, but care must be exercised to exclude from counts extraneous particulate matter always found in semen. Sperm count has been the single most often employed test to assess effects of physical and chemical agents on human spermatogenesis (Wyrobek et al. 1982). There is a disturbing lack of agreement as to what constitutes optimal sperm concentration in good-quality human semen. In all species, sperm counts, like sperm motility, will fluctuate widely because of confounding external factors such as variable continence time prior to sample collection, frequency of sexual contact, and collection of incomplete ejaculates (Amann 1981). Dispite the huge underlying variability and the associated statistical problems in detecting small changes, there are many clear examples of agent-induced reduction in sperm counts (Wyrobek et al. 1983 a,b).

4.4 Sperm Morphology

Microscopic evaluation of sperm for normalcy of shape (seminal cytology) has a long history in the diagnosis of infertility and the prognosis of fertility. Of the three characteristics most commonly used for evaluation of semen quality, i.e., motility, density and morphology, morphology is the most constant in unexposed individuals and is statistically very sensitive to small changes (Wyrobek et al. 1982). Apart from normally shaped sperm, semen from all animals contains a species-characteristic proportion of grossly abnormal and degenerate sperm representing every deviation such as tapering, oval, slab-sided, giant, double, diploid, amorphous and immature. Sperm-head shape is most often emphasized, but some assessments tabulate midpiece and tail abnormalities as well. A high frequency of structural abnormality generally indicates testicular or epididymal dysfunction. In fertile men abnormal sperm can account for 40% or more of the cells in the ejaculate, whereas in bulls and stallions it seldom exceeds 20%. In rams, boars, and dogs as a rule it is much lower and in several strains of hybrid mice it is as low as 2%.

Visual determination of sperm morphology as usually practiced is highly subjective, critically dependent upon the classification scheme used, and subject to much inter-laboratory and interscorer variability. In general there has been little agreement in the definition of normal shapes (Eliasson 1973; Freund 1966). Approaches to objec-tive, quantitative, assessment of morphology and definition of "normal shape" have been sporadic and preliminary in nature (Benaron et al. 1982; David et al. 1975; Eliasson 1971; Moore et al. 1982; Young et al. 1982) but this seems a ripe area for exciting cytometric applications (Pinkel, this volume).

For the determination of the type and percentage of abnormal forms in smears from mammalian semen, the commonly employed staining techniques are Papanicolaou's method, Harris's hematoxylin, periodic acid-Schiff, aqueous Eosin Y, modifications of these and certain other staining procedures. Sperm are examined in a microscope and minimum of 500 are assessed for morphological abnormalities (Wyrobek 1981).

4.4.1 Radiation

Radiation causes well-behaved dose dependent increases in proportion of abnormal sperm. Acute x-irradiation of the mouse testis results in an increase in proportion of malformed sperm heads with maximal expression at approximately 5 weeks after ir-radiation. The response increases as an exponential power function ($D^{1.5}$) of dose and has a doubling dose of between 25 and 70 rads depending method of assessment (Bruce et al. 1974). Stem cell irradiation at 300 rads can induce an apparently permanent ele-vation in sperm shape abnormalities in mice (Wyrobek, personal communication 1983).

4.4.2 Chemical Effects

The test results of 154 chemicals evaluated with the mouse sperm morphology assay have been reviewed and collated for the EPA's Gene-Tox program (Wyrobek et al. 1983a). The effects of radiation were specifically excluded. The Gene-Tox reviewing Committee judged 41 chemical agents to be positive inducers of sperm-head shape ab-normalities, 103 were negative and 10 were inconclusive. The relationship between induction of shape abnormalities and germ-cell mutagenicity also was evaluated by comparing the effects of 41 agents on mouse sperm shape to available data from the specific locus, the heritable translocation and the dominant lethal tests, all established mammalian germ-cell mutational tests. The mouse sperm morphology assay correctly identified known mutagens 100% of the time. Data were insufficient to determine the rate of false positives. The panel found further that a positive response in the mouse sperm test had a 100% concordance with carcinogenic potential but only 50% of the carcinogens were positive in the test (i.e., sensitivity was approximately 50%). Because so many carcinogens did not produce abnormally shaped sperm, even with administra-tion of lethal doses, negative findings with the sperm assay cannot be used to classify agents as noncarcinogens. The panel of reviewers concluded (Wyrobek et al. 1983b) that the mouse sperm morphology test has potential for use in identifying chemicals that induce spermatogenic dysfunction. They thought it might detect heritable mutations, and recommended an increased emphasis on the understanding of the inheritance of abnormal sperm shapes and a better description of the genetic loci involved in sperm-shape control.

Transmission of chemically induced sperm defects to the F_1 progeny of treated mice has been demonstrated in a limited number of studies using the F_1 sperm morphology test. This test involves exposing sexually mature mice of either sex to the test compound, mating treated animals to untreated partners and scoring for abnormal sperm morphology in mature male progeny (Wyrobek et al. 1983a). This test may provide an *in vivo* measure of heritable genetic damage in a mammal but so far only 7 compounds have been evaluated.

The same group of Gene-Tox reviewers (Wyrobek et al. 1983b) also evaluated the utility of sperm tests as indicators of chemical effects on human spermatogenesis. Again radiation effects were excluded by charter of the review. No useful tests have been developed for radiation-induced damage to the germ-cells of men exposed to ionizing radiation (Wyrobek, personal communication 1983). The tests surveyed included sperm count, motility, morphology and double Y-body (a fluorescence-based test thought to detect Y-chromosomal nondisjunction). When carefully controlled, the morphology test is statistically the most sensitive of these human sperm tests. Data sufficient for evaluation came from 132 papers encompassing 89 different chemical exposures. Of the 89 exposures reviewed, 52 adversely affected spermatogenesis (one or more sperm test was positive), 11 were suggestive of improving semen quality, 14 gave inconclusive evidence of adverse effects and 12 showed no significant changes. The review committee emphasized the need for a better understanding of underlying mechanisms and remained uncommitted on the extent to which sperm tests in the human can detect mutagens, carcinogens, or agents that affect fertility. They also emphasized the importance of rigorous study designs in hazard evaluations using sperm tests.

5 Newly Developed Assays of Sperm Chromosome Constitution

Until quite recently the chromosome constitution of sperm had to be infered from indirect studies of somatic tissues, meiotic spreads, spermatogonia, or embryonic tissue. Two new tests designed to reveal the chromosome complement of sperm now are available.

5.1 Double Y-Body Test

Unlike other sperm tests (motility, count and morphology), the double Y-body test can not be used in laboratory or domestic animals. An exceptionally bright fluorescence from the long arm of the human Y chromosome is seen when somatic cells are stained with quinacrine mustard (Caspersson et al. 1971). Observation of a single bright spot in nuclei of less than 50% of human sperm lead to the interpretation that the bright spot is the Y chromosome. The double Y-body test is based on scoring the frequency of such spots in human sperm stained with a quinacrine dye. Sperm with two or more spots are thought to contain two Y chromosomes due to meiotic non-disjunction. There are major unresolved questions about this interpretation (Beatty 1977). The double Y-body test is quite new and has not been used in more than a few populations of exposed men (Kapp 1979; Wyrobek et al. 1981).

One strain of northern field vole, *Microtus oeconomus* has highly heterochromatic sex chromosomes that can be identified in haploid spermatids (Tates 1979) thus providing a test in laboratory animals for nondisjunction similar to the double Y-body test in humans. While several chemical agents have been studied in the vole, the animal is difficult to obtain and the method does not appear to be applicable to mature sperm.

5.2 Direct Chromosomal Analysis of Human Sperm

Haploid sets of human sperm chromosomes in normal healthy men have been analyzed (Brandriff et al. 1982, 1983; Martin et al. 1982) using the human-sperm/hamster-egg *in vitro* fertilization system (Rudak et al. 1978). In this system human sperm chromatin decondenses and progresses to the mitotic chromosome stage in the hamster egg cytoplasm. After quinacrine or other means of banding the haploid sets of human mitotic chromosomes are identified. In more than 900 complete karyotypes from four normal unexposed men examined by Brandriff et al. (1983), equal proportions of X-chromosome and Y-chromosome containing haploid sets of sperm chromosomes were observed. The four donors differed in frequency of aberrations but did not differ in incidence of aneuploidy or gaps. No spontaneously occurring dicentrics were seen. This method offers the exciting possibility of measuring the frequency of chromosomal abnormalities in human sperm of normal, occupationally and therapeutically exposed men.

6 Dosimetry of Femal Germ Cells

While the thrust of this review centers on dosimetry of sperm, the female germ cells must not be ignored. Mouse and monkey primary oocytes are among the most radiosensitive mammalian cells known. Primordial oocytes in juvenile mice have an LD_{50} of only 6–7 rad, and the germ cell pool in squirrel monkeys is destroyed by prenatal exposure of 0.7 rad/day (Dobson and Felton 1983). Ovarian germ cells can be killedl also by chemicals and the killing has marked similarities to radiation effects. Quantitative dose-effect relations are usually determined by tedious microscopic enumeration of germ cells in histologically processed sections of the ovary or by autoradiographic analysis of precursor uptake. Dosimetry of the female germ cell is a ripe area for application of analytical cytology.

7 Concluding Remarks

Male, as well as female, germ cells respond dramatically to a variety of insults and are important reproductive dosimeters. Semen analyses will be very useful in studies on the effects of drugs, chemicals, and environmental hazards on testicular function, male fertility and heritable germinal mutations. The accessability and separability of male cells makes them well suited for analytical cytology. Challenges are abundant. We might automate the process of determining sperm morphology but should not do so soley for increased speed. Rather, richer tangible benefits will derive from cytometric evaluation through increased sensitivity, reduced subjectivity, standardization between investigators and laboratories, enhanced archival systems and the benefits of easily exchanged standardized data. Inroads on the standardization of assays for motility and functional integrity are being made. Altogether, the task is huge but the goal shall be reached.

Disclaimer

This document was prepared as an account of work sponsored by an agency of the United States Government. Neither the United States Government nor the University of California nor any of their employees, makes any warranty, express or implied, or assumes any legal liability or responsibility for the accuracy, completeness, or usefulness of any information, apparatus, product, or process disclosed, or represents that its use would not infringe privately owned rights. Reference herein to any specific commercial products, process, or service by trade name, trademark, manufacturer, or otherwise, does not necessarily constitute or imply its endorsement, recommendation, or favoring by the United States Government or the University of California. The views and opinions of authors expressed herein do not necessarily state or reflect those of the United States Government thereof, and shall not be used for advertising or product endorsement purposes.

References

Amann RP (1981) A critical review of methods for evaluation of spermatogenesis from seminal characteristics. J Androl 2:37–58

Amann RP, Hammerstedt RH (1980) Validation of a system for computerized measurement of spermatozoal velocity and percentage of motile sperm. Biol Reprod 23:647–656

Amelar RD (1966) The semen analysis. In: Infertility in men: Diagnosis and treatment, Davis, Philadelphia, p 30–53

Beatty RA (1977) F-bodies as Y chromosome markers in mature human sperm heads: a quantitative approach. Cytogenet Cell Genet 18:33–49

Benaron DA, Gray JW, Gledhill BL, Lake S, Wyrobek AJ, Young IT (1982) Quantification of mammalian sperm morphology by slit-scan flow cytometry. Cytometry 2:344–349

Brandriff B, Gordon L, Carrano AV, Ashworth L, Watchmaker G, Wyrobek AJ (1983) Direct analysis of human sperm chromosomes: comparison among four individuals. Hum Genet (in press)

Brandriff B, Gordon L, Watchmaker G, Ashworth L, Carrano A, Wyrobek AJ (1982) Analysis of sperm chromosomes in normal healthy men using the human-sperm/hamster-egg in vitro fertilization system. Abstracts Amer Soc Human Genetics 33rd Ann Mtg, Detroit, Michigan, Sept 29–Oct 2, 1982

Bruce WR, Furrer R, Wyrobek AJ (1974) Abnormalities in the shape of murine sperm after acute testicular x-irradiation. Mutat Res 23:381–386

Carr DH (1967) Chromosome anomalies as a cause of spontaneous abortion. Am J Obstet Gynecol 97:283–293

Caspersson T, Lomakka G, Zech L (1971) The 24 fluorescence patterns of the human metaphase chromosomes – distinguishing characters and variability. Hereditas 67:89–102

Clermont Y (1962) Quantitative analysis of spermatogenesis in the rat: a revised model for the renewal of spermatogonia. Am J Anat 111:111–119

Clermont Y, Trott M (1969) Duration of the cycle of the seminiferous epithelium in the mouse and hamster determined by means of ^{3}H-thymidine and radioautography. Fertil Steril 20:805–81

Clermont Y, Leblond CP, Messier B (1959) Durée du cycle de l'épithélium séminal du rat. Arch Anat Microscop Morphol Exptl (Suppl) 48:37–56

David G, Bisson JP, Czyglick F, Jouannet P, Gernigon C (1975) Abnormalités morphologiques due spermatozoïde humain 1) Propositions pour un systéme de classification. J Gyn Obst Biol Repr 4 (suppl 1):17–36

Dobson RL, Felton JS (1983) Female germ cell loss from radiation and chemical exposures. Am J Indust Med 4:175–190

Eliasson R (1971) Standards for investigation of human semen. Andrologie 3:49–64

Eliasson R (1973) Parameters of male fertility. In: Hafez ESE, Evans TN (eds) Human reproduction conception and contraception, Harper and Row, Hagerstown, New York, Evanston, San Francisco, London, p 39–51

Evenson DP, Higgins PJ, Melamed MR (1984) Detection of male reproductive abnormalities by flow cytometry measurements of testicular and ejaculated germ cells. In: Eisert WG, Mendelsohn ML (eds) Biological dosimetry: cytometric approaches to mammalian systems, Springer, Berlin, Heidelberg, New York, Tokyo, p 99–109

Finsy R, Peetermans J, Lekkerkerker H (1979) Motility evauation of human spermatozoa by photon correlation spectroscopy. Biophys J 27:187–192

Freund M (1966) Standards for the rating of human sperm morphology. Intl J Fertil 11:97–180

Furuhjelm M, Jonson B, Lagergren CG (1962) The quality of human semen in spontaneous abortion. Int J Fertil 7:17–21

Hagopian A, Limjuco GA, Jackson JJ, Carlo DJ, Eglar EH (1976) Experimental allergic aspermatogenic orchitis. IV. Chemical properties of sperm glycoproteins isolated from guinea pig testes. Biochim Biophys Acta 434:354–364

Heller CG, Clermont Y (1963) Spermatogenesis in man: An estimate of its duration. Science 140:184–186

Hertig AT, Mansell H (1957) The female genitalia. In: Anderson WAD (ed) Pathology. CV Mosby, St Louis, p 1038

Homonnai ZT, Paz GF, Weiss JN, David MP (1980) Relation between semen quality and fate of pregnancy: retrospective study on 534 pregnancies. Intl J Androl 3:574–584

Huckins C (1978) Behavior of stem cell spermatogonia in the adult rat irradiated testis. Biol Reprod 19:747–760

Jacobs PA (1972) Chromosome abnormalities and fertility in man. In: Beatty RA, Glueckschn-Waelsch S (eds) The genetics of the spermatozoon, Bogtrykeriet Forum, Copenhagen, p 346–358

Jöel CA (1966) New etiologic aspects of habitual abortion and infertility with special reference to the male factor. Fertil Steril 17:374–380

Kapp RW (1979) Detection of aneuploidy in human sperm. Environ Health Prespect 31:27–31

Katz DF, Dott HM (1975) Methods of measuring swimming speed of spermatozoa. J Reprod Fertil 45:263–272

Klinefelter HF, Reifenstein EC, Albright F (1942) Syndrome characterized by gynecomastia, aspermatogenesis without a-leydigism and increased excretion of follicle-stimulation hormone. J Clin Endocrinol 2:615–627

Leblond CP, Clermont Y (1952) Definition of the stages of the cycle of the seminiferous epithelium in the rat. Ann NY Acad Sci 55:548–573

MacLeod J (1971) Recent advances concerning the role of variocele in male infertility. In: Jöel CA (ed) Fertility disturbances in men and women, Karger, Basel, p 268–277

Mann T, Lutwak-Mann C (1981) Male reproductive function and semen. Springer, Berlin, Heidelberg, New York

Martin RH, Lin CC, Balkan W, Burns K (1982) Direct chromosomal analysis of human spermatozoa: preliminary results from 18 normal men. Am J Hum Genet 34:459–468

Monesi V (1974) Nucleoprotein synthesis in spermatogenesis. In: Mancini RE, Martini L (eds) Male fertility and sterility, Academic Press, London, New York, San Francisco, p 59–87

Moore II DH, Bennett DH, Kranzler D, Wyrobek AJ (1982) Quantitative methods of measuring the sensitivity of the mouse sperm morphology assay. Analyt Quant Cytol 4:199–206

Oakberg EF (1956) A description of spermiogenesis in the mouse and its use in analysis of the cycle of the seminiferous epithelium and germ cell renewal. Am J Anat 99:391–413

Oakberg EF (1957) Duration of spermatogenesis in the mouse and timing of stages of the seminiferous epithelium. Am J Anat 99:507–516

Oakberg EF (1975) Effects of radiation on the testis. In: Hamilton DW, Greep RO (eds) Handbook of physiology: Endocrinology, American Physiological Society, Washington, DC, p 233–243

Ortavant R (1956) Autoradiographie des cellules germinales du testicle de Béliver. Durée des phénomènes spermatogénétiques. Arch Anat Microscop Morph Exptl 45:1–10

Overstreet JW, Katz DF, Hanson FW, Fonseca JR (1979) A simple inexpensive method for objective assessment of human sperm movement characteristics. Fertil Steril 31:162–172

Pinkel D (1984) Cytometric analysis of mammalian sperm for induced morphologic and DNA content errors. In: Eisert WG, Mendelson ML (eds) Biological dosimetry: cytometric approaches to mammalian systems, Springer, Berlin, Heidelberg, New York, Tokyo, p 111–126

Rikmenspoel R (1978) Movement of sea urchin sperm flagella. J Cell Biol 76:310–322

Rudak E, Jacobs PA, Yanagimachi R (1978) Direct analysis of the chromosome constitution of human spermatozoa. Nature 274:911–913

Schirren C (1971) Praktische Andrologie. Hartmann, Berlin

Steinberger E (1962) A quantitative study of the effect of an alkylating agent (Triethylenemelamine) on the seminiferous epithelium of rats. J Reprod Fertil 3:250–259

Steinberger E, Steinberger A (1975) Spermatogenic function of th testis. In: Hamilton DW, Greep RO (eds) Handbook of physiology: Endocrinology, American Physiological Society, Washington, DC, p1–19

Suarez SS, Katz DF (1982) Movement characteristics and acrosomal status of populations of rabbit spermatozoa recovered at the site and time of fertilization. Biol Reprod (suppl 1) 26:146A

Swierstra EE, Foote RH (1963) Cytology and kinetics of spermatogenesis in the rabbit. J Reprod Fertil 5:309–322

Tates AD (1979) *Microtus oeconomus* (Rodentia), a useful mammal for studying the induction of sex-chromosome nondisjunction and diploid gametes in male germ cells. Environ Health Perspect 31:151–159

Timourian H, Watchmaker G (1970) Determination of spermatozoan motility. Develop Biol 21:62–72

Wyrobek AJ (1981) Sperm assays in man and other mammals as indicators of chemically induced testicular dysfunction. In: Waters MD, Sandhu SS, Lewtas-Huisingh J, Claxton L, Nesnow S (eds) Short-term bioassay in the analysis of complex environmental mixtures II. Plenum press, p 495–506

Wyrobek AJ, Gordon LA, Burkhart JG, Francis MW, Kapp RW, Letz G, Malling HV, Topham JC, Whorton D (1983a) An evaluation of the mouse sperm morphology test and other sperm tests in nonhuman animals: a report of the U.S. Environmental Protection Agency Gene-Tox Program. Mutat Res 115:1–72

Wyrobek AJ, Gordon LA, Burkhart JG, Francis MW, Kapp RW, Letz G, Malling HV, Topham JC, Whorton D (1983b) An evaluation of human sperm as indicators of chemically induced alterations of spermatogenic function: A report of the U.S. Environmental Protection Agency Gene-Tox Program. Mutat Res 115:73–148

Wyrobek AJ, Gordon LA, Watchmaker G, Moore II DH (1982) Human sperm morphology testing: Description of a reliable method and its statistical power. In: Banbury Report 13: Indicators of Genotoxic Exposure, Cold Spring Harbor Laboratory, p 527–541

Wyrobek AJ, Heddle JA, Bruce WR (1975) Chromosomal abnormalities and the morphology of mouse sperm heads. Can J Genet Cytol 17:675–681

Wyrobek AJ, Watchmaker G, Gordon L, Wong K, Moore II DH, Whorton D (1981) Sperm shape abnormalities in carbaryl-exposed employees. Environ Health Perspect 40:255–265

Young IT, Gledhill BL, Lake S, Wyrobek AJ (1982) Quantitative analysis of radiation induced changes in sperm morphology. Analyt Quant Cytol 4:207–216

Detection of Male Reproductive Abnormalities by Flow Cytometry Measurements of Testicular and Ejaculated Germ Cells

Donald P. Evenson*, Paul J. Higgins, and Myron R. Melamed

Memorial Sloan-Kettering Cancer Center, 1275 York Avenue, New York, New York 10021, USA

1 Introduction

1.1 Male Germ Cells as Targets of Environmental Modifying Agents

The sequence of biochemical and morphological events that characterize the cellular pathway leading from the diploid spermatogonia to mature haploid sperm represents one of the most dramatic differentiation pathways in biology. Table 1 illustrates some of the reproductive tract events within this differentiation pathway that can be modified by environmental agents. Some of the specific events underlying sperm maturation currently under investigation in this laboratory include studies on macromolecular synthesis, meiosis, exchange of nuclear histones for protamines, disulfide bonding and sperm head shaping as they are influenced by mutagens and other environmental factors.

A recent review article [10] suggests that analyses of peripheral blood chromosomes and sperm may serve as useful monitors of long term biological effects due to exposure to physical and chemical agents in the environment. In this regard, our recent studies have centered around the hypothesis that mammalian spermatogenesis and spermiogenesis are sensitive indicators of chromosome/chromatin alterations induced by chemicals, disease, diet, and other agents. Many of these alterations can be readily detected by by flow cytometry and often confirmed by light/electron microscopy.

1.2 Sperm Chromatin Differentiation

The chromatin of maturing sperm cells undergoes a dramatic exchange of somatic like histones for sperm specific basic proteins. Figure 1 shows observed changes in chromatin as cells progress from a round spermatid to mature sperm. Round spermatids are characterized by diffuse chromatin (Fig. 1A) very similar to that seen in somatic cells. As a male germ cell progresses to an elongated spermatid, the chromatin becomes beaded to fibrillar in appearance (Fig. 1B). Chromatin condensation progresses from the anterior portion of the nucleus towards the posterior region; thus, as can be seen in Fig. 1C, the anterior portion of the nucleus is highly condensed the posterior portion is much less condensed. Finally, a mature testicular sperm nucleus contains highly condensed chromatin (Fig. 1D). Obviously, the testis expends a considerable amount of energy in this exchange of histones for sperm basic proteins and the disulfide bonding between adjacent protein molecules. Since a lack of complete condensation and perhaps exchange

*Current address: Chemistry Department, Box 2170, South Dakota State University, Brookings, South Dakota 57007, USA

Biological Dosimetry. Edited by W. G. Eisert and M. L. Mendelsohn
© Springer-Verlag Berlin Heidelberg 1984

Table 1

Reproductive Tract Events	Differentiation Pathway	Modifiers
Macromolecular synthesis	Spermatogonia ↓	Chemicals Radiation
	Spermatocytes	Drugs
Meiosis	↓	
Exchange of nuclear histones for protamines	Spermatids ↓ Sperm	Diet Air pollutants
Sperm specific morphology		

of protein now appears to be correlated with subfertility (4, 7] Evenson, in preparation), those sperm which have failed to undergo such a complete transition are possibly more vulnerable to chromosomal damage from exposure to nucleases, chemicals and other modifying agents in both the male and female reproductive tract thereby giving rise to subfertility and/or early embryo death.

2 FCM Measurements on Male Germ Cells

2.1 Testicular Biopsies

2.2.1 Human

Figure 2 shows an acridine orange (AO) staining pattern for cellular suspensions prepared from testicular biopsies. In this particular case, tetraploid cells have been exluded from the measurement; ejaculated sperm were admixed to provide an enhanced comparison of stainability between mature sperm and precursor cells. Note the three populations of spermatids. Population a is seen only with a two parameter (DNA, RNA) stain such as AO; treatment of the cells with RNAse shifts population a into population b. Thus, flow cytometry can readily detect the known loss of RNA [11] from sperm during differentiation. Population c consists of elongated spermatids as determined by examining flow-sorted samples under the light microscope. Although the population kinetics of these cells would provide useful information in drug/chemical dose-response studies, it is difficult to obtain human surgical specimens and, therefore, these studies are not practical.

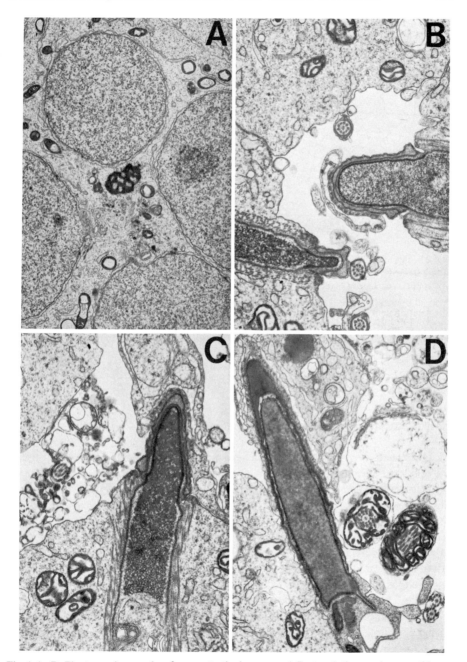

Fig. 1 A–D. Electron micrographs of mouse testicular sperm. **A** Post meiotic round spermatids; x 9,500. **B** Elongating spermatid, x 20,000. **C** Elongated spermatid, x 20,000. **D** Sperm with condensed chromatin, x 20,000. Small biopsies of a mouse testicle were fixed in glutaraldehyde and osmium tetroxide, dehydrated, embedded in Epon 812, thin sectioned, stained and photographed as previously described [8]

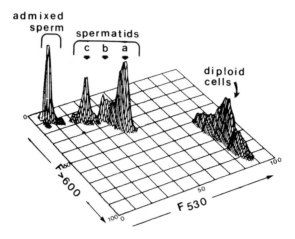

Fig. 2. Computer drawn two-parameter (F_{530} vs. F_{600}) histogram representing the distribution of AO-stained human testicular biopsy cells plus admixed normal human semen. A 0.2 ml aliquot of a cellular suspension made from a scapel-minced biopsy was mixed with 0.4 ml of a detergent solution consisting of 0.1% Triton X-100 in 0.08N HCl, 0.15N NaCl. Thirty seconds later 1.2 ml of a solution containing 0.2M Na_2HPO_4-0.1M citric acid buffer (pH 6.0), 1mM EDTA, 0.15M NaCl and 6μg AO/ml was admixed with the sample as previously described [8]. The sample was immediately measured in a FC 200 Cytofluorograph interfaced to a Nova 1220 minicomputer. The red (F_{600}) and the green (F_{530}) fluorescence emission for each cell was recorded and the integrated values stored in the computer. The data are based on a total of 5000 cells (From Evenson et al. [7], with permission from J Histochem Cytochem)

2.1.2 Mouse Model for Testicular Cell Kinetics

As an alternative to human studies, we have initiated experiments with mice to study the effects of chemicals on testicular cell kinetics. Figure 3 shows preliminary data on the effects of carbon tetrachloride (CCl_4) on mouse testis cell kinetics. The AO staining pattern for mouse vas deferens sperm is shown in the far left panel. As previously shown [5], the AO stainability of mature sperm is reduced approximately five fold relative to round spermatids due to the condensation of the chromatin. A suspension of mouse testicular cells stains similar to those of human [7]. Occasionally the population of elongated spermatids demonstrates two peaks as seen here.

The exposure of mice to CCl_4 has a definite effect on the cell maturation kinetics so that by the second day the major population consists of elongated spermatids. These data suggest that by day three there is a slight recovery with the presence of a higher percentage of round spermatids. We expect that this type of measurement will be useful in future studies of agents that interfere with cell maturation.

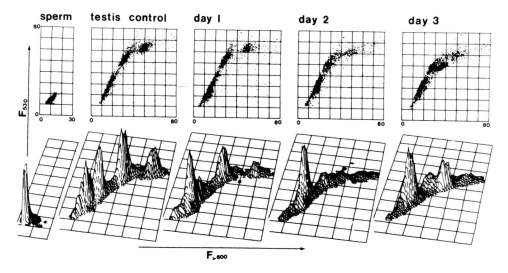

Fig. 3. Computer drawn two-paramter (F_{530} vs F_{600}) histograms representing distributions of AO-stained mouse testicular cells obtained from animals treated with carbon tetrachloride (CCl_4). Mice were injected with 1 ml CCl_4/kg and sacrificed at 0, 1, 2 and 3 days post injection. Samples were prepared and measured as described in the legend for Fig. 2. Note that the diploid and tetraploid cells were excluded from the measurements; also, no mature sperm were added to the testicular samples

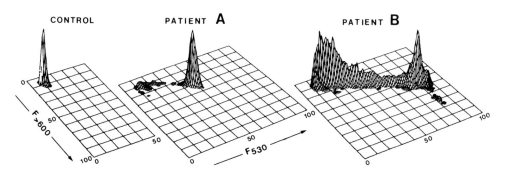

Fig. 4. Computer drawn two-parameter (F_{530} vs F_{600}) histograms representing distributions of AO-stained human semen cells. The control semen was obtained from a fertile donor. Patient A, age 18, had Stage III embryonal cell testicular carcinoma. The sample was obtained 6 months post-unilateral orchiectomy and 3 months post-VAB 6 induction chemotherapy. The sperm concentration was 5×10^4/ml, 8% motility. Patient B, age 28, had Stage I testicular cancer with teratoma, embryonal cell carcinoma and seminoma. The sample, obtained 11 months post orchiectomy contained immature sperm and diploid cells, but no mature normal sperm as verified by light and electron microscopy. (From Evenson et al. [7], with permission from J Histochem Cytochem)

2.2 Semen Samples

2.2.1 Cell Types

An easier method, especially for human studies, for evaluating effects of modifying agents on sperm maturation is to measure various parameters of the sperm in the ejaculate. The left panel in Fig. 4 shows the AO staining pattern of cells present in normal semen (which consists almost entirely of mature sperm). This procedure is very rapid; an aliquot of fresh or frozen (allowing samples to be sent to a laboratory) semen can be stained with AO, measured and the data analyzed by computer assistance within five minutes. Both patient A and B demonstrate a severe testicular dysfunction as an apparent result of cancer or its therapy. The semen cells from patient A are mostly round spermatids as seen by both the flow cytometry staining pattern and light microscopy. The cell types in semen from patient B are very heterogeneous ranging from nearly mature sperm to diploid cells as also confirmed by light/electron microscopy.

Most patients treated with chemotherapy will demonstrate very dramatic alterations of testicular function as seen for patient B; obviously, it would be of interest to follow the dose-effect response with a variety of chemotherapy protocols. Two problems are evident for such experiments, however. First, given the fact that the patients are often faced with a life threatening disease and other complications, it is often difficult to obtain the semen samples. Secondly, cancer itself may cause dramatic changes in testicular function [2, 7] and, therefore, it would be difficult to determine whether the observed effects were due to the disease and/or the treatment.

2.2.2 Sperm Nuclear Chromatin Condensation

Often the effects on sperm differentiation due to disease or other modifying agents are not as dramatic as that seen in Fig. 4. For example, Fig. 5 shows the AO staining pattern of sperm obtained from a patient diagnosed with Hodgkin's disease and prior to any therapy. In this case, the Cytofluorograph photomultiplier gains were set higher than those represented in Fig. 4. Note that a significant number of sperm cells from the patient have an increased stainability with AO most likely due to a lack of complete chromatin condensation. Whereas, 93% of the control sperm stained within three standard deviations of the mean, only 60% of the cells obtained from this patient were within three standard deviations of the mean. Although Figs. 4 and 5 demonstrate a problem in using cancer patients for dose-response studies of the effect of treatment on testicular function, this type of measurement should be useful for both animal and human studies related to normal or altered testicular function. Namely, we hypothesize that agents which cause testicular dysfunction will in many cases produce an interference with normal chromatin condensation which can then be measured by this technique.

AO stained semen cells

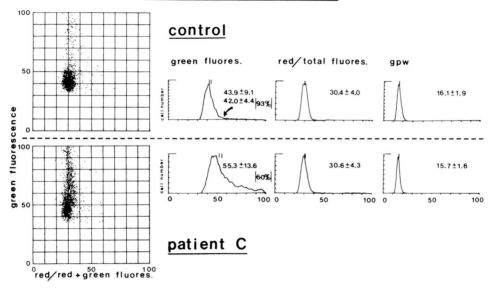

control

green fluores.   red/total fluores.   gpw

43.9 ± 9.1
42.0 ± 4.4 | 93%

30.4 ± 4.0

16.1 ± 1.9

55.3 ± 13.6 | 60%

30.6 ± 4.3

15.7 ± 1.6

patient C

green fluorescence

red/red + green fluores.

Fig. 5. Computer drawn scattergrams of the distributions of semen cells from a fertile, healthy control and patient C according to their green and red/red plus green fluorescence intensities after staining with AO as described in the legend to Fig. 2. The sample from patient C, age 18 and diagnosed with Hodgkin's disease, was obtained prior to any therapy. The sperm count was 30 x 10[6] per ml; sperm motility was 70% 1 hr post-emission. Sperm morphology was normal by light microscopy (From Evenson et al. [7], with permission from J Histochem Cytochem)

2.3 Thermal Denaturation of Sperm Nuclear DNA in situ

A more sensitive technique to measure alterations of chromatin structure is to subject isolated sperm nuclei to thermal stress and then determine the resistance of the DNA to *in situ* denaturation. Since mature sperm nuclei do not contain RNA [11], the red fluorescence of heated, AO stained nuclei corresponds to the proportion of single stranded denatured DNA while the green fluorescence corresponds to the native undenatured DNA. The ratio of red/total fluorescence (termed a_t) provides an index, ranging from 0 to 1, of the extent of denaturation. This technique has previously been used to correlate chromatin structure with fertility [2, 4]. Figure 6 shows an example of this technique using sperm provided by a normal human control and two patients under evaluation for fertility. Note that the staining pattern for the control sample shifts only slightly after heating indicating that little of the DNA is susceptible to thermal denaturation. In sharp contrast, note the high degree of denaturation resulting in high a_t values in the two patient samples. Of particular interest is patient B whose samples was, by the classical criteria of sperm count, morphology and motility, considered to be an excellent specimen. Therefore, the significant point is that this method provides an independent parameter of analysis to detect an abnormality that is not available by any other method. We are currently finding this type of measurement

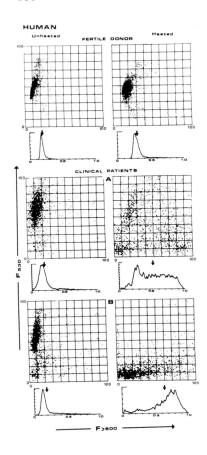

Fig. 6. Computer drawn scattergrams of the AO staining distribution of individual sperm nuclei isolated from cells obtained from a normal, fertile donor and two patients. The sperm nuclei were isolated by sonication, purified through sucrose gradients, fixed and then rehydrated prior to either heating (100 C, 5 min) or not heating and then staining with AO (See Evenson et al. [4] for further details)

useful for analyzing, and perhaps predicting, the recovery of testicular function following cancer and its therapy [2].

Using this latter method, we have begun studies with a mouse model system to detect abnormalities of chromatin structure in sperm obtained from animals exposed to mutagens and other environmental agents. Our hypothesis is that this method will detect alterations of chromatin structure that are not readily detected by other means. Figure 7 shows preliminary data on the effects of diethylnitrosamine on sperm nuclear DNA denaturation *in situ*. Somewhat unexpectedly, the data suggest that the DNA becomes more resistant to denaturation. Additional work will be done to confirm this and establish the biochemical bases for this response, especially since it is of question to what extent DENA enters the testis. In any case, we expect this type of analysis to be useful in dose-response studies on chemicals that affect sperm differentiation.

2.4 Sperm Concentration, Viability and "Motility"

The effect of sperm modifying agents on sperm concentration, viability and "motility" can be monitored easily by flow cytometry. Sperm in an ejaculate are first stained with

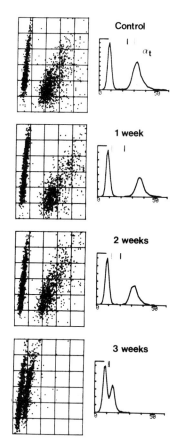

Fig. 7. Computer generated scattergrams of the staining distribution of individual sperm nuclei isolated from cells obtained from mice treated with diethylnitrosamine (DENA); the corresponding α_t frequency histograms are to the right of each scattergram. C57B1/6 mice were treated with DENA (50 ppm in drinking water) and sacrificed over a three week period. The sperm were removed from the vas deferens and the nuclei were isolated and analyzed as described in the legend for Fig. 6 and Evenson et al. [4]. In this case, unheated and heated nuclei were mixed prior to measuring in the flow cytometer

rhodamine 123 (R123) which specifically binds to mitochondria [9] at a level which has been correlated with the vigor of sperm motility [3, 6]. The cells are then counterstained with ethidium bromide which will stain the DNA of nonviable cells. Figure 8 demonstrates the distribution of semen cells stained with R123 and EB. Recently, we have added known numbers of fluorescent beads [6] to the R123 and ethidium bromide stained cells just prior to FCM measurement so as to simultaneously determine sperm concentration, viability and mitochondrial function related to motility.

3 Summary

Flow cytometry of developing and mature sperm from humans and animals with pathological conditions or those exposed to testicular function modifying agents can provide rapidly acquired data that are statistically sound due to the large numbers of randomly measured cells. Of more importance, however, is the fact that we can acquire information on factors such as chromatin structure that cannot be practically obtained in any

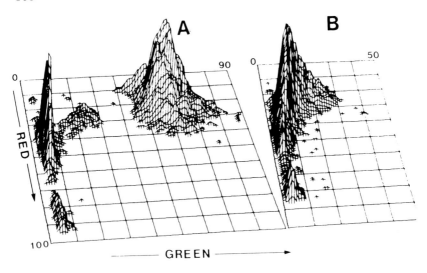

Fig. 8 A,B. Two parameter frequency histograms demonstrating stainability of human sperm cells with rhodamine 123 (R123) and ethidium bromide (EB). A single semen sample from a normal control was stained and measured 4 h after collection (**A**) and 30 h after collection (**B**). The semen was diluted to about $1-2 \times 10^6$ cells per ml in RPMI 1640 media plus 10% fetal calf serum and stained with R123 at 10 μg/ml final concentration for 10 min. The sperm were then pelleted, resuspended in the same fresh medium and counterstained with 10 μg/ml EB prior to analysis in a FC 200 Cytofluorograph equipped with a 100mW argon-ion laser (From Evenson et al. [3], with permission from J Histochem Cytochem). Note the drop in green fluorescence (rhodamine 123) and increase in red fluorescence (ethidium bromide) with time

other manner. This approach, coupled with classical techniques in reproductive biology, including electron microscopy, will provide a powerful methodology to study the response of animals to agents that modify testicular function.

Acknowledgement. This work was supported by grants from the National Institute of Health RO1 HD12196 and RO1 ESHDO3035 and in part by NCI Core Grant CA08148. We thank Ms. Jisoo Lee for expert technical assistance and Ms. Robin Nager for typing the manuscript.

References

1. Darzynkiewicz Z, Traganos F, Sharpless T, Melamed MR (1976) Lymphocyte stimulation: A rapid multiparameter analysis. Proc Natl Acad Sci 73:2881–2884
2. Evenson DP, Arlin Z, Welt S, Claps ML, Melamed MR (1984) Male reproductive capacity mayay recover following drug treatment with the L-10 protocol for acute lymphocytic leukemia (ALL). Cancer 53:30–36
3. Evenson DP, Darzynkiewicz Z, Melamed MR (1982) Simultaneous measurement by flow cytometry of sperm cell viability and mitochondrial membrane potential related to cell motility. J Histochem Cytochem 30:279–280

4. Evenson DP, Darzynkiewicz Z, Melamed MR (1980) Relation of mammalian sperm chromatin heterogeneity to fertility. Science 240:1131–1133

5. Evenson DP, Darzynkiewicz Z, Melamed MR (1980) Comparison of human and mouse sperm chromatin structure by flow cytometry. Chromosoma 78:225–238

6. Evenson DP, Melamed MR (1983) Simultaneous measurment of sperm concentration, viability and mitochondrial function related to motility. Cytometry (Abstract)

7. Evenson DP, Melamed MR (1983) Rapid analysis of normal and abnormal cell types in human semen and testis biopsies by flow cytometry. J Histochem Cytochem 31:248–253

8. Evenson DP, Witkin S, deHarven E, Bendich A (1978) Ultrastructure of partially decondensed human spermatozoal chromatin. J Ultrastruc Res 63:178–187

9. Johnson LV, Walsh ML, Chen LB (1980) Localization of mitochondria in living cells with rhodamine 123. Proc Natl Acad Sci 77:990–994

10. Maugh TH (1982) Biological markers for chemical exposure. Science 215:643–649

11. Monesi V (1971) Chromosome activities during meiosis and spermiogenesis. J Reprod Fertil, Suppl 13:1–14

Cytometric Analysis of Mammalian Sperm for Induced Morphologic and DNA Content Errors*

Daniel Pinkel

Lawrence Livermore National Laboratory, Biomedical Sciences Division, University of California, P.O. Box 5507 L-452, Livermore, California 94550

1 Introduction

Among the signatures of reproductive toxicity of chemical and physical agents are alterations in morphology and/or DNA content of sperm. Sperm based assays for reproductive effects are attractive because of the potential for obtaining samples for monitoring human populations. Visual analysis of sperm morphology has been used for many years as one component of semen assessment. More recently it has been found that shape abnormalities are induced by exposure of animals to a wide range of agents [27, 28] and there is evidence for elevated levels of abnormally shaped sperm in men exposed to drugs or chemical agents in the work place [29]. Experienced observers are consistent in classifying sperm according to shape but classification criteria may differ among them. Additionally there are important aspects of sperm shape which humans can't discriminate well. These considerations have led to efforts to develop quantitative morphologic analysis procedures. These procedures lend themselves to automation, so that once their value is established the time required for large numbers of analyses may be greatly reduced.

DNA content errors in germ cells have been detected by cytological techniques [12, 21, 22] at specific points during spermatogenesis such as meiosis, and the consequences of the errors have been determined by post fertilization karyotype analysis [3, 4, 6, 7, 12, 14]. Recently flow cytometric measurements have been able to detect radiation induced DNA content errors in sperm [19] and spermatids [10, 11], yielding information where it was previously unavailable.

In this article I will review some flow-cytometric and image analysis procedures under development for quantitative analysis of sperm morphology. I will also summarize the results of flow-cytometric DNA-content measurements on sperm from radiation exposed mice, relate these results to the available cytological information, and discuss their potential dosimetric sensitivity.

---

*Work performed under the auspices of the U.S. Department of Energy by the Lawrence Livermore National Laboratory under contract number W-7405-ENG-48

2 Sperm Head Morphology

Mammalian sperm have thin planar heads attached to a long tail. Morphologic abnormalities may occur in both. Procedures for quantitatively classifying sperm shape range from simple measurements of head length [13] to complex descriptions of both head and tail [28]. To illustrate the possibilities and problems of this work I will summarize two image analysis and one flow cytometric study of changes in mouse sperm head shape induced by x-ray exposure.

2.1 Image-Analysis Studies

Quantitative assessment of changes in sperm head shape due to toxic exposure requires establishing of a set of measurements of cell features which are sensitive to the changes. Parallel experiments by Moore et al. [16] (called MBKW in what follows) and Young et al. [30] (called YGLW) illustrate two methods of approaching this problem. Both studies analyzed sperm from the same mice, which had been exposed to testicular x-ray doses of 0, 30, 60, 90, and 120 rad. The three mice in each dose group were killed 5 weeks after exposure and smears of sperm from the cauda epidydimides were prepared. MBKW stained the smears with the protein stain eosin which allowed visualization of the whole sperm head. YGLW used the DNA stain gallocyanin chrome alum which stained the nucleus. Since the sperm head is almost entirely nucleus, the visible sperm shapes were basically the same in these two studies. Five hundred eosin stained sperm from each mouse were visually classified as normal or abnormal [5] so that the image analysis results could be compared with a dose-response curve generated by an experienced human observer. MBKW took photomicrographs (8500x) of 50 randomly selected sperm per mouse and manually measured 8 linear features believed to describe sperm head shape (Fig. 1) as well as area, perimeter, and $perimeter^2/area$. YGLW used a microscope-based, computer image-analysis system in which the outlines of the sperm nuclei were defined using a thresholding technique (Fig. 2). They then computed 10 general descripitons of cell shape. These included area, perimeter, $perimeter^2/area$, and 7 others describing the curvature, roughness, and deviation from circularity of the sperm nuclear outline. No measurements of linear or diagonal dimensions were included. Thus only 3 of the measured features in the two studies were the same.

Visual scoring of sperm heads is based on classifying each as normal or abnormal, and perhaps recognizing several different classes of abnormalities. Quantitative measurements allow the establishment of rigorous criteria for this classification and also make it possible to study changes in each dose group as a whole. The simplest way to describe a group is with the average values and variances for the measured features. The results obtained in the two studies were different. MBKW found no significant changes in the means of their measurements, but did find an increase in the variances in the highest dose groups. On the other hand, YGLW found changes in the mean values of some features but did not report changes in the variances. In the three features measured in common, there is disagreement on their behavior.

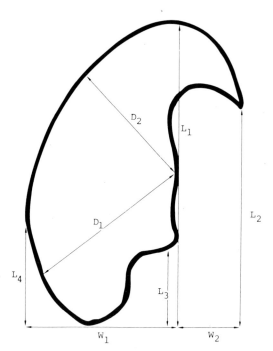

Fig. 1. Definition of measurements used to describe shape of mouse sperm heads [16]. The indicated measurements were manually obtained on enlarged photographs of the sperm heads. In addition to the measurements shown, perimeter, area, and shape factor were also determined

Sensitivity was greatly improved in both studies by combining individual measurements into a statistical measure called the Mahalonobis distance, M. This describes the "distance" in the multidimensional feature space by which a cell or experimental population departs from a reference population. The distance is measured in units related to the variance and covariance of the measurements in the reference population. Explicitly,

$$M = (Y - \bar{Y}_c)' \, S^{-1} \, (Y - \bar{Y}_c) \tag{1}$$

where Y is a vector whose elements are the measured features on a particular cell, $\bar{Y}_c$ is the mean value of the feature vector, for the reference population (the ' indicates the transpose of the vector) and S^{-1} is the inverse of the covariance matrix for the reference population (YGLW calculated S^{-1} by pooling both the exposed and reference populations).

MBKW applied Eq. (1) to each sperm head in each dose group using the zero dose sperm for the reference population. They found that the mean M for each dose group increased with dose but that more sensitivity could be achieved by classifying each sperm as normal or abnormal based on its M value. This was done by defining those with M larger than some critical value as abnormal. Figure 3 shows a series of dose response curves obtained in this way. A choice of ~ 45 for the critical value approximately duplicates the visually obtained dose-response curve. Choosing lower values yields steeper slopes at the expense of classifying more sperm in the unexposed group as abnormal.

The dose-response curves of Fig. 3 were based on all 11 features. MBKW then searched all possible subsets of combinations of 1, 2, ... etc. features to see which yielded the greatest sensitivity to exposure. Searches of this type have to be carried out by directly

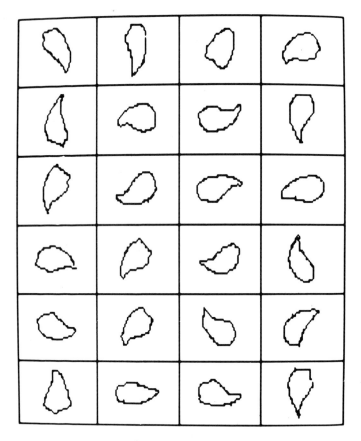

Fig. 2. Sperm head outlines obtained by an automated image analysis system [30]

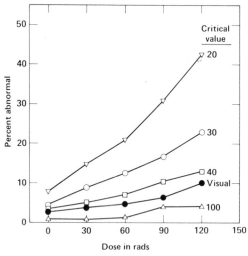

Fig. 3. Dose response curve as a function of the critical value of M [16]. Sperm heads with M greater than the critical value were classified as abnormally shaped

testing all of the possible combinations, in this case 2047. The search was constrained in two ways. First the critical value of M was chosen such that at zero dose 3% of the sperm were classified as abnormal, in agreement with the visual scoring. Second, the resulting dose-response curve was required to be well fitted by a straight line. These subsets were then compared on the basis of the slopes of their dose-response curves. It was found that several subsets yielded greater slopes than the complete set of all 11 features. Thus "noise" in some of the features was obscuring response in the others. The features L_2, L_4, D_2, and area (Fig. 1) appeared in most of the high-slope subsets and thus appear to be more sensitive to radiation exposure than the others. The doubling dose (the dose required to double the percentage of abnormal sperm in the control group) for these subsets was less than 25 rad compared to about 70 rad for visual scoring. Sperm-by-sperm comparisons of visual and quantitative classification indicate that the additional sensitivity of the latter comes from changes in the area and diagnonal measurements, quantities not easily judged by eye.

YGLW applied Eq. (1) to the sperm populations as a whole, using the mean values of the features of the sperm from each mouse and thus calculating a single M for the sperm from that animal. No classification of individual sperm as normal or abnormal was reported. The resulting dose response curve was fitted by an equation of the form $M = a\,D^b + c$, where M is the Mahalonobis distance at dose D, c is the zero dose intercept, a is a constant, and $3 \langle b \langle 4$.

Several issues raised by these studies remain unresolved. Application of the calculation of YGLW to the measurements of MBKW would have found essentially no response to dose since the means of their measured features were very nearly dose-independent. MBKW found it most sensitive to classify individual sperm based on their M value and searched for and found dose-response curves that were well fitted by a straight line. YGLW found a cubic or quadratic behavior of M for the groups and reported no attempt to classify individual sperm. Experience with visual scoring has found dose-response curves increasing approximately as $D^{1.5}$ where D is the dose. The differences in the shape of the curves may be related to the different information captured by the measured features in each case. This issue is important since the sensitivity of the assay at low doses depends on the slope of the curve at D = 0; the larger the exponent the larger the dose required to measure a response.

Methodological improvements might also resolve some of the differences between the quantitative studies. It is important to establish the sampling density required to adequately describe sperm head shape in digital images. It is also important that automatic procedures for finding the sperm head outline be carefully monitored to be sure that the acquired image is that of a single sperm lying flat on the slide and that the computer-found perimeter not contain artifactual deviations due to debris near the cell being measured. This is especially important in rodent sperm because their hooked shape (Fig. 1) results in a normal outline with regions of high curvature. While these studies point up issues that require resolution, they indicate the potential of quantitative sperm head measurements to improve the sensitivity and objectivity of morphological classification. They allow a rigorous definition of classification criteria and the testing for those features most affected by exposure to toxic agents.

2.2 Flow Cytometry

Slit-scan flow cytometry has been used to obtain one-dimensional morphological information about cells and chromosomes. Benaron et al. [2] have applied it to the detection
of x ray induced sperm head shape abnormalities. In their work sperm whose nuclei were
stained with a fluorescent DNA specific dye were scanned as they flowed through an
elliptically focused laser beam whose dimension along the direction of flow (2.5 μm) was
much less than the length ($\sim$7μm) of the nucleus (Fig. 4). The hydrodynamic forces associated with the flow oriented the sperm so that they flowed either head or tail first
through the beam. Because of the optical effects peculiar to sperm (see Sect. 3) data
were only accepted from sperm which were oriented such that their plane was approximately normal to the axis of the fluorescence collection lens. This was judged by the
brightness of the integrated fluorescence signal (cells in the peak of Fig. 6a). Fluorescence intensity was recorded every 20 ns so that about 50 measurements defined the
profile of the fluorescence pulse. The information contained in the pulse profile is the
DNA content in a series of strips across the cell. Its relation to cell shape as seen by a
human observer and the image analysis techniques discussed above is indirect. For example, the slit scan signal is sensitive to thickness variations in the nucleus which were
not determined in the other studies (but could have been).

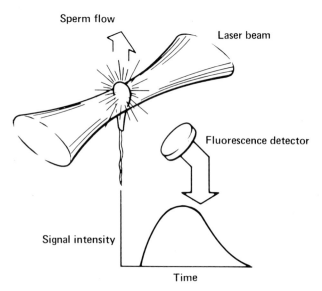

Fig. 4. Sperm morphology measurement by slit scan flow cytometry [2]. Fluorescently stained
sperm heads flow lengthwise through an elliptically focused laser beam. The time course of the
signal intensity gives the distribution of dye along the sperm head and thus an indication of
head shape

A sperm head was classified as normal or abnormal by comparing the shape of its
fluorescence profile, the sequence of 50 or so individual measurements [a (i)], with
a standard profile [b (i)] obtained by averaging the profiles of 100 sperm from a

control population. The departure from standard was quantified by the sum of the squares of the differences (SSD) of the two sets of numbers

$$SSD = \Sigma_i \, [a \, (i) - b \, (i)]^2. \tag{2}$$

Sperm with an SSD value above a chosen threshold were defined as being different from the standard.

A comparison of the flow and visual scoring of sperm from mice exposed to testicular radiation doses between 0 and 900 rad is shown in Fig. 5. The curve depends on the choice of SSD threshold and the one shown corresponds to the value that gives highest correlation with visual scoring. The difference in the background rates of sperm classified as abnormal may be due to a number of factors such as inadequate control of cell orientation during measurement or to actual shape variability, such as nuclear thickness, to which visual observers are not sensitive. Further work is needed to define these factors and perhaps to develop other methods of processing the profiles to define morphological differences. For example, small differences in the length of a sperm nucleus are weighted heavily in the SSD calculation because of the steepness of the fluorescence profile at the ends of the sperm head (Fig. 4). While the information contained in the flow profiles is less complete than that used for visual and quantitative image analysis, the potential for high analysis rates inherent in flow cytometry may compensate for the poorer resolution. Slit scanning may also capture aspects of sperm shape variability missed by other techniques.

3 DNA Content

The incidence of sperm with abnormal DNA content increases after exposure of males to various chemical and physical agents. Two general classes of abnormalities have been of interest. The first consists of polyploid sperm, the most prevalent of which are diploid. The second includes errors of much smaller magnitude such as chromosomal translocations or abnormal chromosome number due to nondisjunction. The size spectrum of the latter results in variability of the DNA content in the sperm population. Diploid sperm occur in mammals with frequencies on the order of 10^{-3} to 10^{-4} and in special cases as high as 10^{-2} [1, 9]. Diploid sperm incidence has been shown to be elevated by exposure of animals to radiation and certain chemicals [23, 25]. Manual scoring of such an assay is laborious because of the low frequency of occurrence. In Sect. 3.1 I will describe one application of flow cytometry to determine the incidence of diploid mouse sperm as a function of x-ray dose.

The smaller DNA content errors can be detected during spermatogenesis by examination of the condensed chromosomes at the cell divisions [12, 21, 22]. Karyotype analysis after *in vitro* or *in vivo* fertilization can be used to determine the incidence of sperm with chromosomal abnormalities which complete spermatogenesis and are capable of fertilization [3, 4, 6, 7, 12, 14]. In special cases particular chromosomes can be visualized in spermatids and sperm by exploiting their unique staining properties

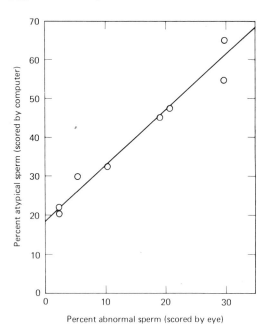

Fig. 5. Comparison of the percent abnormal sperm classified by slit-scan flow
cytometry and by a trained visual observer. The SSD threshold has been
chosen to maximize the slope at the
expense of classifying a larger number
of sperm from unexposed animals as
atypical [2]

and so that errors in their frequency may be recognized [23, 24]. Flow-cytometric
techniques capable of measuring DNA content of sperm and spermatids with sufficient
accuracy to detect these errors have recently been developed [15, 20]. In Sect. 3.2
and 3.3 I will discuss these measurements, compare the information obtained from
them with that obtained from cytologic techniques, and finally speculate on the sensitivity of flow measurements in animals other than the mouse. Measurements of induced DNA content variability in mouse spermatids is discussed by U. Hacker elsewhere in this volume [10] and in Ref. 11.

3.1 Diploid Sperm

Flow-cytometric detection of diploid sperm requires discrimination of true diploid
cells from doublets of normal sperm, whose frequency is difficult to reduce below
1%. Our approach has been to index sort suspected diploid sperm and confirm their
status by quantitative microscopic examination. These measurements, which are still
preliminary, were made on sperm from mice collected 35 days after local testicular
x-ray exposure. Sperm were stained with the acriflavine-Feulgen technique [20] and
each object with a diploid fluorescence intensity was single-drop index sorted in a
rectangular array on an agarose coated microscope slide. The stain stayed in the cell
after the drop dried. Since the sorter was not equipped to fully compensate for the
optical problems of sperm measurement (Fig. 6) the sort windows were chosen to
cover the wide range of expected diploid intensities. Each drop was then examined
on a quantitative fluorescence microscope. If a drop contained a single sperm that
cell was a candidate for being diploid. Its fluorescence intensity was then measured

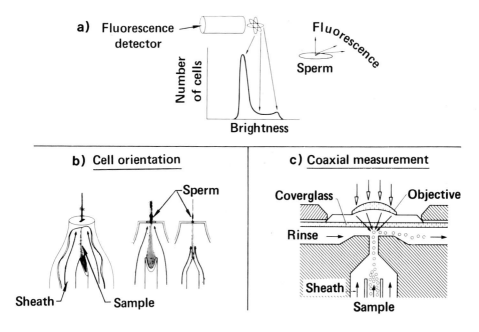

Fig. 6a–c. Accurate DNA content measurements of mammalian sperm require special techniques. **a)** The high index of refraction causes fluorescence to be preferentially emitted in the plane of these flat cells. The measured brightness thus depends on the cell orientation during measurement. Randomly oriented sperm yield a distribution as shown when a flow cytometer with the flow axis orthogonal to the optical axis of the detector is used. **b)** Hydrodynamic orientation of the sperm allows accurate measurements. Sperm are oriented so they are illuminated on one flat face and fluorescence is collected from the other. **c)** Good measurements are also obtained in a system where the cells flow along the optical axis of the cytometer. The rotational symmetry of this geometry makes it insensitive to variations in cell orientation

for confirmation. A preliminary dose response curve is shown in Fig. 7. Since the error bars can be reduced by counting more cells, these measurements can be made very precise. At high doses there is a several order of magnitude increase in the incidence of diploid mouse sperm, roughly in agreement with the spermatid data of Tates [23, 25], but at low doses the response is very flat. Because of the limited number of measurements performed so far, it is not certain this apparent threshold behavior is correct. Confirmation of this dose response relation requires more measurements.

3.2 Sperm DNA Content Variability

Flow-cytometric DNA content measurements of sperm are complicated by the highly condensed state of the nucleus [26]. This makes accurate fluorescent staining for DNA difficult and results in a high index of refraction. The index of refraction coupled with the flatness of the cell makes fluorescence measurements sensitive to the orientation of the cell with respect to the excitation beam and the detectors (Fig. 6a). Thus special staining and optical techniques are required for accurate measurement [20]. Two tech-

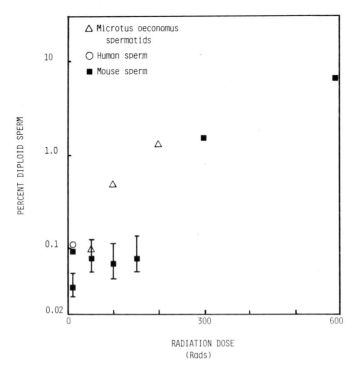

Fig. 7. Incidence of diploid sperm as a function of radiation dose. The three species shown have about the same background incidence. *Microtus* data is from Ref. 23. Mouse and human data are preliminary

niques have demonstrated the capability to overcome the optical problems. One hydro-dynamically orients the sperm heads (Fig. 6b), and the other, for example the commercially available ICP 22 flow cytometer (Ortho Instruments) measures the cells while they are flowing along the optical axis (Fig. 6c). To solve the staining problems, proteolytic decondensation of the nucleus is usually employed. With these techniques the X and Y sperm populations of many species can be clearly resolved [8] and in one case separated [18]. While this resolution is occasionally achieved with human sperm [17] it is very infrequent presumably because the chromatin of human sperm is less uniformly condensed than that of other species.

A flow-cytometrically obtained dose-response curve describing the increase in DNA content variability, CV^2_D, as a function of x-ray dose for mouse sperm is shown in Fig. 8 [19]. CV^2_D is the dose-dependent component of the square of the coefficient of variation of the sperm fluorescence frequency distributions. For this experiment sperm were collected 35 days after testicular exposure. Measurements on sperm collected less than three weeks after exposure show no response, indicating the DNA content errors occur before or during meiosis. Sperm collected 10 weeks or more after exposure also showed no increase in CV^2_D. Measurements with a variety of staining protocols and two types of flow cytometers were compared to be certain that artifacts due to the difficulties of sperm measurement were adequately overcome.

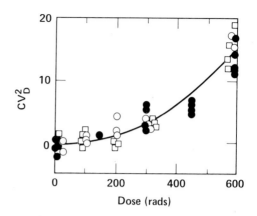

Fig. 8. Increased DNA content variability as a function of radiation dose. The square of the dose dependent part of the coefficient of variation, CV^2_D, is shown as a function of radiation exposure as determined by several combinations of staining and flow cytometric techniques: *Solid circles - -* acriflavine-Feulgen stain and orienting flow cytometer. *Squares - -* ethidium bromide-mithramycin double stain and orienting flow cytometer. *Open circles* 4–6 diamidino-2-phenylindole (DAPI) stain and ICP22 flow cytometer. The *solid line* is a fit of an equation of the form $CV^2_D = BX + CX^2$ where X is the dose and B and C are coefficients whose values are given in the text

The dose response curve is well described by the equation

$$CV^2_D = BX + CX^2 \tag{3}$$

where X is the dose in rad. The coefficients B and C found by fitting the data are $B \leq 0.23 \times 10^{-2}$, $C = (0.44 \pm 0.06) \times 10^{-4}$. Note that the slope of the curve is small at low doses and may even be zero. Thus this technique is not sensitive to detecting low radiation exposures. As shown below, this is consistent with other estimates of radiation induced aneuploidy in sperm. Modeling of the shapes of the fluorescence distributions indicates that errors equivalent in magnitude to two whole chromosomes occur at the higher doses and that by 600 rad 30 to 40% of the sperm have abnormal DNA content. Exposure to benzo(a)pyrene and mitomycin-C (MMC) show no increase of CV^2_D up to lethal doses. The MMC result is consistent with fetal karyotype determinations [6].

Assuming that at low doses the DNA content errors result from single chromosome nondisjunction, that the response is linear with dose, and that the probability of non-disjunction is the same for each chromosome, these measurements allow putting an upper bound on the aneuploidy induction rate in sperm which reach maturity. Due to cell death during spermiogenesis this should be smaller than or equal to the probability of nondisjunction determined immediately after meiosis (sperm count was reduced about 30% by a 100 rad exposure). For single nondisjunction, $CV^2_D = fCV^2_A$ where f is the fraction of aneuploid sperm and CV^2_A is the square of the coefficient of variation of the DNA distribution of all sperm with a single chromosome aneuploidy. For the assumed linear response,

$$f = 2\bar{p} \; X N \tag{4}$$

where X is the dose in rads, N is the haploid number of chromosomes and $\bar{p}$ is the probability of a nondisjunction per chromosome per rad in sperm which reach maturity. Using the linear term of Eq. 3,

$$CV^2_D = 2\,\bar{p}\,X\,N\,CV^2_A \leq 0.23 \times 10^{-2}\,X. \tag{5}$$

For the mouse N = 20 and CV^2_A = 27 in units where CV is specified in percent (Table 1). Thus $\bar{p} \leq 0.2 \times 10^{-5}$.

For comparison we can calculate meiotic nondisjunction probabilities from cytological measurements and postfertilization probabilities from analysis of early embryos. Again assuming a linear response, the fraction of aneuploid cells f is given by

$$f = 2pNX \tag{6}$$

where p is the probability of nondisjunction per chromosome per rad, N is the haploid number of chromosomes, and X is the dose in rads. From the data of Szemere and Chandley [22], p $\sim 10^{-5}$ for meiotic metaphase II at 100 rad in the mouse; and from Tates et al. [23, 25] $0.25 \times 10^{-5} \leq p \leq 10^{-5}$ for the sex chromosomes in spermatids of the vole *M. oeconomous* at 100 rads. The frequency of aneuploidy in early mouse embryos whose fathers were exposed to x rays and mitomycin-C was barely measureable [6]. For the radiation exposure p $\leq 10^{-6}$. Thus the flow cytometric results are consistent with other determinations of the nondisjunction rate. The fact that the sensitivity of all these detection methods is low is fundamentally due to the low probability of radiation induced nondisjunction. Cytological methods, which allow direct visualization of the chromosomes and individual counting of the errors, have the greatest potential sensitivity but require high effort.

Table 1. Comparison of factors related to dose response sensitivity

	ΔXY[a]	CV^2_A[b]	N[c]	NCV^2_A[d]
Human	3.0	24	23	552
Mouse	3.2	27	20	540
Microtus montanus	2.5	100	12	1200
Microtus oregoni	9.2	180	9	1620

[a]DNA content difference of X- and Y-chromosome bearing sperm, percent

[b]Square of coefficient of variation of DNA distribution of sperm with all possible single chromosome aneuploidies. Probability is assumed equal for each chromosome. The number encompases both X and Y subpopulations. Units are percent squared

[c]Haploid number of chromosomes

[d]Expected dose response given by Eq. (5) is $CV^2_D = 2\,\bar{p}\,X\,NCV^2_A$, where $\bar{p}$ is the probability of nondisjunction per chromosome per rad and X is the dose in rads

The above comparison has been predicated on the assumption that the DNA content errors are due to whole chromosome aneuploidies. The flow cytometric results shed no light on this assumption since all that is measured is the total sperm DNA content. Cytologic data [22] indicates that at doses up to 200 rad single chromosome aneuploidy occurs, while at higher doses a more complex fragmentation of the genome may dominate. I have also assumed a linear response up to 100 rad in extracting the values of p from the cytologic data. The flow measurements indicate that by 100 rad the response is dominated by the quadratic term (the C coefficient of Eq. (3) even assuming the maximum value for the linear term. Given the experimental uncertainties both in the cytologic and flow data a more rigorous comparison is not warranted. The one presented above serves to demonstrate the rough agreement of the results.

3.3 Comparative Sensitivity of Animal Systems

The relative ability to detect sperm with abnormal DNA content in different species depends both on the aneuploidy probability and the spectrum of the DNA content errors. In the model calculations performed above, this depends on $\bar{p}$ and the distribution of chromosome sizes, CV^2_A. Applying Eq. (6), $CV^2_D = 2 \bar{p} N \times CV^2_A$. In Fig. 9 the DNA content distributions of normal sperm and those with a single chromosome error are shown for the mouse, *Microtus montanus,* and *Microtus oregoni.* The separation of the X and Y sperm peaks, N, CV^2_A and N CV^2_A are shown for these as well as the human in Table 1. Note that if the value of $\bar{p}$ is comparable for all of these species, detection of aneuploidy in the human should be as sensitive as in the mouse, while the two types of Microtus may offer a doubling or tripling of the sensitivity. *M. montanus* is attractive because the smaller split between the X and Y populations of normal sperm may make it easier to detect cells with slightly abnormal DNA content. An ideal species would have X and Y chromosomes of equal size.

The distributions of Fig. 9 suggest an alternative detection strategy. Rather than trying to measure all aneuploid sperm, one could just detect those containing errors of the larger chromosomes. For example, both *M. montanus* and *M. oregoni* contain 8 or 9 errors that would result (if they are compatible with cell survival) in sperm departing by 10% or more from the mean DNA content. Thus one would expect, using $\bar{p} = 10^{-6}$, a frequency of about 10^{-5} per rad for aneuploid sperm with large DNA content errors. The rarity of these events makes their detection difficult in the face of the normally expected fluorescent debris. Measurements of the debris frequency present in bull sperm samples indicate that a sensitivity of a few tens of rads might be possible if the assumption of $\bar{p} \sim 10^{-6}$ is correct. Sorting of these suspected aneuploid sperm for a confirmatory analysis would be required, at least until confidence in the procedure was established.

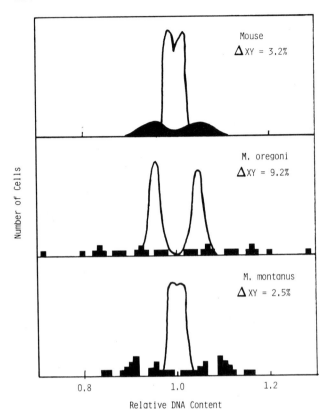

Fig. 9. The DNA distributions of normal sperm (*line*) and sperm with single chromosome aneuploidies (*solid area*) in sperm from three species. The distributions for the normal sperm are shown broadened as they appear in flow cytometric histograms. ΔXY is the percentage DNA content difference between X and Y sperm. In *M. oregoni* ΔXY refers to the difference between the Y and "O" sperm populations [18]

4 Conclusion

Both image analysis and slit-scan flow cytometry have the ability to detect head shape abnormalities in sperm. Their eventual application is unclear since results to date have been restricted to training sets of data. In the current state of development, a specifically tailored set of shape parameters in conjunction with computer image processing promises the most sensitivity.

The incidence of diploid sperm may be a sensitive biological dosimeter for x rays but uncertainties in the low dose region of the response curve need to be resolved. Measurement of total DNA content can detect induced aneuploidy in sperm. However, the sensitivity to x ray exposure is low because of the small size and low frequency of the DNA content errors.

References

1. Beaty RA, Fechheimer NS (1972) Diploid spermatozoa in rabbit semen and their experimental separation from haploid spermatozoa. Bio of Reprod 7:267–277
2. Benaron DA, Gray JW, Gledhill BL, Lake S, Wyrobek AJ, Young IT (1982) Quantification of mammalian sperm morphology by slit-scan flow cytometry. Cytometry 2:344–349
3. Brandriff B, Gordon L, Watchmaker G, Ashworth L, Carrano A, Wyrobek AJ (1982) Analysis of sperm chromosomes in normal healthy men using the human-sperm/hamster-egg *in vitro* fertilization system. Abstracts Amer Soc Human Genet 33rd Ann Mtg, Detroit, Mich., Sept. 29-Oct. 2, 1982
4. Brandriff B, Gordon L, Carrano AV, Ashworth L, Watchmaker G, Wyrobek AJ (1983) Direct analysis of human sperm chromosomes: comparison among four individuals (in preparation)
5. Bruce WR, Furrer R, Wyrobek AJ (1974) Abnormalities in the shape of murine sperm after acute testicular x-irradiation. Mutat Res 23:381–386
6. Chandley AC, Speed RM (1979) Testing for nondisjunction in the mouse. Environ Health Perspect 31:123–124
7. Fraser LR, Maudlin I (1979) Analysis of aneuploidy in first-cleavage mouse embryos fertilized *in vitro* and *in vivo*. Environ Health Perspect 31:141–149
8. Garner DL, Gledhill BL, Pinkel D, Lake S, Stephenson D, Van Dilla MA, Johnson LA (1983) Quantification of the X- and Y-chromosome-bearing spermatozoa of domestic animals by flow cytometry. Biol Reprod 28:312–321
9. Gledhill BL (1964) Cytophotometry of presumed diploid bull spermatozoa. Nord Vet-Med 17: 328–335
10. Hacker-Klom U (1984) In: Eisert WG, Mendelsohn ML (eds) Biological Dosimetry. Cytometric Approaches to Mammalian Systems. Springer, Berlin Heidelberg New York Tokyo p 127
11. Hacker U, Schumann J, Göhde W, Müller K (1981) Mammalian spermatogenesis as a biological dosimeter for radiation. Acta Radiol Oncology 20:279–282
12. Hansmann I, Probeck HD (1979) Detection of nondisjunction in mammals. Environ Health Perspect 31:161–165
13. Illisson L (1969) Spermatozoal head dimensions in two inbred strains of mice and their F_1 and F_2 progenies. Australian J Biol Sci 22:947–963
14. Martin RH, Lin CC, Balkan W, Burns K (1982) Direct chromosomal analysis of human spermatozoa: preliminary results from 18 normal men. Am J Hum Genet 34:459–468
15. Meistrich ML, Göhde W, White RA (1978) Resolution of X and Y spermatids by pulse cytophotometry. Nature 274:821–823
16. Moore II DH, Bennett DH, Kranzler D, Wyrobek AJ (1982) Quantitative methods of measuring the sensitivity of the mouse sperm morphology assay. Analyt and Quant Cytol 4:199–206
17. Otto FJ, Hacker U, Zante J, Schumann J, Göhde W, Meistrich ML (1979) Flow cytometry of human spermatozoa. Histochemistry 62:249–254
18. Pinkel D, Gledhill BL, Lake S, Stephenson D, Van Dilla MA (1982) Sex preselection in mammals? Separation of sperm bearing Y and "O" chromosomes in the vole *Microtus oregoni.* Science 218:904–906
19. Pinkel D, Gledhill BL, Van Dilla MA, Lake S, Wyrobek AJ (1983) Radiation-induced DNA content variability in mouse sperm. Radiat Res 95:550–565
20. Pinkel D, Lake S, Gledhill BL, Van Dilla MA, Stephenson D, Watchmaker G (1982) High resolution DNA content measurements of mammalian sperm. Cytometry 3:1–9
21. Searle AG, Beechey CV, Green D, Humphreys ER (1976) Cytogenetic effects of protracted exposures to alpha-particles from plutonium-239 and to gamma-rays from cobalt-60 compared in male mice. Mutat Res 41:297–310
22. Szmere G, Chandley AC (1975) Trisomy and triploidy induced by x-irradiation of mouse spermatocytes. Mutat Res 33:229–238
23. Tates AD (1979) *Microtus oeconomus* (Rodentia), a useful mammal for studying the induction of sex-chromosome nondisjunction and diploid gametes in male sperm cells. Environ Health Perspect 31:151–159

24. Tates AD, Pearson PL, Geraedts JPM (1975) Identification of X and Y spermatozoa in the northern vole *Microtus oeconomus.* J Reprod Fertil 42:195–198

25. Tates AD, Pearson PL, Ploeg Mvd, de Vogel N (1979) The induction of sex-chromosomal non-disjunction and diploid spermatids following x-irradiation of pre-spermatid stages in the northern vole *Microtus oeconomus.* Mutat Res 61:87–101

26. Van Dilla MA, Gledhill BL, Lake S, Dean PN, Gray JW, Kacher V, Barlogie B, Göhde W (1977) Measurement of mammalian sperm deoxyribonucleic acid by flow cytometry – problems and approaches. J Histochem Cytochem 25:763–773

27. Wyrobek AJ, Bruce WR (1978) The induction of sperm shape abnormalities in mice and humans. In: Hollaender A, de Serres F (eds) Chemical mutagens, Vol 5. Plenum Publishing Corp, New York, p. 257

28. Wyrobek AJ, Gordon LA, Burkhart JG, Francis MW, Kapp RW, Letz G, Malling HV, Topham JC, Whorton MD (1983) An evaluation of the mouse sperm morphology test and other sperm tests in nonhuman mammals. Mutat Res 115:1–72

29. Wyrobek AJ, Gordon LA, Burkhart JG, Francis MW, Kapp RW, Letz G, Malling HV, Topham JC, Whorton MD (1983) An evaluation of human sperm as indicators of chemically induced alterations of spermatogenic function. Mutat Res 115:73–148

30. Young IT, Gledhill BL, Lake S, Wyrobek AJ (1982) Quantitative analysis of radiation induced changes in sperm morphology. Analyt and Quant Cytol 4:207–216

Mammalian Spermatogenesis as a Biological Dosimeter for Ionizing Radiation

U. Hacker-Klom,[1] W. Göhde,[1] and J. Schumann[2]

[1]Institut für Strahlenbiologie, Universität Münster, Hittorfstr. 17, D-4400 Münster
[2]Fachklinik Hornheide, Universität Münster, Dorbaumstr. 44–48, D-4400 Münster

Summary

Mammalian spermatogenesis is an extremely sensitive *in vivo* system for biologic dosimetry of ionizing radiation. Differentiated spermatogonia which are responsible for the radiation-induced reduction of DNA-synthesizing testicular cells are the most sensitive cells we know in adult mammals. The D_{50}-dose of murine S-phase diminution is about 0.25 Gy. The dose-response relationship of S-phase reduction with 200 kV X-rays is shown in the range from 2.5 Gy down to 0.1 Gy using data from flow cytophotometric analysis of ethidium bromide/mithramycin stained cells. In addition to the inactivation of spermatogonia the value of mutagenic actions of ionizing radiation on spermatogenic cells (the increase of the coefficient of variation and of the frequency of diploid spermatids) for biologic dosimetry is investigated. These data are compared with other models of biologic dosimetry using spermatogenic cells. The application of this system to man is discussed as well.

1 Introduction

In case of an accidental radiation exposure it is of importance for genetic counseling to be able to estimate the gonad dose to an exposed person. The ovaries cannot be used as a model of quantitative biologic dosimetry because of the low number of oocytes in a woman. The testes, however, are a steady state system with high proliferative activity and thus are expected to be very sensitive to the cytotoxic action of ionizing radiation (cf. Bergonié & Tribondeau 1906). Actually, mammalian spermatogenetic cells proved to be most radiosensitive (cf. e.g. Hacker et al. 1981).

The classic method of biologic dosimetry by counting the radiation-induced chromosome aberrations in lymphocytes of the peripheral blood (Bender & Gooch 1962) only renders a rough estimate of the gonad dose. The cytogenetic analysis of lymphocytes is only an indicator model, not a risk model. A further disadvantage of this classic method is that it is very timeconsuming. The analysis of 100 to 200 metaphases requires one-man-day. In the low dose range (below 0.3 Gy) 1500 to 2 000 metaphases should be analyzed and this requires about 10 man-days (G. Stephan 1982). Thus, the necessity

Biological Dosimetry. Edited by W. G. Eisert and M. L. Mendelsohn
© Springer-Verlag Berlin Heidelberg 1984

of counting a large number of cells limits the value of this model of biologic dosimetry in the low dose range. Methods designed for use in radiation accidents where physical dosimeters were not worn must be sufficiently sensitive and rapid. Their sensitivity should allow detection within the official radiation-exposure limits. Acceptable time limits will depend in part on the number of individuals exposed. Thus, the technique of cytogenetic dosimetry cannot be used for accidental radiation exposures of large scale (e.g. accidents in nuclear power plants) because the facilities for analyzing large quantities of blood samples do not exist. Other models of biologic dosimetry of similar or even greater sensitivity should be available. Here a method is proposed using the *in vivo* system of spermatogenesis in combination with flow cytometry.

2 Material and Methods

Male NMRI mice aged 9 to 12 weeks were irradiated with X-rays (200 kV, 0.5 mm Cu, 1.29×10^{-2} C/kg/min). Only the testis region was irradiated while the body was shielded with lead (0.2 mm thick). During the irradiation the mice were rotated along their longitudinal axis in order to provide a homogeneous dose-distribution. Doses ranged from 0.1 to 15 Gy. 3 mice per dose and time point (2 to 70 days following irradiation) were sacrificed; at doses of 0.5 Gy and below 5 or 10 mice were analysed. The data of the irradiated mice and the 18 sham-irradiated control mice were pooled.

Testes were prepared using pepsin-HCl for cell separation and ethidium-bromide/mithramycin for DNA-staining according to the method described earlier (Zante et al. 1976, 1977; Hacker et al. 1981). Flow cytometry was performed with a pulse cytophotometer developed by Göhde et al. (1979), which had a special device to count the number of cells in a certain volume of suspension (0.2 ml) in order to calculate the total number of the different cell types in the testis.

3 Results

3.1 Composition of a Testis DNA Histogram:

Figure 1 demonstrates the cellular DNA content changes during spermatogenesis which are reflected in a DNA histogram. Peak no. I represents elongated spermatids (step 10 or 11 to 16 of murine spermiohistogenesis), and peak no. II contains the DNA values of round spermatids (step 1 to 9 or 10 of murine spermiohistogenesis) (Zante et al. 1977; Hacker et al. 1981). G_1-spermatogonia, G_1-preleptotene spermatocytes and secondary spermatocytes are registered in peak no. III in the 2c-region. Spermatogonia and primary spermatocytes synthesizing DNA are registered in the S-phase region between 2c and 4c. In peak no. IV in the 4c-region primary spermatocytes and (G_2+M)-spermatogonia are recorded.

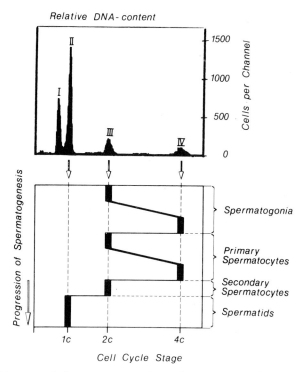

Fig. 1. Changes of the cellular DNA content during spermatogenesis as reflected in a DNA frequency distribution of an untreated control mouse

Around 90% of all testicular cells are germ-cells and only 10% are non-germ cells (Hacker 1981). All non-germ cells are registered in the 2c-peak and comprise about 80% of this peak. Among the 2c-germ cells there are about 60% secondary spermatocytes, 26% spermatogonia and 14% primary spermatocytes. About 48% of the S-phase cells are spermatogonia, 52% are spermatocytes. (G_2+M)-spermatogonia contribute only to about 2% of the 4c-peak; 98% of the DNA values in the 4c-region are formed by primary spermatocytes which mainly are in pachytene (Hacker 1981). The calculations of the different testicular cell-types were performed using cumulative frequency distributions (Göhde et al. 1979). The composition of a testis DNA histogram described above is typical not only for the mouse but also for all other mammals investigated up to now including man.

3.2 Radiation Effects

The mutagenic radiation effects that can be detected by flow cytometry (induction of diploid elongated spermatids and the increase of the variability of the cellular DNA content) as well as the cytotoxic radiation effects as measured by a reduction of elongated spermatids have been described recently (Hacker et al. 1981). Here only the induction of diploid elongated spermatids and the reduction of the number of DNA-synthesizing cells shall be demonstrated.

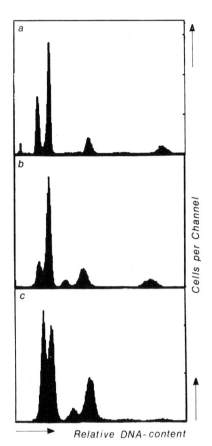

Fig. 2a–c. The histogram shows a normal DNA frequency distribution of unirradiated mouse testis. **b** This histogram demonstrates a high frequency of diploid elongated spermatids in a mouse. The mouse was untreated. I do not have any explanation for the high rate of genome mutations that occurred in that animal. **c** A histogram 14 days after irradiation of a mouse testis demonstrates that a lot of the elongated spermatids are diploid

3.2.1 Induction of Diploid Spermatids

In control histograms of unirradiated mice there is a certain number of DNA values registered between the 1c- and the 2c-region. These signals form a normal distribution with a maximum at twice the modal channel number of peak no. I (Fig. 2). This small peak represents diploid elongated spermatids (Hacker 1981). The control value of unirradiated mice is 4.1 ± 0.82 per cent of all testicular cells or 7.8 ± 1.74 x 10^5 per testis.

After irradiation increases of the number of diploid spermatids are seen. Figure 3 shows oversized polyploid round spermatids in a histologic cross-section 14 days following an irradiation with 10 Gy. The occurrence of these oversized spermatids in histologic cross-sections was followed by an increase of the number of DNA values in the peak between 1c and 2c representing diploid elongated spermatids. These cells were sorted with a self-developed flow-sorter attached to our self-developed flow-cytometer (Fig. 4). Under the fluorescence microscope these cells looked either oversized or two-headed. Obviously the meiotic divisions were not completed properly in these cells. The origin of diploid meiotic products appears to be a failure of the second meiotic division (Sora and Magni 1982), at least in *Saccharomyces cerevisiae.* Figure 5 shows the radiation-induced changes of the number of diploid elongated spermatids at different

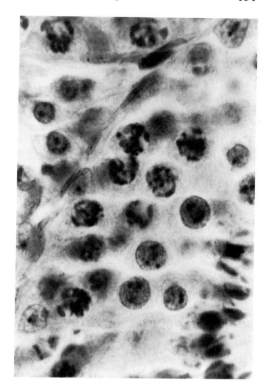

Fig. 3. This histologic cross-section demonstrates that 14 days following an irradiation with 10 Gy there are oversized diploid round spermatids in the mouse testis

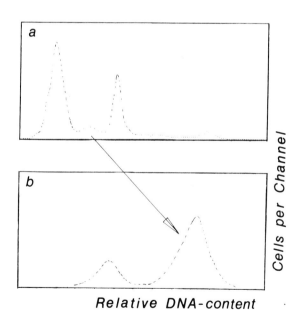

Fig. 4. a A histogram of testicular cells of a mouse irradiated with 6.75 Gy 21 days before is shown. An increase of the number of diploid elongated spermatids is reflected by the hight of the peak between 1c and 2c. **b** The diploid elongated spermatids of the sample represented in **a** were sorted and measured with the fluid switch cell sorter (for a detailed description, see Dühnen et al. 1982). The first peak represents haploid elongated spermatids, the second peak diploid elongated spermatids

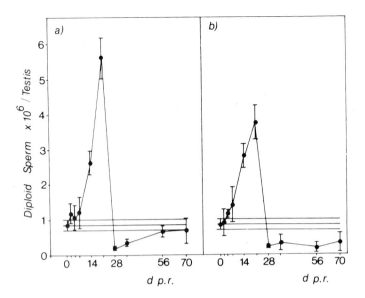

Fig. 5a,b. Time-dependent changes of the number of diploid elongated spermatids 2 to 70 days following radiation exposures with **a** 5 Gy and **b** 10 Gy. The arithmetic means and the 95% confidence limits are indicated. The *straight lines* show the mean control values with the 95% confidence limits

intervals following irradiation with 5 Gy and 10 Gy. 21 days following a radiation with 5 Gy the number of diploid elongated spermatids increased to a maximum at about the sixfold of the control value. Figure 6 demonstrates the dose-dependent changes 14 and 21 days following radiation.

3.2.2 Reduction of the Number of DNA-synthesizing Cells

The most sensitive criterion of radiation action in the low dose range and thus the most interesting feature for biologic dosimetry is, however, the reduction of the number of DNA-synthesizing cells. All those cells contributing to the decrease of the number of DNA-synthesizing cells were irradiated as spermatogonia. Figure 7 shows the time-dependent reduction of the number of S-phase cells following an irradiation with 2.5 Gy 2 to 70 days after the irradiation. This reduction shows a maximum 2 to 7 days after the radiation exposure. A similar timing of the reduction of the S-phase cells is found following radiation with other doses. The decrease of S-phase cells was parallelled by a reduction of spermatogonia as seen in the histologic cross-section (Hacker 1981). Figure 8 demonstrates the dose-response relationship 2 days after irradiation: The D_{50} value of this curve is 0.25 Gy. Even a dose of 0.1 Gy leads to a significant ($p < 0.05$) reduction of S-phase cells. There is no shoulder in this dose-response curve (for more details cf. Hacker 1981).

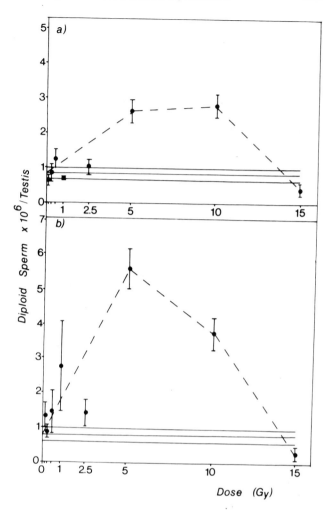

Fig. 6a,b. Dose-dependent changes of the number of diploid elongated spermatids **a)** 14 and **b)** 21 days following irradiations with 0.1 to 15 Gy. The arithmetic means and the 95% confidence limits are indicated. The straight lines show the mean control values with the 95% confidence limits

4 Discussion

The premeiotic stages, especially the differentiated spermatogonia, are most sensitive against the radiation-induced cell-inactivation which leads to a decrease of the number of haploid germ-cells (Hacker 1981). The D_{50}- or LD_{50}- dose für the reduction of S-phase cells and elongated spermatids was 0.26 Gy and 0.75 Gy respectively (Hacker 1981). Oakberg (1957) described an LD_{50}-dose of 0.20 to 0.24 Gy for differentiated spermatogonia.

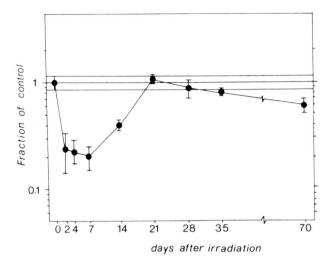

Fig. 7. Time-dependent changes of the number of DNA-synthesizing cells 2 to 70 days after an irradiation with 2.5 Gy. The arithmetic means and the 95% confidence limits are indicated. The *straight lines* show the mean control values with the 95% confidence limits

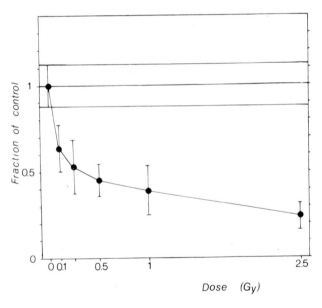

Fig. 8. Dose-dependent reduction of the number of S-phase cells 2 days after irradiations with different doses (0.1 to 2.5 Gy). The arithmetic means and the 95% confidence limits are indicated. The *straight lines* show the mean control values with the 95% confidence limits. The half-logarithmic plot looks like the linear one

The meiotic stages, mainly the spermatocytes, are most sensitive against the mutagenic action of ionizing radiation. This was shown previously. The spermatocyte was the only male germ-cell in which genome mutations and detectable aneuploidies as measured by a broadening of the DNA-histogram could be induced. Spermatocytes are known to be most sensitive against the radiation-induction of aneuploid sperm (Tates et al. 1979), sperm head abnormalities (Wyrobek 1979), and unscheduled DNA synthesis (Sotomajor et al. 1979). The doubling dose of the induction of diploid spermatids was about 1 Gy, the doubling dose of the increase of the genetic variability as measured by the coefficient of variation was about 5 Gy (Hacker 1981). Thus, the radiation-induced mutations do not render a suffiently sensitive criterion for biologic dosimetry. They are, however, indicators of the mutagenic actions of irradiation and chemical noxae (Hacker 1977) thus providing a mutagenicity test system. The biologic relevance of these mutations is not known. Up to now it has not been possible to measure a genetic response to radiation in man (Neel et al. 1980), but man is espected to be like all other organisms we know in this respect since Muller's work on Drosophila (1927).

The model proposed here using the reduction of the number of DNA-synthesizing testicular cells is the most sensitive and can detect a radiation exposure of only 0.1 Gy with 95% confidence. The usefulness of this model of biologic dosimetry in the low dose range relates especially to the absence of a shoulder in the initial part of the dose-response curve. The major advantage of the system proposed here is its speed and correspondingly its cheapness: The analysis of one testicular sample only takes about 15 min. Thus, in the low dose range, this technique is about 300 times less time-consuming than the conventional method of biologic dosimetry, the chromosome analysis. The large number of cells analyzed (about 30 000 cells per histogram) minimizes the statistical error. The analysis of the incorporation changes of radioactive thymidine into DNA of testicular cells has been proposed as an *in vivo* short-term test for the identification of chemical mutagens by Friedman and Staub (1976), Seiler (1977) and Lambert and Eriksson (1979), and for the intercomparison of neutron beams by Geraci et al. (1977). Labelling techniques, however, have two disadvantages: First, more time is required until results can be seen, and second, labelling techniques cannot be applied in man.

The method proposed here might be an additional way to detect or exclude suspected accidental radiation overexposures in man. Two-parameter measurements of DNA and protein allow to discriminate between germ-cells with a 2c-DNA content and non-germ cells thus enabling a further improvement in the precise analysis of radiation-induced damage of testicular cells (Hacker 1981). It is of great importance that this model allows the direct estimation of the gonad dose. The most important theme is the risk estimation for man. It can only be performed with human germ-cells. Lymphocytes are not relevant for estimating the genetic risk. The general concept of germ-cell sensitivity of the mouse can be applied to man as well (Oakberg and Lorenz 1972). Spermatogonia which are responsible for the radiation-induced S-phase reduction described here seem to be even more sensitive in man than in the mouse if cell-inactivation is the criterion. This can be concluded from the fact that an irradiation of only 0.15 Gy leads to a long-lasting reduction of sperm in man to about 25% of the control value (Fig. 9). The aspiration biopsy technique using a fine needle renders a sufficient num-

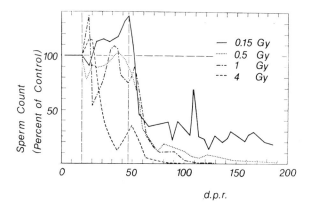

Fig. 9. Time-dependent radiation-induced reduction of human sperm count after irradiation with different doses (from: Heller 1967)

ber of cells for flow cytometry (Clausen and Åbyholm 1980). This technique might by used in special cases to evaluate radiation exposures. Or, in case of a longer interval between irradiation and analysis, evaluation of sperm counts in the ejaculate might be applied.

6 References

Bender MA, Gooch PC (1962) Types and rates of X-ray induced chromosome aberrations in human blood irradiaten in vitro. Proceedings of the National Academy of Sciences 48:522–532

Clausen OPE, Åbyholm T (1980) Desoxyribonucleic acid flow cytometry of germ cells in the investigation of male infertility. Fertil Steril 34:369–374

Dühnen J, Stegemann J, Wiezorek C, Mertens H (1983) A new fluid switching flow sorter. Histochemistry 77:117–121

Friedman MA, Staub J (1967) Inhibition of mouse testicular DNA synthesis by mutagens and carcinogens as a potential simple mammalian assay for mutagenesis. Mutation Research 37:67–76

Geraci JP, Jackson KL, Christensen GM, Thrower PD, Weyer BJ (1977) Mouse testes as a biological test system for intercomparison of fast neutron therapy beams. Radiation Research 71:377–386

Göhde W, Schumann J, Büchner T, Otto F, Barlogie B (1979) Pulse cytophotometry. Application in tumor cell biology and clinical oncology. In: Melamed MR, Mullaney PF, Mendelsohn ML (eds) Flow cytometry and sorting. John Wiley & Sons Inc., New York, Chichester, Brisbane, Toronto, p 599

Hacker U (1977) Der Einfluß physikalischer und chemischer Noxen auf die Spermatogenese der Maus. Staatsexamensarbeit, University of Münster

Hacker U (1981) Biologische Dosimetrie ionisierender Strahlung am Modell der Spermatogenese der Maus. Ph. D. Dissertation, University of Münster

Hacker U, Schumann J, Göhde W (1981) Mammalian spermatogenesis as a biologic dosimeter for radiation. Acta Radiologica Oncology 20:279–282

Heller CG (1967) Effects on the germinal epithelium. In: Langham H (ed) Radiobiological factors in manned space flight. NRC Publication 1487, National Research Council, Washington DC, p 124

Lambert B, Eriksson G (1979) Effects of cancer chemotherapeutic agents on testicular DNA synthesis in the rat. Evaluation of a short-term test for studies of the genetic toxicity of chemicals and drugs *in vivo*. Mutation Research 68:275–289

Luckey TD (1982) Hormesis with ionizing radiation. CRC Press Inc., Boca Raton, Florida, p 46

Muller HJ (1927) Artificial transmutation of the gene. Science 66:84–87

Neel JV, Satch C, Hamilton HB et al. (1980) Search for mutations affecting protein structure in children of atomic bomb survivors: Preliminary report. Proc Natl Acad Sci USA 77:4221–4225

Oakberg EF (1957) Gamma-ray sensitivity of spermatogonia of the mouse. J Exptl Zool 134:343–356

Oakberg EF, Lorenz EC (1972) Irradiation of generative organs. In: Hug O, Zuppinger A (eds) Handbuch der Med. Radiologie II, Springer-Verlag, Berlin, Heidelberg, New York, p 217

Seiler JP (1977) Inhibition of testicular DNA-synthesis by chemical mutagens and carcinogens. Preliminary results in the validation of a novel short-term test. Mutation Research 46:305–310

Sora S, Magni GE (1982) Meiotic diploid progeny and meiotic non-disjunction in *Saccharomyces cerevisae*. Genetics 101:17–33

Sotomajor RE, Sega GA, Cumming RE (1979) An autoradiographic study of unscheduled DNA synthesis in the germ cells of male mice treated with X-rays and methyl methansulfonate. Mutation Research 62:293–309

Tates AD, Pearson PL, van der Ploeg M, de Vogel N (1979) The induction of sex-chromosomal nondisjunction and diploid spermatids following X-irradiation of the pre-spermatid stages in the northern vole *Microtus oeconomus*. Mutation Research 61:87–101

Zante J, Schumann J, Barlogie B, Göhde W, Büchner T (1976) New preparation and staining procedures for specific and rapid analysis of DNA distributions. In: Göhde W, Schumann J, Büchner T (eds) Second International Symposium on Pulse Cytophotometry. European Press, Ghent, p 97

Zante J, Schumann J, Göhde W, Hacker U (1977) DNA-fluorometry of mammalian sperm. Histochemistry 54:1–7

Mutagenic Effects

Biological Dosimetry of Mutagenesis: Principles, Methods, and Cytometric Prospects*

Mortimer L. Mendelsohn

Lawrence Livermore National Laboratory, Biomedical Sciences Division, University of California, P.O. Box 5507 L-452, Livermore, California 94550

1 Introduction

Since the beginning of biological time the alteration of genetic information through mutation has been a threat and a salvation: a threat to survival because of loss of essential function; and a salvation by providing the degrees of freedom with which selection caused genetic diversity and evolution. Geneticists and more recently molecular biologists have been fascinated by the mechanisms of mutation and repair. Now regulators and public health authorities have been drawn into the subject out of concern that contemporary chemically-oriented societies may be genetically toxic to somatic cells thereby causing cancer or to germinal cells thereby causing heritable defects and a lessening in quality of the human gene pool.

Society needs three kinds of information about mutagenesis:
- screening information for identification of mutagenic agents.
- mammalian and human dosimetric information for species extrapolation, hazard analysis and identification of hypersensitive individuals.
- sufficient knowledge of mechanism to understand the consequences of mutation, to evaluate risk, and to make the appropriate short- and long-term regulatory decisions.

Thousands of chemical mutagens have been identified in the screening process, even though only a small fraction of chemicals have yet to be tested. Potency and issues of specie and individuals-within-specie differences in sensitivity are poorly understood. Few methods are available in mammals and in the human, and we have yet to measure unequivocally an induced change in either somatic or heritable mutation rates in any exposed human population. The medical and social cost of mutagenesis could be huge but have yet to be properly documented. With the rapid increase in ability to test mutagenicity of chemicals, we face the difficult responsibility of deciding what to do about those found to be mutagenic. The temptation to regulate strenuously is difficult to resist even though the cost-benefit equation has too many unknowns to be compelling.

Improved, practical, definitive mutagenic biologic dosimetry would go a long way to solving the above problems. Since mutagenic events are cellular and rare, and can be highly specific, they are prime targets for measurement by cytometric automation. It is

---

*Work performed under the auspices of the U.S. Department of Energy by the Lawrence Livermore National Laboratory under contract number W-7405-ENG-48

Biological Dosimetry. Edited by W. G. Eisert and M. L. Mendelsohn
© Springer-Verlag Berlin Heidelberg 1984

Fig. 1. The major steps in chemical mutagenesis

the purpose of this overview to show briefly: how mutation is thought to occur; how mutations and related events are currently detected; and where there may be targets of opportunity for cytometric automation leading to improved biologic dosimetry of mutagenesis in the human and other species.

2 The Mutagenesis Process

The general scheme for chemical mutagenesis as presently understood is shown in Figure 1. The typical chemical mutagen is inhaled, ingested or synthesized as a promutagen, i.e., as a genetically inactive, precursor molecule. It is converted to the active mutagen by enzymatic and other chemical changes which are frequently a side reaction of the normal detoxification mechanisms in the organism. Once activated, the mutagen is generally a reactive electrophile and therefore is attracted to and covalently reactive with DNA as well as with any other negatively charged molecule or molecular region. DNA adducts might involve the bulky addition of an entire mutagen, or the transfer of a simple group such as the methyl or ethyl group in alkylation. Some chemicals interact with DNA by intercalation between the stacked bases of the DNA double helix, and can exert their effect without the formation of covalent bonds. Ionizing radiation, through direct hits or indirect water and other chemical radicals, can break DNA strands or chemically change the sugars and bases. Ultraviolet radiation, largely by direct absorption in the DNA, modifies the bases and produces unique lesions such as thymine dimers, which covalently link adjacent thymines on a single strand of DNA.

The lesions in DNA can be recognized by sensitive chemical analysis and by monoclonal antibodies (Adamkiewicz et al., this volume). This provides a form of dose (Ehrenberg et al., in press) which when related to subsequent mutation eliminates the vagaries of metabolic activation and pharmacologic behavior of the promutagen and mutagen. Detection of the weakly fluorescent signals from monoclonal antibodies in nuclei or chromosomes is well suited to image and flow cytometry and should become an important tool in the near future of such DNA dosimetry.

The lesions in DNA can also be recognized by a remarkable ensemble of repair mechanisms which can excise the adduction region (single base or often a large region of single stranded DNA), resynthesize a normal patch from the complementary DNA strand, and ligate the patch in place. Single strand breaks are ligated directly. The result is a high liklihood that the DNA will be returned to its original state with no memory of the lesion.

However, some repair mechanisms are error prone (i.e., produce incorrect DNA sequences) and some lesions are "fixated" into abnormal sequences by normal DNA replication before repair takes place. These altered DNA sequences are chemically stable, and, depending on location, represent permanent, potentially heritable, informational changes in the genetic code. These are mutations.

The change of a single base (a base substitution) is the smallest mutational event. A base substitution in the exon region of a gene is likely to express as a single amino acid change in the corresponding gene product. This may or may not change the protein's ability to function. If it occurs in a control region of the gene, a base substitution could prevent gene expression.

Intercalation of a mutagen and other types of damage can cause the addition or deletion of a base in a sequence. This results in a frame-shift away from the normal triplet coding of the DNA and generally scrambles the gene product from the point of the lesion to the end of the molecule. Frameshifts also cause misreading of the termination triplet and hence render gene products that run on until the next termination accidentally occurs. Frame shift abnormalities thus are considerably larger than base substitutions, and they often result in inactive or badly compromised gene products.

Deletion is the result of strand breakage and misrepair. Such lesions can range from a few nucleotides, through a few genes, to the macroscopic lesions scored at the chromosomal level by the cytogeneticist. As seen in mutation testing, deletion is generally a lesion that irreversibly wipes out one or a few contiguous genes.

Table 1 summarizes the above mutations in terms of functional effect, size and revertability. TS refers to temperature sensitive product, a common type of degraded but still functioning protein. Forward mutation is used in the sense that normal DNA is being changed to a new or abnormal state. For those examples in the Table where sufficient information is preserved to allow for reversal of the effect, the return toward the normal state (revertability) would be via a reverse mutation.

3 Illustrative Methods in Mutagenesis Testing

The Ames Test is presently the most commonly used method to identify mutagenic chemicals (McCann et al. 1975). Special strains of Salmonella typhimurium have been made repair defective and membrane permeable to increase their sensitivity to mutation. A specific mutation (a base substitution or frame shift at a G-C specific or A-T specific site) causing histidine dependence has been built into each strain. The bacteria are plated in large numbers on agar with the chemical being tested, and with enough histidine to allow a few bacterial divisions for expression of the revertant phenotype. Two days later colonies are counted. Each colony represents a revertant bacterium that is able to grow indefinitely in the absence of histidine because the chemical (or a random background event) reversed the specific mutation built into that strain. If the revertants are significantly above background, the chemical is a mutagen and can be roughly classified on the basis of which strains responded. Metabolic activation is

Table 1

Forward mutation	Mechanism	Size	Revertability
Loss of linked gene products	Deletion	Very Large	No
Loss of gene product	Deletion	Variable	No
Loss of gene function	Frame shift	Variable	Yes
Degraded gene function, TS	Substitution	Small	Yes
Modified gene product			
Specific amino acid substitution	Substitution	Small	Yes

generally provided by adding microsomes from supernatent of rat liver homogenates to the culture at the same time as the chemical being tested.

The Ames test is simple enough to be done manually, although some laboratories use commercial automated colony counters. There are many other bacterial methods for mutagenicity testing, and some have been combined with time-lapse photography and image analysis to identify mutants by changes in growth rate with time or temperature, or by colony morphology (Busch et al., 1980).

Almost all tests involving mammalian cells, mammals or humans are based on forward mutation since arranging for specific testor stock in which to test for revertants in such material ranges from difficult to impossible. Drug resistance offers a good way to test for rare, forward mutations. A prototypic and perhaps commonest example is the hypoxanthine-guanine phosphoribosyl transferase (HGPRT) assay. HGPRT mutants can be identified by resistance to 6-thioguanine or 8-azaadenine, two toxic base analogues that require HGPRT to become activated. When used with cultured cell lines (Hsie et al. 1975), the chemical, with or without metabolic activator, is added to a monolayer of cells, after a suitable induction period one of the analogues is also added, and a week or two later the plates are scored for colonies. The absence or reduction of HGPRT enzyme can be confirmed in the surviving, drug-resistant colonies. The corresponding *in vivo* version for human cells, as developed by Strauss and Albertini (1979), identifies 6-thioguanine resistance in PHA-treated lymphocytes by autoradiographic detection of DNA synthesis. Unfortunately, it is not possible to check further that cells with overlying silver grains are indeed rare mutants of HGPRT. The early results with the human test were heavily compromised by so-called phenocopies, that is cells which were drug resistant for reasons other than a mutation of HGPRT. Albertini (1982) has since developed a way to reduce the phenocopies, as well as a second method based on cloning in which the genetic mechanism can be confirmed. As examples of cytometric approaches, the detection of rare autoradiographically positive cells is an obvious application of image analysis (see Perry et al., in this volume), concentrating cells in S prior to autoradiography is a time saving role for flow sorting (Amneus et al, 1982) and direct counting of DNA-synthesizing cells should soon be possible using BrdU detected by an anti-BrdU fluorescent monoclonal antibody in flow or image analysis (Dolbeare et al., 1983).

A somatic mutation test of base substitution in circulating human erythrocytes has been introduced by Papayannopoulou et al. (1976). The method uses a fluorescent antibody to sickle hemoglobin to detect single red cells that make substantial amounts of sickle hemoglobin in otherwise normal people. The change of glutamic acid to valine at the 6th position of β globin causes sickle hemoglobin and is due to the change GAG to GTG in the corresponding codon of the DNA. Hemoglobin alleles are codominant hence this codon change in an otherwise normal erythrocyte precursor results in red cell descendants with sickle trait (ie, with 30 to 50% sickle hemoglobin). The background frequency of fluorescent cells in the slide-based microscopic assay is around 10^{-7} and it takes one-person-month to score a sample.

Two different approaches toward automation of this test are being taken: one by Jensen et al. (this volume) uses flow cytometry, and one by Tanke et al. of Leiden uses image analysis. Neither has yet surmounted the enormity of counting a one-in-10^7 event at the cellular level. The sickle hemoglobin method is prototypic of many possible specific locus somatic cell assays based on immunologic detection of deviant proteins (Jensen et al., 1980).

A variant of the sickle method has also been used for a germinal mutation assay, the sperm LDH-X test, as developed by Ansari et al. (1980). Lactate dehydrogenase X is a germinal-specific enzyme that appears in the mid-piece of mature sperm. Rat and mouse have almost identical LDH-X enzymes and what differences there are are presumably due to genetic drift caused by single base substitutions leading to nondescript single amino acid changes. Mice injected with rat sperm develop antibodies to the deviant regions of the rat enzyme; these antibodies then can be used to recognize a subset of mutation-induced amino acid changes in the mouse enzyme. Using fluorescent enzyme and microscopic detection, Ansari et al. (1980) have obtained in mice a well-behaved dose response to procarbazine, a potent germ cell mutagen. This method would be improved by monoclonal antibodies and by an automated method of detecting the mutant sperm. Detection of such rare events in the mid-piece of sperm will be challenging to flow and to image analysis.

The classical heritable specific locus test measures mutations in offspring at seven loci controlling non-lethal easily detected morphologic changes such as coat color and tail shape (Russell and Matter 1980; Ehling and Neuhäuser; 1979). Like the cellular method where millions of cells are processed to find a few mutations, these experiments have background rates of 10^{-6} mutations per locus per mouse and are literally megamouse in size. They are straightforward to do, and are the standard for identifying heritable mutagens and doing risk analysis. Apart from expense, the major problem with these tests is the question of how typical the seven morphological genes are of the remaining genotype. So far it has proven difficult to predict heritable mutagens from bacterial and somatic tests, since many conventional mutagens fail either to reach the gonads or to produce lesions that survive the meiotic filter, fertilization, early embryonic development or repair. A cytometric approach to this test makes little sense because every cell in the offspring carries the mutation. It is not a problem of rare cells, but of rare offspring.

A heritable specific locus test suitable for human or mouse application has been based on electrophoretic detection of variant proteins in offspring. The test identifies the subset of mutations which cause a change in charge in an easily sampled gene product such as kidney enzymes in the mouse or serum proteins in the human. It is one of the tests that Neel and colleagues have used to evaluate heritable genetic damage in the children of a-bomb survivors from Hiroshima and Nagasaki (Schull et al., 1981, also see Awa, this volume). By examining up to 30 gene products per child, these studies (according to the latest count, Feb. 1983) have examined 543,664 loci in children of exposed parents and 386, 706 loci in control children. The finding of 3 probable mutations in the exposed and 2 probable mutations in the controls leads to identical mutation rates for each group of 5×10^{-6} per locus per child. Thus this heroic study has lacked the sensitivity to resolve a human radiation effect from an average parental radiation exposure of 60 rems. To increase the sensitivity requires more loci, and one possible approach is to go from one-dimensional electrophoresis to a two-dimensional gel electrophoresis based on charge and molecular weight (Anderson and Anderson, 1982). In such a system, the mutant child displays a spot not shown in either of the parental patterns. Successful deployment of these methods will require computer techniques including image reconstruction of gels, image analysis for identification and quantitation of spots and for discovery of non-congruency of parent and child, data basing, etc. (Skolnick et al., 1982). Beyond the protein approach, there is also the exciting possibility of mutant detection by direct analysis of DNA using various aspects of the rapidly emerging restriction enzyme and recombinant DNA technologies.

To complete this survey, brief comment should be made about biological dosimetry per se. From a mechanistic and public health view, it is crucial to understand the dose-response relationships in human mutagenesis. Extrapolation across species, such as from mouse to human, may work to some extent, but well-demonstrated human dose-responses will be needed for validation of extrapolation and if this fails for case by case analysis of mutagens. The core process of mutagenesis is likely to be inherently non-thresholded, but will be modulated, at times extensively, by the complexities of metabolic activation and repair (Ehrenberg et al., in press; Ehling et al, in press). A variety of studies (eg., Jenssen and Ramel, 1980; Maher and McCormick, 1980) have shown contrasting dose-response curves depending on either the repairability of the adducts or the repair capacity of the cells. In the absence of repair, these *in vitro* studies give linear, non-thresholded dose-responses for induced gene mutation. In the presence of repair the responses become threshold-like (ie., hockey-stick shaped) suggesting complete or near complete prevention of mutation induction at low doses. Such contrasting dynamics will exist between species and between individuals within species. Thus biological dosimetry will be important for humans in general in the quantification of genetic hazard, as well as on a person by person basis in the identification of repair defectives or of individuals with metabolic peculiarities that make them vulnerable to particular agents. The likelihood is that methods will be needed for somatic, germinal and heritable studies of human populations. The technology for such measurements is too young to see clearly but the prospect that automated cytometry will be involved seems highly likely.

4 References

Adamkiewicz J, Ahrens O, Rajewsky MF (1984) High-affinity monoclonal antibodies specific for deoxynucleosides structurally modified by alkylating agents: applications for immunoanalysis. In: Eisert W, Mendelsohn ML (eds) Biological Dosimetry: Cytometric Approaches to Mammalian Systems. Springer, Berlin Heidelberg New York, Tokyo p 325

Albertini RJ (1982) Studies with T-Lymphocytes: An Approach to Human Mutagenicity Monitoring. In: Bridges BA, Butterworth BE, Weinstein IB (eds) Banbury Report 13, Indicators of Genotoxic Exposure. Cold Spring Harbor Laboratory, 393–411

Amneus H, Matsson P, Zetterberg G (1982) Human lymphocytes resistant to 6-thioguanine: Restrictions in the use of a test for somatic mutations arising *in vivo* studied by flow-cytometric enrichment of resistant cell nuclei. Mutation Res 106:163–178

Anderson NG, Anderson NL (1982) The human protein index. Clin Chem 28:739–748

Ansari AA, Baig MA, Malling HV (1980) *In vivo* germinal mutation detection with "monospecific" antibody against lactate dehydrogenase-X. Proc Natl Acad Sci 77:7352–7356

Awa AA, Sofuni T, Honda T, Hamilton HB, Fujita S (1984) Preliminary reanalysis of radiation-induced chromosome aberrations in relation to past and newly revised dose estimates for Hiroshima and Nagasaki A-bomb survivors. In: Eisert W, Mendelsohn ML (eds) Biological Dosimetry: Cytometric Approaches to Mammalian Systems. Springer, Berlin Heidelberg New York, Tokyo p 77

Busch DB, Cleaver JE, Glaser DA (1980) Large-scale isolation of UV-sensitive clones of CHO cells. Somat Cell Genet 6:407–418

Dolbeare F, Gratzner H, Pallavicini M, Gray JW (1983) Flow cytometric measurement of total DNA content and incorporated bromodeoxyuridine. Proc Natl Acad Sci USA, In Press

Ehling UH, Neuhauser A (1979) Procarbazine-induced specific-locus mutations in male mice. Mutation Res 59:245–256

Ehrenberg LE, Moustacchi E, Osterman-Golkar S (1983) Dosimetry of genotoxic agents and dose-response relationships of their effects. Mutation Res (In press)

Hsie AW, Brimer PA, Mitchell TJ, Gosslee DG (1975) The dose-response relationship for ethyl methanesulfate-induced mutations at the hypoxanthine-guanine phosphoribosyl transferase locus in chinese hamster ovary cells. Somat Cell Genet 1:247–261

Jensen RH, Bigbee W, Branscomb EW (1984) Somatic mutations detected by immunofluorescence and flow cytometry. In: Eisert W, Mendelsohn ML (eds) Biological Dosimetry: Cytometric Approaches to Mammalian Systems. Springer, Berlin Heidelberg New York, Tokyo p 161

Jenssen D, Ramel C (1980) Relationship between chemical damage of DNA and mutations in mammalian cells, I. Dose-response curves for the induction of 6-thioguanine-resistant mutants by low doses of monofunctional alkylating agents, x-rays and UV radiation in V79 Chinese hamster cells. Mutation Res 73:339–347

Maher VM, McCormick JJ (1980) Comparison of the mutagenic effect of ultraviolet radiation and chemicals in normal and DNA-repair-deficient human cells in culture. In: F.J. de Serres, A. Hollaender (eds) Chemical Mutagens, Vol 6, Plenum Publ. Corp., New York-London, 309–329

McCann J, Choi E, Yamasaki E, Ames BN (1975) Detection of carcinogens as mutagens in the Salmonella/microsome test: Assay of 300 chemicals. Proc Natl Acad Sci, USA, 72:5135–5139

Papayannopoulou TH, McGuire TC, Lim G, Garzel E, Nute PE, Stamatoyannopoulos G (1976) Identification of haemoglobin S in red cells and normoblasts using fluorescent anti-Hb S antibodies. Br J Haematol 34:25–31

Perry PE, Thomson EJ, Stark MH, Tucker JH (1983) Detection of HGPRT-variant lymphocytes using the FIP high-speed image processor. Presented at Interntl Symposium "Biological Dosimetry: Cytometric Approaches to Mammalian Systems", München/Neuherberg, Germany, October 14–16, 1982

Russell LB, Matter BE (1980) Whole-mammal mutagenicity tests: Evaluation of five methods. Mutation Res 75:279–302

Schull WJ, Otake M, Neel JV (1981) Genetic effects of the atomic bombs: A reappraisal. Science 213:1220–1227

Skolnick MM, Sternberg SR, Neel JV (1982) Computer programs for adapting two-dimensional gels to the study of mutation. Clin Chem 28:4:969–978

Strauss GH, Albertini RJ (1979) Enumeration of 6-thioguanine-resistant peripheral blood lymphocytes in man as a potential test for somatic cell mutations arising *in vivo.* Mutation Res 65:353–379

Detection of HGPRT-Variant Lymphocytes Using the FIP High-Speed Image Processor

P.E. Perry, E.J. Thomson, M.H. Stark, and J.H. Tucker

MRC Clinical and Population Cytogenetics Unit, Western General Hospital, Crewe Road, Edinburgh, EH4 2XU, UK

Introduction

The central issue in genetic toxicology is the assessment of the contribution of chemical and physical agents to genetic ill-health in Man. The major approach to this problem has been to assay the effects of suspect environmental agents upon laboratory animals or cultured prokaryotic or mammalian cells. On the basis of these experiments many carcinogens and mutagens have been identified and steps have been taken to eliminate some of these from our environment, but attempts to use the data so obtained to calculate the risks to Man have proved less than satisfactory. There is thus a clear requirement for a method to estimate directly the mutation rate in some readily available human tissue such as peripheral lymphocytes and such a technique would be of inestimable value in the screening of individuals exposed, through accident or occupation, to genotoxic agents.

Strauss and Albertini (1979) have described a method for the identification in the human peripheral blood lymphocyte population of cells that are variant at the hypoxanthine guanine phosphoribosyl transferase (HGPRT) locus. Although this "salvage" enzyme is not necessary for cell proliferation, its absence or deficiency is the fundamental biochemical defect in individuals suffering from the Lesch-Nyhan syndrome, which is characterised by accelerated *de novo* purine biosynthesis. HGPRT is the enzyme responsible for the conversion of the purines hypoxanthine and guanine to their respective nucleotides during DNA synthesis. Purine analogues such as 6-thioguanine (6-TG) and 8-azaguanine (8-AG) are also substrates for the enzyme, but as the phosphorylated analogues are cytotoxic normal cells cannot survive in the presence of these analogues. On the other hand, mutant cells that have little or no HGPRT activity will survive and make DNA because they are unable to synthesise the toxic nucleotides.

Briefly, the method of Strauss and Albertini (1979) involved the *in vitro* culture of freshly drawn separated white blood cells in the presence of a mitogen, phytohaemagglutinin, which stimulates a proportion of the resting white cell population to synthesise DNA. Some of these cultures received 6-TG at initiation and after a suitable duration of culture the cells in DNA synthesis were labelled with tritiated thymidine and slide preparations were made. Cells engaged in DNA synthesis were identified by visual scanning of autoradiographs by the presence of an overlayer of silver grains in the nuclear emulsion. From a comparison of the labelling index in 6-TG cultures with

Biological Dosimetry. Edited by W.G. Eisert and M.L. Mendelsohn
© Springer-Verlag Berlin Heidelberg 1984

the labelling index in control (untreated) cultures the variant frequency (Vf) could be calculated (see Materials and Methods). Strauss and Albertini reported that the frequency of 6-TG resistant lymphocytes in normal healthy individuals was in the order of 1.3×10^{-4} at 2×10^{-4} M 6-TG, while cancer patients undergoing cytotoxic drug chemotherapy displayed variant frequencies as much as 20 times this value. More recently, this method was adopted by Evans and Vijayalaxmi (1981) who found that the variant frequency increased linearly with the age of the donor and could be increased by *in vitro* exposure of the cells to the alkylating and cross-linking agent, mitomycin-C (MMC).

Despite the surprisingly high variant frequency found by this method (compared with that found in other mutation assay systems) the frequency of labelled cells in azaguanine treated cultures may be less than one in every ten thousand cells. To obtain a statistically adequate number of labelled cells therefore entails the tedious microscopical examination of tens of thousands of cells at a high power of magnification, while direct estimation of the total population of cells is beyond the scope of visual microscopy. The latter value was obtained by Coulter counting methods prior to slide preparation in the work of Strauss and Albertini (1979), while Evans and Vijayalaxmi (1981) made use of an image analysis system (Tucker 1976). However in both studies the count of labelled cells had to be obtained by visual screening.

In the experiments described in this paper the measurement of labelled and unlabelled cells was carried out using FIP (Fast Interval Processor — see Fig. 1), a multipurpose high speed computer analyser which provides accurate morphological and densitometric measurements on cells and cell-like objects, and is capable of high scanning rates ($\rangle$ 500 cells/s). Briefly, FIP consists of a microscope with motorised stage/focus movement, a charge coupled linear diode array scanner, a hardware preprocessor unit for data reduction and a microprocessor which carries out the cell measurement and classification processes required for each individual application. A detailed description of the FIP system is given elsewhere (Shippey et al. 1980).

With the use of a stoichiometric DNA staining technique such as the Feulgen reaction, the integrated optical density (IOD) of all nuclei before DNA synthesis (i.e. with the 2C DNA value) is approximately equal, resulting in a peak in the IOD-frequency distribution (see Fig. 2). The apparent IOD of autoradiographically labelled nuclei is substantially higher than this modal value due, primarily, to the high opacity of the silver grains overlying the nuclei (see Fig. 3 and Table 1). Thus by measuring the IOD of each object and applying classification limits as shown in Fig. 2, it is possible to distinguish small debris particles, unlabelled cells with the modal 2C DNA value and suspect objects with a high IOD which will include labelled cells and artifacts such as overlapping pairs of nuclei and large debris particles. To eliminate these "false positive" signals, the system incorporates an "operator review" facility which allows the operator to relocate and examine each suspect object and to then assign a code designating the object as a labelled cell, unlabelled cell or non-nuclear object. The labelling index of the specimen is then calculated from the total cell count and the labelled cells identified by the operator. Experiments to compare labelling indices from various specimens measured visually and by FIP have shown excellent correlation (Stark et al. 1984).

Fig. 1. The Fast Interval Processor (FIP)

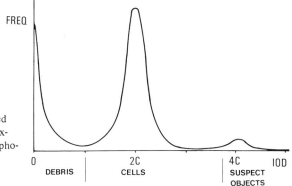

Fig. 2. Distribution of Integrated Optical Density (IOD) Values expected for a 8-AG treated Lymphocyte Population stained by the Feulgen reaction

In this paper we describe experiments to measure the variant frequencies obtained in lymphocytes cultured in the presence of different concentrations of MMC using automated scanning techniques.

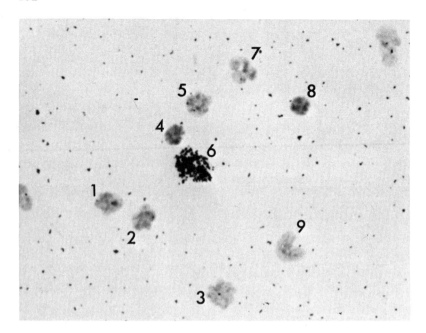

Fig. 3. Typical field containing autoradiographically labelled and unlabelled lymphocytes stained by the Feulgen reaction. IOD values for the various cells are shown in Table 1

Table 1. Measured integrated optical density values for nuclei in Fig. 3

Cell nucleus number	IOD (FIP Units)
1	5611
2	5615
3	5949
4	5725
5	5561
6	24166
7	5779
8	5184
9	4454

Materials and Methods

1. Cell Culture Method

Samples of whole blood were defibrinated immediately after withdrawal to remove polymorphonuclear cells and platelets. The defibrinated blood was then spun down in a bench centrifuge at 1500 x G for 15 min, the serum was removed and the cells were made up to their original volume with serum free medium. The cells were then carefully layered onto freshly prepared ficoll/hypaque (1 part water, 2.5 parts 34% hypaque and 5 parts 12% ficoll) and spun at 1500G for 15 min. The interface layer containing the mononuclear cells was removed carefully by pipette, and the cells were washed twice with RPMI 1640 medium containing 20% dialysed foetal calf serum (FCS). On average, haemocytometer counts indicated a yield of approximately 10^6 mononuclear cells per ml of whole blood.

Two ml cultures were set up in plastic bijoux bottles with 1 x 10^6 cells per ml in RPMI 1640 medium supplemented with 15% dialysed foetal calf serum, phytohaemagglutinin (PHA) and, for some cultures, 2 x 10^{-4}M 8-azaguanine (8-AG). The cells were harvested 40 hours after the addition of PHA and tritiated thymidine (1 μCi/ml) was added for the final 12 h of culture. Exposure of cells to MMC was for two hours immediately before the addition of PHA. The cultures were washed twice with phosphate buffered saline after chemical treatment.

Despite the higher affinity of 6-TG for the enzyme HGPRT we decided to use 8-AG because it inhibits RNA synthesis (Nelson et al. 1975) and therefore prevents decondensation of the nucleus during blast transformation. The higher stain intensity obtained from condensed nuclei considerably simplifies the thresholding process during image analysis, and so leads to more accurate nuclear IOD measurements.

For optimum performance from the image analysis system, the slide preparations must be reasonably dense (~200 cells/mm^2) with as few overlapping nuclei as possible. To produce a monodispersed cell suspension the cell aggregates were dispersed by vigorous vortex mixing followed by washing in 0.2% EDTA. At all stages of preparation the cell suspension was mixed thoroughly. The cells were finally fixed in 5:1 methanol acetic acid, centrifuged and resuspended in a suitable volume of fix. Slides were cleaned thoroughly by soaking and wiping with 70% alcohol and the slide preparations were made in a horizontal laminar flow hood to avoid contamination by dust particles. Slides were stored in a dust free box until required for staining.

The cells were stained using the Feulgen reaction. The preparations were fixed overnight with paraformaldehyde vapour at 50°C in a sealed container, hydrolysed in 5N hydrochloric acid at room temperature for 1 h, rinsed, stained for 90 min at room temperature in Schiff's reagent and finally washed in running water for at least 20 min. To ensure good stain density the Schiff's reagent was never more than 3 months old when used and was membrane filtered to remove charcoal and other debris particles immediately before each use.

Autoradiographs were made by standard methods using Ilford G5 nuclear emulsion. After 3 days exposure at -18°C the autoradiographs were developed in undiluted D19 developer for 4 min fixed in Kodafix (diluted 1+3 with water) and rinsed in slow running water for at least 20 min. The slides were then dried, soaked in xylene and mounted in DePeX.

2. Image Analysis Method

Automatic scanning on FIP was carried out at a scan speed of 1mm/s using a broad-band green filter (500–600 nm), and the "0.7" density range on the FIP scanner (1 grey level= 0.003 odu). Segmentation was carried out at a fixed threshold of 14 grey-levels (0.042 odu). Focus was adjusted automatically during the scan using the FIP autofocus system (Shippey et al. 1980). The specimen was scanned in a series of 400 μ wide swathes, with an additional 56 μ guard region at either side of the swathes to eliminate errors due to edge-bisecting objects (Stark et al. 1984). Object classification was automatically referred during the scan to the modal value for 2C nuclei (PK). Objects in the range 0.5PK–1.8PK were classified as unlabelled cells, and objects with IOD $\rangle$ 1.8PK were recorded as suspect objects requiring subsequent manual review. A total slide area of 2.5cm x 1.5cm (63 swathes) was scanned in the case of 8-AG treated specimens, while 2mm x 1.5cm (4 swathes) was scanned in control (without 8-AG) specimens.

Review of the suspect objects was carried out using a program which automatically centres each object in the microscope field, allowing the operator to inspect the object and type in a code classifying the object-class as a labelled nucleus, unlabelled nucleus, or artifact. On completion of the review, a report giving the labelled cell count, unlabelled cell count and labelling index for the scanned area was printed.

The variant frequency was calculated from the labelling indices as follows:

$$Vf = \frac{LI \text{ in presence of 8-AG}}{LI \text{ without 8-AG}}$$

3. Counting of Cells in Culture Using Flow Cytometry

The method of Darzynkiewicz et al. (1976) was used. 0.2 ml aliquots of cell suspension were mixed thoroughly and added to 0.4 ml solution containing 0.1% (v/v) Triton X-100, 0.2M sucrose, 10^{-4} EDTA, and 2 x 10^{-2} citrate phosphate buffer at pH 3.0. After 1 min 0.8 ml solution containing 0.002% acridine orange (Gurr), 0.1M NaCl, and 10^{-2} citrate phosphate buffer at pH 3.8 was added. After 5 min equilibration at room temperature the red fluorescence ($\rangle$ 600 nm) emission of each cell was recorded as the cells passed through a 488 nm argon-ion laser beam in an in-air flow cytometer (Green and Fantes 1983).

Results and Discussion

Our preliminary experiments were designed to compare the results obtained by FIP with previously published results from this laboratory (Evans and Vijayalaxmi 1981). The results are presented in Table 2 and show a Vf of 6.3×10^{-4} in this individual together with a dose related increase in the Vf with increasing in vitro concentrations of MMC. Although these results are within the ranges reported by Strauss and Albertini (1979) and Evans and Vijayalaxmi (1981) they are nevertheless rather higher than other mutation studies would predict and our current experiments are aimed at identifying the factors that may be responsible for this.

Table 2. Cell counts and variant frequencies at different MMC concentrations

MMC Conc	Total Cells	AG+ Lab. Cells	Lab. Index $\times 10^{-4}$	Total Cells	AG− Lab. Cells	Lab. Index	Var Freq $\times 10^{-4}$
0	32650	6	1.84	756	215	0.28	6.6
10^{-8}M	19500	6	3.07	1221	272	0.22	13.9
10^{-7}M	32082	7	2.18	1324	48	0.036	60.6
10^{-6}M	16900	8	4.73	1065	123	0.12	39.4

FIP analysis proved to be considerably faster than visual microscopy. The overall analysis time for a typical preparation from an 8-AG treated culture was approximately 45 min using the FIP system, compared with 1.5 days by visual scanning. This was determined largely by the time taken to review the large numbers of "suspect" objects found by the scanner (typically 1000−1500 per 40000 cells in the 8-AG treated specimens). Most of these were either debris particles or pairs of touching or overlapping cells. Further improvements in specimen preparation and/or artefact rejection techniques during scanning should further improve the efficiency of the method.

Certain precautions in the preparation of specimens were necessary to achieve optimum results. Tests previously carried out (Stark et al. 1984) indicated that accurate counts of the labelled and unlabelled cells in Feulgen-stained autoradiographs required the use of freshly-prepared Schiff's reagent for the Feulgen staining reaction (to ensure sufficient stain density for good thresholding during scanning) and a large-grain highly sensitised nuclear emulsion such as Ilford G5 to give adequate discrimination of lightly-labelled cells. In addition, the number of "false positive" signals due to overlapping nuclei or to carbon particles from the Feulgen reagents was minimised by the use of EDTA to reduce clumping, and by membrane filtration of the stain reagents to eliminate debris particles.

Close comparison of the mutagen induced variant frequencies reveals that the relative contribution of the control and 8-AG labelling indices from which the variant frequencies are calculated differed between the present experiments and those of Evans and Vijayalaxmi. In the present experiments the labelling indices in the control cultures decreased with increasing MMC concentration, due presumably to delay of cells reaching DNA synthesis or to inhibition of DNA synthesis itself. Cells resistant to 8-AG would be similarly affected by MMC induced cytotoxicity in the drug treated cultures, although mutagen induced variant cells would contribute to any dose related increase in the labelling index. However, several lines of evidence suggested to us that other factors may have a considerable impact on the labelling index observed in preparations from 8-AG treated cultures and would thus affect the final variant frequency. We observed that after methanol/acetic fixation by routine cytogenetic methods, the yield of cells from 8-AG cultures was considerably lower than that from control cultures. Furthermore, the incidence of nuclei with irregular morphology or diffuse Feulgen staining reaction and the amount of cellular debris on the slides was frequently high in preparations made from 8-AG treated cultures. These visual observations are supported by a comparison of the FIP DNA/frequency histograms shown in Fig. 4a,b which indicate a considerable build-up of debris particles with an IOD value intermediate between the 2C DNA value and the background grain labelling in cultures treated with the selective agent.

The method of harvest used, which involves several centrifugation and washing steps, inevitably results in the loss of a considerable proportion of the original cell population and also in the selective removal of cellular debris. While cell loss due to fixation and harvest should not affect the labelling index observed, it will prevent direct assessment from slide preparations of the extent of cell loss as a result of the cytotoxic effects of 8-AG. We therefore attempted to estimate the extent of cell loss in cultures before fixation and harvest by using Coulter counting techniques to measure the object size distribution. Unlike the experiments of Albertini and Strauss (1979) in which Coulter counts showed no loss of cells due to 6-TG treatment after periods of up to 96 hours of culture, we found that any reduction in the total population of cells could not be estimated because the debris and 2C DNA peaks merged at high 8-AG concentrations. Since, in the present study, the nuclei with the 2C DNA content are identified by FIP by their IOD after staining with a quantitative DNA stain, it seemed more relevant to us to identify and count the number of surviving intact cells by their DNA content.

To investigate this a Flow cytometer (Green et al. 1980) was used to assess the concentration of 2C DNA nuclei remaining in the cultures after various treatments. To ensure minimal cell or debris loss, an acridine orange staining technique based on a method by Darzynkiewicz et al. (1976) was used in which living cells were stained directly without centrifugation steps. The volume of culture medium measured was accurately monitored using a syringe drive attachment to the Flow machine. Preliminary results are given in Table 3 and it is evident that in comparison with control cultures 8-AG, at the concentrations usually employed in these experiments, resulted in the loss of 59% of the cell population, while the combined effects of tritiated thymidine and 8-AG resulted in a total loss of approximately 64% of the cell population. Thus if the labelling index were calculated in this experiment using estimated total cell counts from slides, the final variant frequency would be approximately three times too high due to cell disintegration alone.

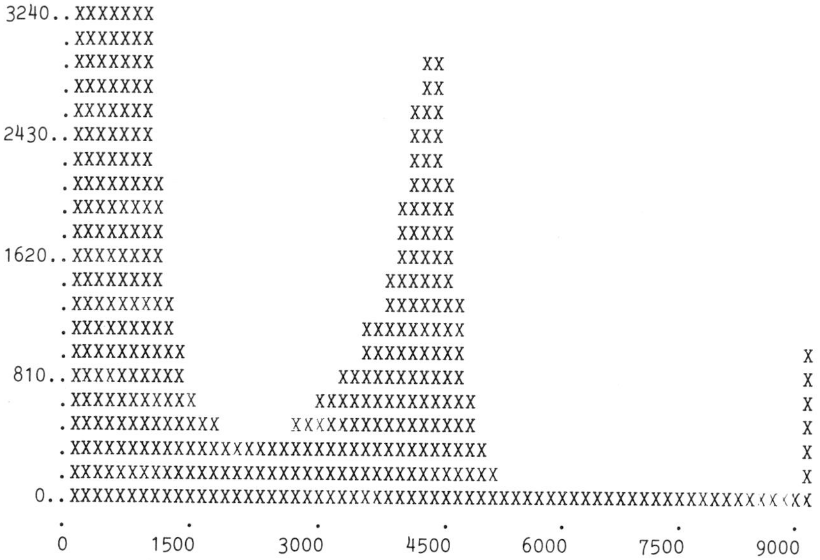

```
3240..XXXXXXX
    .XXXXXXX
    .XXXXXXX                            XX
    .XXXXXXX                            XX
    .XXXXXXX                           XXX
2430..XXXXXXX                           XXX
    .XXXXXXX                           XXX
    .XXXXXXXX                         XXXX
    .XXXXXXXX                        XXXXX
    .XXXXXXXX                        XXXXX
1620..XXXXXXXX                        XXXXX
    .XXXXXXXX                       XXXXXX
    .XXXXXXXXX                      XXXXXX
    .XXXXXXXXX                    XXXXXXXXX
    .XXXXXXXXXX                   XXXXXXXXX                                    X
 810..XXXXXXXXXX                 XXXXXXXXXXX                                   X
    .XXXXXXXXXXX              XXXXXXXXXXXXX                                    X
    .XXXXXXXXXXXXX          XXXXXXXXXXXXXXX                                    X
    .XXXXXXXXXXXXXXXXXXXXXXXXXXXXXXXXXXXXXXX                                   X
    .XXXXXXXXXXXXXXXXXXXXXXXXXXXXXXXXXXXXX                                     X
   0..XXXXXXXXXXXXXXXXXXXXXXXXXXXXXXXXXXXXXXXXXXXXXXXXXXXXXXXXXXXXXXX<X<XX
      .         .          .           .          .          .          .
      0        1500       3000        4500       6000       7500       9000
```

AREA	OBJ	DEB	CELLS	LOBJ	SUSP	OVE	PK
407	197914	169160	27504	1251	1240	0	4575

a   CELLS/SQMM   67

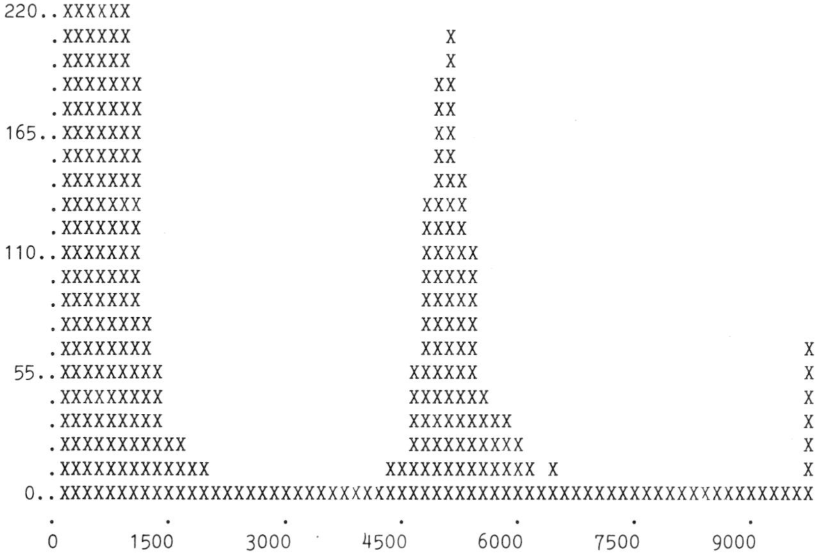

```
220..XXXXXX
   .XXXXXX                              X
   .XXXXXX                              X
   .XXXXXXX                            XX
   .XXXXXXX                            XX
165..XXXXXXX                            XX
   .XXXXXXX                            XX
   .XXXXXXX                           XXX
   .XXXXXXX                          XXXX
   .XXXXXXX                          XXXX
110..XXXXXXX                          XXXXX
   .XXXXXXX                          XXXXX
   .XXXXXXX                          XXXXX
   .XXXXXXXX                         XXXXX
   .XXXXXXXX                         XXXXX                                    X
55..XXXXXXXXX                         XXXXX                                    X
   .XXXXXXXXX                        XXXXXX                                    X
   .XXXXXXXXX                       XXXXXXXX                                   X
   .XXXXXXXXXX                      XXXXXXXXX                                  X
   .XXXXXXXXXXXXX                  XXXXXXXXXXXX X                              X
  0..XXXXXXXXXXXXXXXXXXXXXXXXXXXXXXXXXXXXXXXXXXXXXXXXXXXXXXXXXXXXXXXXXXXXXXXXXXXXX
     .         .          .           .          .          .          .
     0        1500       3000        4500       6000       7500       9000
```

AREA	OBJ	DEB	CELLS	LOBJ	SUSP	OVE	PK
14	15416	13978	1276	163	163	0	4875

b CELLS/SQMM   91

Fig. 4a,b. IOD Histograms for **a)** an 8-AG treated specimen, and **b)** its untreated control

Table 3. Cell loss due to azaguanine and tritiated thymidine

Condition	Cell Count
NoAG	0.928×10^6
2×10^{-5}M AG	0.617×10^6
2×10^{-4}M AG	0.384×10^6
2×10^{-4}M AG $1\ \mu Ci/ml\ ^3H$-Tdr	0.328×10^6

In the experiments described above dialysed foetal calf serum was used because the presence of exogenous purines in non-dialysed serum is known to reduce the selection pressure exerted by 8-AG (Peterson et al. 1976; Van Zeeland and Simons 1975). However, dialysed serum increases the cytotoxic effects of 8-AG (Peterson et al. 1976), and this may be responsible for the high level of cell disintegration observed in our experiments. The cytotoxic effects of tritiated thymidine are not surprising in the light of previous findings that cells in DNA synthesis are particularly susceptible to DNA damage by incorporated isotopes (Burki et al. 1975), and that exposure of stimulated lymphocytes to tritiated thymidine of high specific activity for periods longer than 6 h results in a non-linear uptake (Bain 1970), while a significant delay in cell cycle progression occurs with chronic exposures at levels as low as 0.1 $\mu Ci/ml$ (Pollack et al. 1979). In our current experiments we have obtained satisfactory labelling levels by reducing the period of exposure from 12 h to 2 h and increasing the specific activity of tritiated thymidine from 0.5 $\mu Ci/ml$ to 2.0 $\mu Ci/ml$.

Another factor which we are currently studying is the possible influence of cycling cells present in fresh blood samples upon the labelling index after 8-AG treatment. If a proportion of these cells were to escape the cytotoxic effects of the selective agent, the variant frequency would be artificially increased. We have separated mononuclear lymphocytes into subpopulations using sheep red blood cell rosetting methods in an attempt to isolate a fraction of the population free of cycling cells. However, our preliminary results from several individuals have indicated that a low labelling index occurs in all separated cell subpopulations.

Conclusions

We have demonstrated that automated methods can be used to measure the HGPRT variant frequency in Man. Our FIP method gives results that are within the ranges reported by other workers using similar, but non-automated techniques. However, other factors such as cell disintegration caused by the selective agent, and the presence of cycling cells in the fresh blood sample, may significantly affect the observed variant frequency, and further experiments are required to investigate this.

References

Bain B (1970) Tritiated-thymidine uptake in mixed leucocyte cultures: Effect of specific activity and exposure time. Clin exp Immunol 6:255–262

Burki HJ, Bunker S, Ritter M, Cleaver JE (1975) DNA damage from incorporated isotopes: Influence of the ^{3}H location in the cell. Radiation Research 62:299–312

Darzynkiewicz Z, Traganos F, Sharpless T, Melamed MR (1976) Lymphocyte stimulation: A rapid multiparameter analysis. Proc Natl Acad Sci USA 73:2881–2884

Evans HJ, Vijayalaxmi (1981) Induction of 8-azaguanine resistance and sister chromatid exchange in human lymphocytes exposed to mitomycin-C and X-rays *in vitro*. Nature 292:601–605

Green DK, Fantes JA (1983) Improved accuracy of in-flow chromosome fluorescence measurements by digital processing of multi-parameter flow data. Signal Processing. 5(2):175–186

Morley AA, Cox S, Wigmore D, Seshadri R, Dempsey JL (1982) Enumeration of thioguanine-resistant lymphocytes using autoradiography. Mutat Res 95:363–375

Nelson JA, Carpenter JW, Rose LM, Adamson DJ (1975) Mechanism of action of 6-thioguanine, 6-mercaptopurine, and 8-azaguanine. Cancer Res 35:2872–2878

Peterson AR, Krahn DF, Peterson H, Heidelberger C, Bhyuan BK, Li LH (1976) The influence of serum components on the growth and mutation of Chinese Hamster Cells in medium containing 8-azaguanine. Mutat Res 36:345–356

Pollack A, Bagwell CB, Irvin GL (1979) Radiation from tritiated thymidine perturbs the cell cycle progression of stimulated lymphocytes. Science 203:1025–1027

Shippey G, Bayley RJH, Farrow ASJ, Rutovitz DR, Tucker JH (1981) A fast interval processor. Pattern Recognition 14:345–365

Stark MH, Tucker JH, Thomson EJ, Perry PE (1984) The human lymphocyte HGPRT assay: Development of an automated image analysis system for the detection of rare autoradiographically labelled cells. Cytometry 5(4)

Strauss GH, Albertini RJ (1979) Enumeration of 6-thioguanine-resistant peripheral blood lymphocytes in man as a potential test for somatic cell mutations arising *in vivo*. Mutat Res 61:353–379

Tucker JH (1976) Cerviscan: An analysis system for experiments in automatic cervical smear screening. Comp Biomed Res 9:93–107

Van Zeeland AA, Simons JWIM (1975) The effects of calf serum on the toxicity of 8-azaguanine. Mutat Res 27:135–138

List of Abbreviations

HGPRT	hypoxanthine-guanine phosphoribosyltransferase	FCS	foetal calf serum
		PHA	phytohaemagglutinin
6-TG	6-thioguanine	FIP	fast interval processor
8-AG	8-azaguanine	PK	peak
Vf	variant frequency	LI	labelling index
MMC	mitomycin C	odu	optical density units
IOD	integrated optical density		

Somatic Mutations Detected by Immunofluorescence and Flow Cytometry*

R.H. Jensen, W. Bigbee, and E.W. Branscomb

Lawrence Livermore National Laboratory, Biomedical Sciences Division, University of California, P.O. Box 5507 L-452, Livermore, California 94550

1 Introduction

In order to detect the genotoxic effects of living in today's complex industrialized society, we must devise sensitive ways to measure early subtle changes in human beings that may lead to more pathological effects in longer periods of time. The detection of mutational changes in somatic cells is a possible way of monitoring for effects that could ultimately lead to carcinogenic or heritable lesions. Depending on the lifespan of the somatic cells analyzed, the frequency of mutated somatic cells in an individual may be a measurement of the effects of recent toxic exposure or an indication of cumulative genetic insult.

One means of detecting cells that have received mutational damage is to use flow analysis to identify those cells that contain specific variant proteins. We have devised ways to label red blood cells that contain variant hemoglobin that is mutant by a single amino acid substitution by exposing the blood to antibodies directed against specific variant hemoglobin, and analyzing the sample for labeled cells using flow cytometry and cell sorting. Our efforts have included using polyclonal horse antibodies against human hemoglobin variants, immunopurified to recognize only the variant hemoglobin and not normal hemoglobin. These purified sera have been used to determine the frequency of putative mutant red blood cells that are labeled with the antibody and therefore presumably contain the variant hemoglobin. In order to acquire greater specificity and reproducibility in labeling, we are searching for monoclonal antibodies which would be specific for detecting variant hemoglobin.

A second technique for detecting mutant erythrocytes is to use flow cytometry to search for cells that have lost the expression of one allele of a cellular protein gene. This 'gene loss' appraoch has the advantage that many different kinds of mutagenic damage to a gene can result in a gene product being synthesized incorrectly or not at all. Thus the genetic target size of the possible damage is many times larger than the small genetic target that leads to a particular single amino acid substitution. The gene loss approach is being pursued using monoclonal antibodies to the two allelic forms of the red cell surface glycoprotein, glycophorin A. Dual parameter flow cytometric ana-

* Work performed under the auspices of the U.S. Department of Energy by the Lawrence Livermore National Laboratory under contract number W-7405-ENG-48, EPA Grant R808642-01, and USPHS Grant No. RO1 CA 31549-01 awarded by the National Cancer Institute, DHHS

lysis of red blood cells labeled with two monoclonal antibodies, one that is specific to each of the allelic forms of the protein, should lead to an accurate assessment of the loss of one of the forms in the presence of full expression of the other form. Thus any cells detected should be true somatic mutants of the glycophorin gene, not phenocopies that could be formed by the loss of metabolic, membranous, or transport functions that are required to express the surface glycoprotein.

2 Antibodies Against Variant Hemoglobins to Detect Single Amino Acid Substitutions

In collaboration with Dr. G. Stamatoyannapoulos we developed a means for analyzing the red cells from humans of normal hemoglobin phenotype to detect any red blood cells that bind antibody specific for the single amino acid substitution variant hemoglobin S or alternatively an antibody monospecific for another single amino acid substitution variant, hemoglobin C [1-3]. These two hemoglobin variants each contains an amino acid substitution at the sixth position in the beta globin chain as compared to hemoglobin A. Either can result from a single base substitution in the gene that codes for the normal hemoglobin A beta chain. Thus, hemoglobin S or hemoglobin C carriers have only single base differences from people that carry the normal gene codes for hemoglobin A. If a stem cell or a committed erythroid cell in a normal, hemoglobin A individual was mutated in exactly the same way in the same codon, that cell and its daughters would be committed to producing variant hemoglobin S or hemoglobin C. Therefore we have postulated that a small fraction of the circulating red cells in all normal people should be somatic mutant cells and contain hemoglobin S or alternatively hemoglobin C. Using manual fluorescence microscopic analysis of samples labeled with the monospecific anti-hemoglobin S, Dr. Stamatoyannopoulos determined that antibody labeled cells do occur at a frequency of about one cell in 10 million [4].

To increase the speed of analysis for such cells we devised a means for preparing red cells that could be immunofluorescently labeled and analyzed by flow cytometry. A fixation procedure was developed that depended on the methods of Wang and Richards [5], in which the membrane permeable cross-linker dimethylsuberimidate was used to study the proteins in close proximity to the red cell membrane. We used it to cross-link hemoglobin to the cellular membranes, so that when the cells were hypotonically treated, they formed ghosts that retained about one percent of the cellular hemoglobin attached covalently to the membrane. This resulted in ghosts that could be exposed to antibodies against hemoglobins and retain their integrity as cell sized particles and as specific antigens [3]. Thus, cells that contain hemoglobin S would be labeled with antibody specific to that variant hemoglobin whereas cells that contain hemoglobin A would not be labeled by such an antibody. An example of the distinction of these two red cell types is shown in a histogram generated by flow analysis of a mixture of red cells from a normal hemoglobin A individual and one tenth as many red cells from a hemoglobin AS heterozygous individual (Fig. 1).

The labeling procedure required to produce bright red cell ghosts was such that a number of difficulties have interfered with successful automated analysis using the

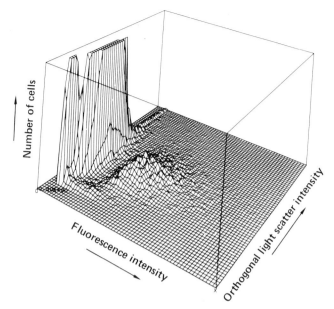

Fig. 1. Flow histogram of a 10:1 mixture of red cell ghosts from a homozygous hemoglobin AA individual and a heterozygous hemoglobin AS invididual. The mixture was labeled with purified horse anti-hemoglobin S antibody that was conjugated with FITC. Ghosts were prepared and stained as described previously [3]

high speed flow cytometric method [6]. A major difficulty has been that a small but significant number of fluorescent particles that are not red blood cells appear to be formed during the antibody labeling procedures. While these particles are easily differentiated from red cells under microscopic examination, the flow cytometer cannot distinguish them from fluorescently labeled red cells. The result has been that we have been unable to obtain direct cytometric enumeration of putative mutant red cells in human blood. Instead it has been necessary to perform cell sorting operations at high speed and to then microscopically analyze the sorted, enriched material to determine the frequency of labeled red cells in the blood using this combination approach. The results of our preliminary study using these techniques as compared with the more tedious direct counting results obtained by Dr. Stamatoyannopoulos are shown in Table 1. The two approaches give results that are in general agreement considering the large variation in measurement reproducibility in each measurement.

In addition we found that the very high flow rates necessary to process a large number of cells was not detrimental to analysis on the flow cytometer nor was it detrimental to flow sorting. Since the frequency of putative mutant cells is about 10^{-7}, at least 10^9 cells should be processed to obtain statistically acceptable numbers of labeled cells (about 100). To perform such a task in a reasonable time (say 10 min), requires a sorter throughput rate of about 10^6 cells per second. This is roughly a thousand time faster than flow sorters have been used in the past. However, we have found that this high rate of analysis has not degenerated the histogram analysis nor the flow sorting to a significant extent [6].

Table 1. Red cells that exhibit fluorescence with anti-variant hemoglobin antibodies

	Frequency of Mutant cells	Number of Samples Analyzed
	Hb S	
LLNL (Sorter + Microscope)	$1.1 \times 10^{-8} - 1.1 \times 10^{-7}$	5
U. of Wash. (Microscope)	$4 \times 10^{-8} - 3 \times 10^{-7}$	15
	Hb C	
LLNL (Sorter + Microscope)	$6.7 \times 10^{-8} - 2.6 \times 10^{-7}$	3

In order to pursue this method of immunofluorescent labeling of cells and enumerating the rare events occurring in the search for mutant cells, we determined that monoclonal antibodies against hemoglobin variants are the necessary labeling reagents. A comprehensive search of the availability of variant hemoglobins from heritable carriers of genetic changes has resulted in a listing of variants that should be usable for our somatic cell mutation analysis. In addition to the availability of these hemoglobins, a number of other criteria used to determine the utility of each of the hemoglobins for our purposes are listed below:

- the variant can result from a single base change in the hemoglobin A gene;
- the variant is stable, so that variant somatic cells with remain intact in the circulatory system;
- the single amino acid substitution in the variant is a molecule surface change, so that it is detectable by immunolabeling;
- the single amino acid substitution in the variant is different from the sequence in mice, so that it will serve as an immunogen in mice;
- the single amino acid substitution is relatively well separated from other differences between mouse and human hemoglobin so that antibodies will not be found that recognize two amino acid differences simultaneously;
- a variety of sites and of kinds of amino acid substitutions are represented so as to detect changes in several parts of the hemoglobin molecule caused by several different kinds of mutagenic changes to the hemoglobin genes;

Application of these criteria results in a list of 21 hemoglobin changes that would serve as good sites for detecting single amino acid substitution mutations in red cells from normal hemoglobin A individuals. Table 2 shows the amino acid and codon base substitutions that are required to produce the mutant changes, and also lists 38 different inherited variant hemoglobin types that display changes at the listed hemoglobin sites. We have obtained variant hemoglobins that represent members of seven of the different useful sites for our application, and are in the process of attempting to isolate hybridomas that produce monoclonal antibodies against these variant hemoglobins. Ultimately, if our approach is successful, we should generate a library of monoclonal antibodies

Table 2. Hemoglobin variants acceptable for generating monoclonal antibodies

Position	Hemoglobin Variant	Amino Acid Substitution	Base Change
α-Chain			
54	Mexico, J, J-Paris-II,		
	Uppsalà	Gln → Glu	C → G
	Shimonoseki, Hiroshima	Gln → Arg	A → G
56	Thailand	Lys → Thr	A → C
	Shaare, Zedek	Lys → Glu	A → G
57	L-Persian Gulf	Gly → Arg	G → A
	J-Norfolk, Kagoshima,		
	Nishik-I, II, III	Gly → Asp	G → A
60	Zambia	Lys → Asn	G → T or C
	Dagestan	Lys → Glu	A → G
61	J-Buda	Lys → Asn	G → T or C
85	G-Norfolk	Asp → Asn	G → A
	Atago	Asp → Tyr	G → T
	Inkster	Asp → Val	A → T
90	J-Broussais, Tagawa-I	Lys → Asn	G → T or C
	Rajappen	Lys → Thr	A → C
120	J-Meerut, J-Birmingham	Ala → Glu	C → A
β-Chain			
20	Olympia	Val → Met	G → A
43	G-Galveston, G-Port Arthur,		
	G-Texas	Glu → Ala	A → C
	Hoshida, Chaya	Glu → Gln	G → C
87	D-Ibadan	Thr → Lys	C → A
95	N-Baltimore, Hopkins-I, Jenkins,		
	N-Memphis Kenwood	Lys → Glu	A → G
	Detroit	Lys → Asn	G → T or C

each of which is specific for a single amino acid substitution on hemoglobin A. A battery of such antibodies mixed together could then be used to detect a variety of different somatic mutant red cells in blood samples.

3 Antibodies Against Glycophorin A to Detect Gene Loss Mutants

In parallel with the variant hemoglobin-based assay we are developing a second independent system based on the biochemically well-studied protein glycophorin A [7].

Glycophorin A is a glycosylated red cell membrane protein present at about 5–10 x 10^5 copies per cell [8]. Its 131 amino acid residues span the membrane with the amino-terminal portion presented on the red cell surface. The utility of this protein as a basis for a somatic cell mutation marker was suggested by the work of Furthmayer [9] which showed that this protein was responsible for the M and N blood group determinants and that these determinants were defined by a polymorphism in the amino acid sequence of the protein coded for by a pair of co-dominantly expressed alleles. The polymorphic sequence at the amino-terminus is shown below:

Glycophorin A(M) Ser-Ser(*)–Thr(*)–Thr(*)–Gly–Val–. . .
Glycophorin A(N) Leu-Ser(*)–Thr(*)–Thr(*)–Glu–Val–. . .
(*) indicates a glycosylated amino acid

Except for the amino acid substitutions at positions one and five of the sequence, the two proteins are identical, both in amino acid sequence and sites and structures of glycosylation. Individual humans that are homozygous for the M or N allele synthesize only the A(M) or A(N) sequence respectively, while heterozygotes present equal numbers of the two proteins on the surface of their erythrocytes [9]. The assay that we are developing using the glycophorin A system is called a "gene expression loss" or "null mutation" assay. In this approach, we attempt to detect rare erythrocytes in the blood of glycophorin A heterozygotes which fail to express one or the other of the two allelic forms of the protein. Such an approach has been described for an *in vitro* system using human cells heterozygous for the multi-allelic HLA determinants [10, 11]. Using immunologic selection, these researchers have demonstrated that lymphoid cells can lose expression of one or more polymorphic HLA cell surface antigens as a result of spontaneous mutation. Exposure to radiation or chemical mutagens increases the mutation rate by greater than two orders of magnitude. Analysis of these variants showed the majority to be single HLA gene mutants.

The glycophorin A "gene expression loss" approach has several inherent practical and biological advantages over the the hemoglobin-based amino acid substitution system. First, this antigen is presented on the surface of the red cell and is firmly anchored in the membrane; thus, cell preparation and antibody labeling procedures are simple and straightforward. Second, since the mutant phenotype detected can result from a variety of mutational lesions, e.g., single nucleotide changes, deletions, or frameshifts occurring either in the glycophorin A structural gene or its control elements, the frequency of variant cells should be much higher (perhaps 100–1000 times the frequency seen for a single amino acid substitution at a single site). Hence such cells should be easier to detect and the frequency of such cells, representing the sum of all of these mutational mechanisms, may more accurately reflect the integrated genetic damage in that individual.

To detect the presence of these functionally hemizygous cells we are at present generating mouse monoclonal antibodies which differentiate the M and N forms of the protein (manuscript in preparation). Mouse monoclonal antibodies recognizing glycophorin A have been produced by Edwards [12] and we have adopted a variation of his immunization protocol. First, mice were injected with a equal mixture of homozygous MM and NN red cells, then boosted with purified glycophorin A(M) and A(N). The serum was then assayed for the presence of anti-red cell antibodies. The spleens from responding mice were then fused with SP2/0 mouse myeloma cells, and anti-red cell producing clones were selected using a red cell enzyme-linked immunosorbant assay (ELISA). Positive clones were then assayed using homozygous MM and NN cells, and those showing specificity for either cell type were selected, sub-cloned and expanded.

Using this procedure, we have isolated four clones, two of which are specific for glycophorin A(M), one specific for A(N) and one which recognizes a shared determinant. Purified A(M)- and A(N)-specific monoclonal antibodies will be labeled with green and red fluorophors, e.g., fluorescein isothiocyanate (FITC) and a derivative of rhodamine like Texas Red [13] and incubated simultaneously with erythrocytes from MN heterozygotes. Variant cells, defined by binding of only one of the antibodies and hence fluorescing only green or red, will be enumerated as shown in Fig. 2 using the LLNL two-color dual-beam flow sorter [14]. The variant frequency will simply be the number of green- or red-only cells divided by the total number of cells processed (the sum of the signals in all three peaks). Because we expect the frequency of these variant cells to be as much as a 1000-fold higher than the frequency of single amino acid substitution-variant cells, direct flow cytometric quantitation should be possible. Also adequate numbers of variant cells should be obtainable by sorting for biochemical analysis.

To determine the sensitivity and specificity of our labeling procedures, we have performed preliminary dual laser analyses using only one of the two immunological labels, monoclonal anti-glycophorin A(M) labeled with fluorescein. Figure 3 shows a histogram generated from the flow analysis of an artificial mixture of different red cells in which there is an equal mix of cells from the blood of a homozygous A(MM) individual and a homozygous A(NN) individual. The histogram is a two parameter display, but the Y-axis plots the intensity of light scatter for easy identification of cells compared to debris. Thus, it is not analogous to the histogram shown in Fig. 2, where both axes are fluorescence from immunofluorescent probes. It does demonstrate the separation of fluorescently labeled cells from unlabeled cells in one dimension, and from these data we can estimate the possible errors that could occur using this technique. Note that the labeled cells are roughly eight times as bright as the unlabeled cells. In our proposed experiment heterozygotes would be analyzed, so we expect that the separation of positives from negatives will be roughly a factor of 4. In enumerating the frequency of mutant cells, the negatively stained cells must be counted carefully. Separation of signals from positive cells from those from negatives depends on the breadth of the histogram peaks as well as their separation. To estimate the potential error rate for different separations between normal and variant cells, we have assumed that a gaussian distribution of the cells will occur. In the cases that we have measured, the coefficient of variation for the peaks is around 17% (as in the histogram in Fig. 3). For the factor of four separation that we might expect for variants and normal cells as labeled in this experiment, we calculate that the frequency of occurrence of normal cells at the position of the variant cells would be 10^{-5}. If our estimates of the frequency of real somatic mutant cells is about 100 times the frequency that we measured for single amino acid mutants in the hemoglobin assays, we would expect to have a signal to noise ratio of about one. To improve this signal to noise we should label our antibodies more brightly to obtain a separation of normal from variant cells by a factor of about 10. Then we calculate that a false mutant cell frequency of 10^{-7}

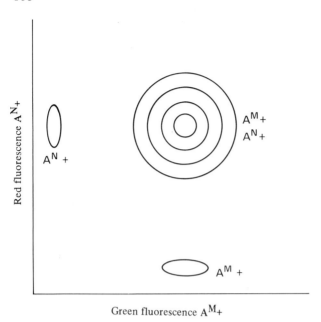

Fig. 2. Dual beam flow cytometric detection of glycophorin A "null" variant cells in a preparation from a heterozygous A(MN) individual. A hypothetical two parameter histogram of a cell population stained with two different monoclonal antibodies, FITC labeled anti-glycophorin A(M) and Texas Red labeled anti-glycophorin A(N). The green fluorescence is to be induced by excitation at 488 nm and the red fluorescence is to be induced by excitation at 568 nm. Normal cells will bind both antibodies and exhibit both green and red fluorescence. Variant cells would be lacking the expression of the glycophorin A(M) and thus fluoresce only red, or would be lacking the expression of the glycophorin A(N) allele and thus fluorescence only green

should occur and the signal to noise ratio for enumerating gene loss mutant cells would be 100. Our present efforts are aimed at this objective, and we have recently obtained intensity differences of 112-fold using secondary antibody labeling of the cells that are labeled with the primary monoclonal anti-glycophorin A(M). Although secondary antibody labeling or biotin avidin labeling would be complex and cumbersome, it will probably be adequate to allow flow cytometric enumeration of somatic gene loss mutants.

Since this assay is based on the detection of cells which fail to express a gene product, it is difficult to be sure that the counted variant cells are true glycophorin A structural gene mutants. This is important since there are both genetic and non-genetic mechanisms which could cause the protein to fail to appear on the red cell membrane. For example, mutations outside the glycophorin A locus leading to loss of function of proteins necessary for processing, transporting, glycosylating or inserting glycophorin A into the membrane could produce apparent glycophorin A "null" cells. Non-genetic events include loss of membrane integrity or insufficient levels of substrate sugars for glycosylating enzymes due to metabolic anomalies. Nevertheless, we believe that this assay is strongly protected against such false "phenocopies" since it requires antibody binding to one of the glycophorin A types. The proper cell surface presentation of the

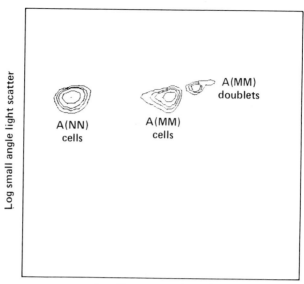

Fig. 3. Flow histogram of a 1:1 mixture of cells from two individuals, a homozygous glycophorin A(MM) and a homozygous glycophorin A(NN), labeled with fluoresceinated monoclonal antibody against glycophorin A(MM). This represents half of the labeling that will be performed for the detection of glycophorin A "null" variant cells. The ordinate plots the intensity of light scattered, in order to separate signals from cells from signals from debris. The logarithmic scales are such that the range of fluorescence and light scatter displayed is roughly a factor of 500 on each axis

glycophorin A product of the unaffected allele insures that the rest of the cell apparatus necessary for the expression of the protein is intact. Finally, we can be assured that the variant cells will not be selected against *in vivo* since erythrocytes from genetically homozygous glycophorin A "null" individuals, completely lacking expression of the protein, appear to exhibit normal viability [15].

4 Summary

We conclude that we have at hand the tools for development of both a single amino acid substitution and a gene loss type assay for somatic cell mutations in human red cells. We are close to being able to assess these assays as possible means for screening the human population for individuals who may be at risk of acquiring abnormally high frequencies of mutant cells. The cause for such a high frequency of mutant cells could be a prior exposure of the individual to large amount of mutagen, or it could indicate a member of a vulnerable subpopulation that may have poor DNA repair systems. Either of these cases could well be an indication that the particular individual is at high risk that exposure to mutagens could lead to potential pathology. In the near future we expect to determine the reliability of the assays and to perform experiments to confirm the validity of flow cytometry in measuring real somatic mutagenic events that occur *in vivo*.

Acknowlegdment. The authors wish to thank Drs. B. Lubin, T.H.J. Huisman, W. MooPenn, G. Zetterberg, D. Rucknagel, and J. Szelenyi for donating hemoglobin or blood from genetically variant individuals.

References

1. Papayannopoulou Th, McGuire TC, Lim G, Garzel E, Nute PE, Stamatoyannopoulos G (1976) Identification of Haemoglobin S in red cells and normoblasts, using fluorescent anti-Hb S antibodies. Brit J Haemat 34:25−31
2. Papayannopoulou Th, Lim G, Mcguire TC, Ahern V, Nute PE, Stamatoyannopoulos G (1977) Use of specific fluorescent antibodies for the identification of Hemoglobin C in erythrocytes. Amer J Hemat 2:105−115
3. Bigbee WL, Branscomb EW, Weintraub HB, Papayannopoulou Th, Stamatoyannopoulos G (1981) Cell sorter immunofluorescence detection of human erythrocytes labeled in suspension with antibodies specific for Hemoglobins S and C. J Immunol Meth 45:117−127
4. Stamatoyannopoulos G (1979) Possibilities for demonstration point mutations in somatic cells, as illustrated by studies of mutant hemoglobins. In: Berg K (ed) Genetic Damage in Man Caused by Environmental Agents, Academic Press, New York, p 49−62
5. Wang K, Richards FM (1975) Reaction of dimethyl-3, 3-dithiobispropionimidate with intact human erythrocytes. J Biol Chem 250:6622−6626
6. Bigbee WL, Branscomb EW, Jensen RH (1983) Detection of mutated erythrocytes in man. In: Individual Susceptibility to Genotoxic Agents in the Human Population. NIEHS (in Press)
7. Furthmayr H, Metaxas MN, Metaxas-Bühler M (1981) Mutations within the amino-terminal region of glycophorin A. Proc Natl Acad Sci USA 78:631−635
8. Dahr W, Kordowicz M, Beyreuther K, Krüger J (1981) The amino-acid sequence of the M(c)-specific major red cell membrane sialoglyprotein- an intermediate of the blood group M- and N-active molecules. Hoppe-Seyler's Z Physiol Chem 362:363−366
9. Furthmayer H (1978) Structural comparison of glycophorins and immunochemical analysis of genetic variants. Nature 271:519−524
10. Pious D, Soderland C (1977) HLA variants of cultured human lymphoid cells: Evidence for mutational origin and estimation of mutation rate. Science 197:769−771
11. Kvathas P, Bach FH, DeMars R (1980) Gamma ray-induced loss of expression of HLA and glyoxalase I alleles in lymphoblastoid cells. Proc Natl Acad Sci USA 77:4251−4255
12. Edwards PAW (1980) Monoclonal antibodies that bind to the human erythrocyte-membrane glycoproteins Glycophorin A and Band 3. Biochem Soc Trans 8:334−335
13. Titus JA, Haugland R, Sharrow SQ, Segal DM (1982) Texas red, a hydrophilic red-emitting fluorophore for use with fluorescein in dual parameter flow microfluorometric and fluorescence microscopic studies. J Immunol Meth 50:193−204
14. Dean PN, Pinkel D (1978) High resolution dual laser flow cytometry. J Histochem Cytochem 26:622−627
15. Tanner MJA, Anstee DJ (1976) The membrane change in En(a-) human erythrocytes. Biochem J 153:217−277

Computer Scoring of Micronuclei in Human Lymphocytes

H. Callisen, A. Norman, and M. Pincu

Department of Radiological Sciences, University of California, School of Medicine,
900 Veteran Avenue, Los Angeles, CA 90029, USA

Summary

Micronuclei in cells arise from chromosome fragments or from whole chromosomes
that lag behind at anaphase. Their frequency in human peripheral blood lymphocytes
provides a convenient measure of chromosome damage in man. In order to improve
the assay and to study the mechanism of micronuclear formation, we built an image
analysis system consisting of a light microscope coupled to a low-light-level TV camera,
a video image digitizer and two microcomputers for image acquisition and processing.

Our initial results on computer aided identification of micronuclei on microscope
slides are gratifying. They demonstrate that automation of the micronucleus assay is
feasable. Moreover, DNA content distributions of micronuclei exposed to graded doses
of ionizing radiation show variations that may be interpreted in terms of different
mechanisms for radiation induced chromosome damage.

Introduction

The chromosomes are the primary target of physical and chemical mutagens and carcinogens in our environment. Many of these agents produce chromosome aberrations
in the circulating peripheral blood lymphocytes. The aberrations, in turn, give rise
to micronuclei when lymphocytes proliferate in culture (Fig. 1). The assay of micronuclei provides, therefore, a direct method for monitoring chromosome damage in
man [1]. Figure 2 shows a direct relationship between radiation exposure and manual
scoring of micronuclei for a dose range between 20R and 200R.

The micronucleus assay is much easier and more rapid than the conventional cytogenetic scoring of damage in metaphase chromosomes. Typically an observer is able
to score the micronuclei in several thousand interphase cells in the time that a trained
cytogeneticist needs to analyse a hundered metaphase cells. Additional improvement
may be possible if the micronucleus test is automated. Automation promises three
distinct benefits over the current manual practice. It promises to
1. improve the counting statistics since a greater number of cells can be scored without system fatigue,

Biological Dosimetry. Edited by W. G. Eisert and M. L. Mendelsohn
© Springer-Verlag Berlin Heidelberg 1984

Fig. 1. Digitized image of a human peripheral blood lymphocyte showing a nucleus with two micronuclei. The lymphocyte was stained with crystal violet, digitized into sixteen levels of gray, and measures approximately 40 by 30 pixels

2. provide more objective and consistent criteria for the identification of a micronucleus than used by the human observer, and
3. yield quantitative information on nuclear size, shape, texture, and DNA content, useful to better understand the micronucleus life cycle.

It is tempting, therefore, to consider an image analysis system for the automated or semi-automated detection of micronuclei based on nuclear area, DNA content, and the relative distance to their associated nuclei. Such a system may be useful not only for scoring micronuclei relatively quickly, but also for determining whether they originate from chromosome fragments, from whole chromosomes, or from dicentrics. Furthermore, quantitative data on lymphocyte nuclei yield parameters which may indicate

MICRONUCLEATED CELLS / 3000 CELLS

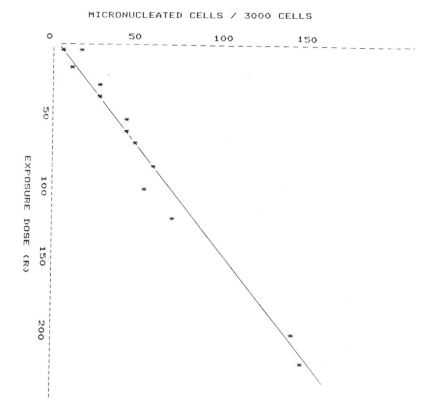

Fig. 2. Scoring results of micronuclei from samples of peripheral blood lymphocytes exposed to graded doses of X-radiation. A linear relationship is shown within a 20R to 200R exposure range

the extent of cell proliferation. The latter measure is needed to compute the extent of biological damage equivalent to the ratio of the number of micronucleated cells to the number of proliferating cells.

System Description and Methods

In order to study cellular and nuclear properties quantitatively, we have assembled a low-cost video/microscope densitometric system. The system consists of a conventional light microscope (optionally equiped with epi-luminescence fluorescence) coupled to a low-light-level closed-circuit TV camera whose output is sequentially processed by two microcomputers, one for image acquisition and one for data analysis. This high-resolution optical low-light-level micrometry expert system (Holmes) allows us to (a) acquire and store microscope images, (b) take spatial measurements such as size and shape, and (c) make grayness measurements needed to compute optical density and texture descriptors. A more detailed description of detective Holmes has been presented in the literature by Callisen et al. [2].

Image processing software algorithms necessary to identify micronuclei vary with the type of stain and microscope illumination used. Below, we describe one protocol for sample preparation and image processing used to develop and verify the software needed to identify and study the micronucleus. The process is a sequence of steps to prepare the microscope slide, digitize a field, correct the digitized image, extract objects, identify micronuclei, and verify their identity.

First, whole blood samples were exposed to ionizing radiation. Lymphocytes in the sample were stimulated to divide and were cultured for 76 h. Microscope slides were prepared and stained specifically for DNA with Azure-A using the Feulgen method.

Second, the slides were manually scanned under the microscope using bright field transmission light. Scenes with at least one micronucleus identified visually by the observer were digitized 16 times, summed, and stored on a computer diskette.

Third, the digitized images were corrected for non-uniform microscope field illumination, for temporal variations in the amplitude of the composite video signal, and for the camera's gamma response at each pixel. Our method for non-uniform illumination correction was to calculate the optical density for each pixel using the pixel values of a flood image as the incident illumination intensities. These optical density values were corrected for the camera's non-linear response to light (gamma correction). The gamma correction was particularly necessary in order to measure DNA content of objects in poorly illuminated regions of the field.

Fourth, all objects were extracted from the digitized image by histogram thresholding, recorded in a line-segment tabular format, and screened from obvious debris. The extraction process consisted of screening each line of the digital image, by collecting all line segments due to the objects in the field, and by associating these segments to a particular object by comparing the position of their centers. The screening procedure involved the following tests:

1. the objects must be relatively round, i.e. the ratio of pixel area to the area calculated from the diameter along either X or Y axis must approach unity,
2. the objects must be uniquely identified within the image, i.e. the center coordinate must be the same whether the object is extracted by scanning along the X or the Y axis, and
3. the objects must be whole and not sliced by the edge of the image or by the object extraction algorithm, i.e. the object's contour must not have a straight edge of significant length.

Fifth, the resulting "good" objects were automatically screened for lymphocyte nuclei and micronuclei combinations based on measurements of relative position and size. The screening was done by calculating an "affinity coefficient" which is greatest for a small object (micronucleus) next to, but not touching, a large object (associated nucleus) and which significantly drops in value for combinations of either two distant objects or two objects of equivalent size. The affinity algorithm was defined as the ratio of the difference in radii between two objects divided by the distance between their centers. If this ratio is greater than a given constant and less than unity, then the two objects were assumed to be associated. This algorithm mimics one criterion used in the manual assay. That is, the micronucleus must be found within the cell's cytoplasm and its size must be small relative to the associated nucleus.

And sixth, suspect micronuclei were verified to be within acceptable absolute quantitative limits. These conditions include

1. the optical density at the center of the micronucleus must be less than or equivalent to that of the associated nucleus but not less than half of the density at the nucleus' center, and

2. the combined area and/or DNA content of the micronucleus and nucleus must not significantly exceed that of a typical proliferating lymphocyte.

Preliminary Results

To evaluate Holmes' ability to automatically detect and uniquely identify micronuclei, we compared Holmes' results with observer collected data. Table 1 shows these preliminary results. Listed in the table are (a) the reference count obtained manually by directly viewing the microscope fields, (b) the number of micronuclei that were automatically extracted from the digitized fields, and (c) the number of micronuclei that were identified and labeled as such by Holmes. The number of "not found" micronuclei are typically those which were faint and were not resolved from the image's background or those which appeared attached to the nucleus due to insufficient spatial resolution of the digital image. The "artifacts" are those objects that were mistakenly identified as micronuclei by the software algorithms. The number of micronuclei which were "not labeled" are those which the software algorithms failed to identify as a micronucleus. Our preliminary results show that approximately 70% of the micronuclei were found and of those found, 90% were properly identified.

In order to gain more information and to determine the range of size and DNA content of micronuclei that would be acceptable for the purpose of verification, we have compiled frequency distributions of area and integrated optical density of known nuclei. The bimodal distribution shown in Fig. 3 was used to verify that optical density measurements can yield DNA content values and to calibrate our percent DNA content units relative to the diploid amount.

Figure 4 shows distributions of micronuclear DNA content from lymphocytes exposed to x-radiation doses of 0R and 200R. Cells exposed to 200R show several micronuclei with DNA content greater than the content found in the largest chromosome. Since these distributions provide boundaries for the DNA content of chromosome fragments, whole chromosomes, and dicentrics, such data may be useful to determine micronuclear origin. Additional experimentation is needed to validate this finding.

Table 1. Preliminary results of computer aided micronuclei scoring. Error rates due to the extraction and identification software algorithms are shown

Category	Number of micronuclei	Relative success
Visual count (ref.)	166	– –
Holmes: extracted ok	117	70% of visual
.. not found	−29	
.. artifacts	+ 9	
Holmes: identified ok	104	90% of extracted
.. not labeled	−13	
.. debris, etc.	+18	
Total count by Holmes	131	80% of visual

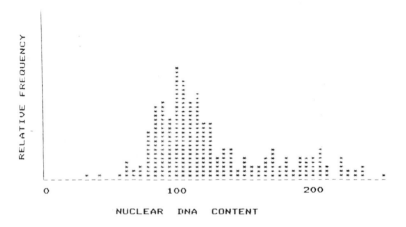

Fig. 3. Frequency distribution of DNA content in lymphocyte nuclei exposed to 200R x-radiation and cultured for 72 h after PHA stimulation

Discussion

These initial steps on automating the micronuclear assay are gratifying. They have demonstrated that it is feasable to identify nuclei and micronuclei. Our preliminary results show a counting efficiency of approximately 80%. It is not necessary to improve on this value as long as it remains constant accross different sets of data obtained from new samples and prepared with new culture and staining solutions.

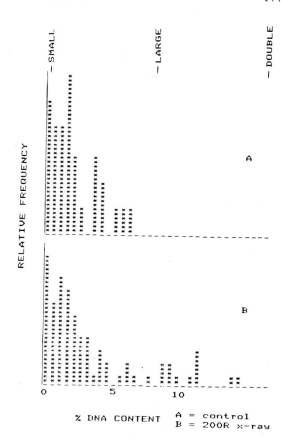

Fig. 4A,B. Variation in frequency distributions of micronuclear DNA content from (**A**) unexposed lymphocytes and (**B**) lymphocytes exposed to 200R x-ray. The *arrows* indicate the relative size of the smallest chromosome, largest chromosome, and double the largest chromosome

Furthermore, quantitative information on micronuclei leads not only to objective identification, but also gives insight into the mechanism of micronuclei formation which, in turn, may reveal additional criteria to verify more precisely the presence of a micronucleus in a digitized image.

It is tempting to speculate that the different mechanisms of micronuclei formation, such as from chromosome breakage due to ionizing radiation or from chemicals that interfere with spindle fiber formation, will result in different micronuclear distributions of DNA content. The micronucleus assay would then serve not only as a measure of the extent of chromosome damage as determined from micronuclear frequency, but also as a means to identify the type of chromosome damage that had been incurred.

Our current implementation of the automated micronucleus assay needs to be refined. It is desireable, for example, to analyse microscope slides collected and prepared for cytogenetic analysis other than for micronuclei identification. An enormous amount of these slides exist from previous studies. These slides, however, often include staining of other cells or of cytoplasmic regions, some of which our current algorithms cannot discriminate from true micronuclei. Figure 1 shows a region in the cytoplasm (near the top) which is as optically dense as the micronucleus to the lower left.

For the purpose of speeding up the identification and scoring process, we are currently investigating various types of microscope illumination and lymphocyte staining techniques. Fluorescent staining has the inherent advantage of a high object-to-background contrast allowing us to use simple thresholding algorithms for object extraction. However, fluorescent fields are slightly more difficult to calibrate for quantitative measurements than fields illuminated with bright field transmission light. We are also considering counter staining and image subtraction techniques in order to select smaller regions of interest equivalent to the cytoplasm as well as to aid in the classification of lymphocytes on the basis of nuclear and cytoplasmic parameters.

Lymphocyte classification into proliferating and non-proliferating populations is necessary in order to validate the micronuclear measure of chromosome damage. This measure is a function of the number of micronucleated cells divided by the number of cells that have proliferated in culture. Consequently, it is important to determine the extent of cell proliferation of the digitized lymphocytes as well as to study and understand the fate of micronuclei during cell division. A model of lymphocyte cell cycle kinetics based on microscopic measurements would be desireable.

Last but not least, the micronuclei finding and quantitation algorithms have room for improvement in terms of spatial and grayness measurement accuracy and in terms of computer memory usage and program execution time. Our difficulties in obtaining quantitative information from digitized images were primarily due to (a) critical measurement tolerances on micronuclear parameters because of their small size (e.g. 5—20 pixels) resulting in significant variations in area measurements when the object segmentation threshold was offset by only one level of gray, and (b) a non-uniform image background and sometimes low gray level resolution (proportional to contrast) surrounding the micronucleus' contour. More complex and conditional algorithms may need to be developed as long as computer memory and execution time are kept within reasonable limits. These latter constraints are essential if one wishes to apply the micronucleus assay to population type surveys, such as to study the effect of low radiation dose to human population.

Conclusion

Our preliminary results on computerized micronuclei detection and analysis lead us to three encouraging conclusions. First, scoring of micronuclear frequency in peripheral blood lymphocytes is a useful method for measuring chromosome damage and may be developed into a powerful assay for timely estimating mutagen and carcinogen damage in human population surveys. Second, automation of the assay holds the promise for improving it in terms of (a) statistically more significant results, (b) more objective identification of micronuclei, and (c) obtaining quantitative spatial and grayness measurements on nuclei. And third, the additional information from quantitative measurements gives us insight into the mechanism of micronucleus formation as well as a means to identify the proliferating cells needed to compute the extent of chromosome damage.

References

1. Norman A, Adams HA, Riley RF (1978) Cytogenetic Effects of Contrast Media and Triiodobenzoic Acid Derivatives in Human Lymphocytes. Radiology 129(1):199–203
2. Callisen HH, Pincu M, Norman A (1982) Video Imaging System for Microscopy: Technical Aspects. IEEE Proceedings on the International Workshop on Physics and Engineering in Medical Imaging, Asylomar

Hematopoietic and Immunologic Effects

Review of Biological Dosimetry by Conventional Methods in Haematopoiesis

Jan W.M. Visser

Radiobiological Institute TNO, 151 Lange Kleiweg, 2288 GJ Rijswijk, The Netherlands

1 Introduction

The adult human body contains on the average five liters of blood and three liters of bone marrow. About half of these volumes are occupied by haematopoietic cells which total about 4×10^{13} in number (10^{13} in the bone marrow and 3×10^{13} in the circulation). The most predominant blood cells are the erythrocytes ($4.5-5.5 \times 10^6$ per μl) which normally have a life span of 110–135 days. A variety of other cell types can be recognized in the blood. The life span of most of the other cell types is much shorter than that of the red cells and, mainly because of that, their incidence is lower. Some cell types (e.g., monocytes; normally 375 per μl blood) migrate through the blood to extravascular areas with one day transit times in the circulation. Others (e.g., reticulocytes; 6×10^5 per μl) are the immediate precursors of mature blood cells and may have turn over times of less than one day. The daily loss of blood cells in the adult is about 5×10^{11} cells. The same number is normally released from the bone marrow, which contains graded numbers of progenitor cells of each of the blood cell types. The production of blood cells in the bone marrow is in sensitive equilibrium with the demands elsewhere in the body.

The blood cell formation is one of the most sensitive equilibria in the mammalian body. Exposures to drugs or ionizing irradiation easily disturb the haematopoietic system. The bone marrow syndrome is the prime lethal effect of ionizing radiation at dosages below 10 Gy (Fig. 1). At higher dosages, the intestinal syndrome (6–30 Gy) and the brain syndrome (20–100 Gy) cause the mortality. While the brain syndrome becomes apparent within hours after exposure to irradiation, the bone marrow syndrome may not appear for several weeks. This indicates that the mortality resulting from the bone marrow syndrome is not due to malfunctioning of the mature or almost mature blood cells but that it is due to damage at the source of the blood cells, viz., the haematopoietic stem cells and their regulation. The incidence of the stem cells is very low, even in haematopoietic organs. It has been estimated that 0.1 to 0.7% of all mouse bone marrow cells — a relatively rich source of stem cells — are pluripotent haematopoietic stem cells (Van Bekkum et al. 1979). The incidence of stem cells in larger mammalian species is even lower (Vriesendorp and Van Bekkum, 1980). Therefore, these cells cannot be studied by direct microscopical and cytochemical techniques. A number of indirect techniques has been developed to analyse these progenitor cells and their regulation. Most of these methods are based on the capability of each stem cell to produce large numbers of mature blood cells. Both *in vitro* and *in vivo*

Biological Dosimetry. Edited by W. G. Eisert and M. L. Mendelsohn
© Springer-Verlag Berlin Heidelberg 1984

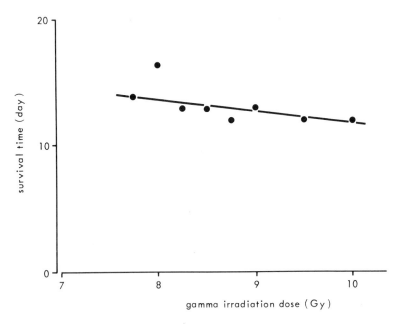

Fig. 1. Survival time of male BC3 mice *versus* dose of gamma radiation (total body ^{137}Cs; 1 Gy/min)

colony formation can be achieved with most hematopoietic stem cells. Since many different cell types play a role in haematopoiesis, techniques for the distinction between these cell types and especially for the recognition of the earliest progenitor cells are of importance in the quantitation of treatments. Cell separation techniques are useful for this purpose.

If effects of treatments such as exposure to drugs or irradiation are studied with the haematopoietic system as the dosimeter, it is of importance to identify the target cells of the treatment which are hidden within all other haematopoietic cells and to take the kinetics of this dynamic production system into account. The radiosensitivity of the pluripotent haematopoietic stem cell will be presented as an example for biological dosimetry.

2 Conventional Methods for Studying Haematopoiesis

2.1 Microscopy

As early as 1665 Marcello Malpighi described "microscope" observations of red spheres in tubes in the body. Some years later Jan Swammerdam confirmed this observation and in 1674 Leeuwenhoek (Fig. 2) accurately determined the size and shape of erythrocytes. Only at the end of the last century could other cell types be recognized in the blood, as the result of the development of staining techniques. This discovery

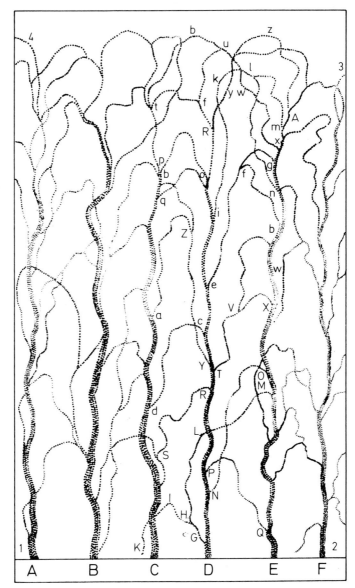

Fig. 2. Blood vessels containing cells as depicted by Van Leeuwenhoek in 1698. Courtesy of Boerhaave Museum, Leiden

initiated studies to elucidate the behaviour and function of the different blood cell types. Five main categories of blood cell types can be distinguished on the basis of apparent differences in shape (Fig. 3; Table 1). The blood cell types can be further subdivided if other properties are taken into account, e.g., the presence of certain enzymes, capacities to take up particles or the presence of specific membrane structures. Such properties correlate with the subdivision of tasks among the blood cells.

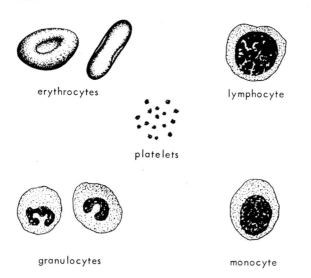

erythrocytes   lymphocyte

platelets

granulocytes   monocyte

Fig. 3. Schematic representation of the morphological appearance of the five main categories of circulating blood cells

Table 1. Circulating Blood Cell Types

Cell type	Incidence (per μl blood)	Average life span in circulation
erythrocytes	$4.5-5.5 \times 10^6$	110–135 days
reticulocytes	6×10^5	1–2 days
platelets	2.5×10^5	8–12 days
leukocytes:		
basophils	25	–
eosinophils	200	–
juvenile neutrophils	300	–
segmented neutrophils	4000	1 day
monocytes	375	–
lymphocytes	2100	hours to years

Each type has a specialized function, e.g., the production of antibodies, transport of gases. The microscopical examination of blood samples is general practice in the clinic because the composition of the blood cells reflects reactions of the body to challenges and treatments.

The extreme specialization of the blood cells occurs at the expense of their capability to divide. They function for a limited time period and are then removed from the blood. The human red blood cells remain in the circulation for about 120 days, the polymorphonuclear white blood cells (the granulocytes) have a mean life span of only a few days. Most blood cells can therefore be regarded as disposable entities. About 5×10^{11} blood cells are removed from the circulation in the adult human body each day. It is of importance that there is a continuous production of new blood cells and that this production is adequately and quickly adapted to the demands for various cell types. The production of the various blood cell types normally takes place in the bone marrow. The blood cells originate there by division of closely packed immature cells. Microscope slide preparations of the bone marrow show a large variety of cell types of differing maturity. Many of these cannot be directly related to the mature blood cell types. Only the immediate precursors of the mature cells can be recognized by microscopical examination. Since the production of blood cells requires cell division, the incidence of the progenitor cells is lower the earlier the progenitor cell type (Fig. 4). Since the typical structures of the specialized blood cells are also not yet developed in the earlier cell types, a microscopical examination is not useful under normal conditions. If the bone marrow composition is severely altered, e.g., by leukemia, microscopical examination is, of course, the method of choice for a preliminary quantitation.

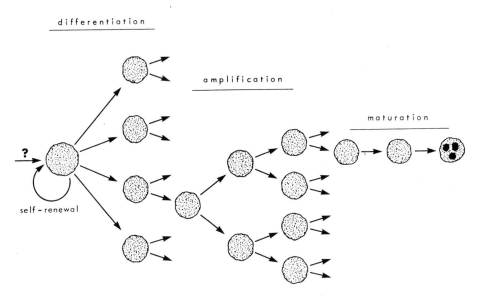

Fig. 4. Schematic representation of blood cell formation

2.2 Stem Cell Cultures

Studies on the relation between bone marrow cells and the mature blood cells were made possible by the development of laboratory techniques for culturing of the progenitor cells. It can be shown that in spite of their many differences, all blood cells are continuously formed from one cell type, the pluripotent haematopietic stem cell, which also produces itself. The evidence for the existence of one stem cell for all blood cells comes from experiments with mice. The blood forming cells are very sensitive to ionizing radiation. It is therefore possible to irradiate mice such that the haematopoietic organs are fully destroyed without serious short term damage to the functioning of the other organs (e.g., Jacobson et al., 1949). The irradiation damage can then be fully repaired by transplantation of the bone marrow cells of a healthy donor mouse. The stem cells which are amongst the transplanted cells home in the haematopoietic organs and repopulate these with their offspring. The blood cell formation can be fully restored by this procedure. Similar bone marrow transplantations have been successfully performed in other species, including man. A portion of the stem cells homes in the spleen which is a blood cell forming organ under conditions of haematopoietic stress. At eight to ten days after transplantation, the mouse spleen contains macroscopically visible nodules which consist of dividing blood forming cells (Till and McCulloch, 1961; Fowler et al., 1967). These nodules are called spleen colonies (Fig. 5). It has been shown that all cells within one colony arise from a single cell (Becker et al. 1963; Chen and Schooley, 1968). This cell is called the colony forming cell. It was also shown that a spleen colony may be of mixed composition, containing erythrocytes, granulocytes and megakaryocytes, which are blood cell types of three different lineages (Curry and Trentin, 1967; Fowler et al., 1967; Silini et al., 1968). The lymphocytes have also been shown to arise from cells with the capability of forming spleen colonies (Micklem et al., 1966; Wu et al., 1968). In addition, the colonies contain cells which can form spleen colonies upon retransplantation (McCulloch and Till; 1964; Lajtha, 1965; Siminovitch et al., 1967; Matioli et al., 1969; Lahiri and Van Putten, 1969). The synthesis of these observations leads to the hypothesis of a pluripotent haematopoietic stem cell which forms all blood cells as well as itself.

Two methods are used to enumerate and analyse the pluripotent stem cells in mice. The first one makes use of the capacity to form spleen colonies. The number of spleen colonies is linearly proportional to the number of transplanted bone marrow cells (Fig. 6) if the spleen contains less than 20 colonies. At higher numbers overlapping colonies cannot be distinguished by eye. Only a portion of the transplanted stem cells homes in the spleen to form colonies there. This fraction can be determined by retransplantation of spleen cells shortly after transplantation. The results of such experiments indicate that about 5% of the transplanted colony forming cells actually produce a spleen colony (Matioli et al., 1969; Lahiri et al., 1970). Therefore, the number of pluripotent stem cells in a graft can be calculated by multiplying the number of spleen colonies (CFU-S; colony forming units spleen) by 20. The normal incidence of CFU-S in mouse bone marrow is 10 to 30 per 10^5 nucleated cells; this means that the incidence of pluripotent stem cells is 0.2 to 0.6%.

There are a number of reports which describe in vitro culture of mixed or immature colonies (Fauser and Messner, 1978; Humphries et al., 1981; Neumann et al., 1981; Nakahata and Ogawa, 1982a; 1982b). These indicate that the pluripotent stem cell can also be studied in vitro. The plating efficiency of these assays is not known so that these assays cannot be directly used to enumerate the stem cells.

The second method for analysing the pluripotent stem cells in a graft makes use of the capacity of these cells to restore the haematopoietic system of lethally irradiated mice. One of the endpoints of this method is the 30-day survival of lethally irradiated recipients of transplanted cells. If less than 10^4 bone marrow cells are transplanted

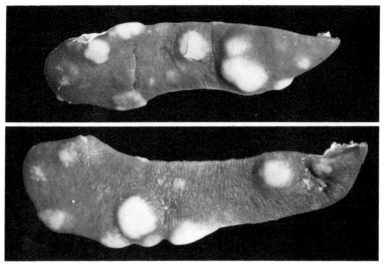

Fig. 5. Picture of mouse spleen containing nodules of blood cells

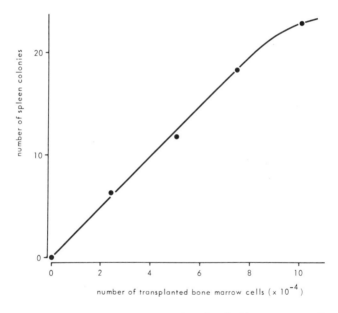

number of spleen colonies

number of transplanted bone marrow cells ($\times 10^{-4}$)

Fig. 6. Number of macroscopic visible spleen colonies *versus* number of grafted bone marrow cells 10 days after lethal irradiation (9.5 Gy ^{137}Cs total body) and transplantation

in mice most recipients die within three weeks (Fig. 7) due to disorders of the haematopoietic system. All recipients survive for 30 days or longer if more than 2×10^4 normal syngeneic bone marrow cells are given. A graft of 10^4 normal bone marrow cells contains 2.5 CFU-S. Therefore, 20 x 2.5 or 50 stem cells per mouse are sufficient to protect 50% of the lethally irradiated animals. A normal adult mouse contains about

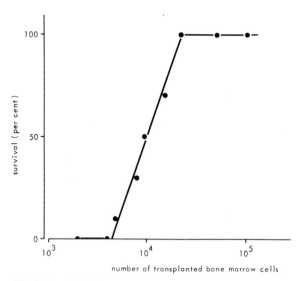

Fig. 7. Survival of lethally irradiated (BC3) mice *versus* the number of transplanted bone marrow cells. The percentage surviving animals is determined at 30 days after irradiation and transplantation

6 x 10⁴ CFU-S or 1.2 x 10⁶ pluripotent stem cells. The presence of pluripotent stem cells capable of this restoration in grafts of different origin and composition can be compared by this assay. This method can also be applied to species other than the mouse (Table 2). The number of bone marrow cells needed to protect 50% of the lethally irradiated recipients increases with the size of the species, also on a per kg body weight basis (Vriesendorp and Van Bekkum, 1980). Other endpoints of the second method are the reappearance of the different blood cell types at various time points after transplantation (Fig. 8). Such results give a good impression of the kinetics of blood cell formation. Shortly after irradiation some cells crucial for survival disappear from the blood. The half times of the loss of granulocytes and platelets are 1 and 3 days, respectively. The progeny of the transplanted bone marrow cells starts to become detectable at about one week after transplantation. The rate of blood cell production seems to be similar in most species (Vriesendorp and Van Bekkum, 1980). The mature cell types in the blood reappear earlier than the progenitor cells in the bone marrow. The thymocytes appear somewhat later than other maturated blood cells (cfr., Kadish and Basch, 1976). The self renewal of the murine pluripotent stem cells in the bone marrow requires several weeks before normal CFU-S numbers are present (Fig. 9).

Burton et al. (1982) recently demonstrated that the alloenzyme content of mouse blood cells shows large, and frequently sudden, fluctuations which may be taken to indicate that under normal equilibrium conditions only a few clones of cells contribute to the blood cell formation. Computer simulations indicated that three clones produce red blood cells for 14 days and then other clones take over. These clones each produce 2 x 10⁹ erythrocytes. This requires 31 doublings, whereas the production of erythrocytes from spleen colony forming cells requires only 12 to 14 doublings. Therefore, Burton et al. (1982) concluded that the spleen-colony-forming cells are intermediate cells separated by 17–19 doublings from the clonogenic cell under

Table 2. Graft size for radioprotection in different species

Number of isogeneic bone marrow cells needed to protect 50% of supralethally irradiated animals are given in absolute numbers and as fraction of all bone marrow cells.

species	number of bone marrow cells needed for 50% protection		literature references
	absolute number	fraction of all bone marrow cells	
mouse	10^4	4×10^{-5}	this paper
rat	10^6	15×10^{-5}	van Bekkum, 1977
rhesus monkey	2×10^7	3×10^{-4}	Vriesendorp and Van Bekkum, 1980
dog	2.4×10^7	17×10^{-4}	Vriesendorp and Van Bekkum, 1980

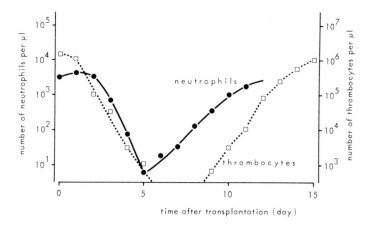

Fig. 8. Reappearance of platelets (*open squares*) and granulocytes (*closed circles*) after irradiation and bone marrow transplantation

normal steady state conditions. This would indicate that the effects of low doses of drugs or radiation may become expressed and apparent at the mature blood cell level only months or years after exposure. The late appearance of leukemia after irradiation may be one of these delayed expressions of damage to the precursors of the stem cells.

2.3 In vitro Cultures

Since almost 20 years, it has been possible to culture bone marrow cells outside the body. A variety of culture techniques for producing colonies of erythrocytes, platelets, granulocytes, monocytes, lymphocytes, immature cells, and new colony forming cells

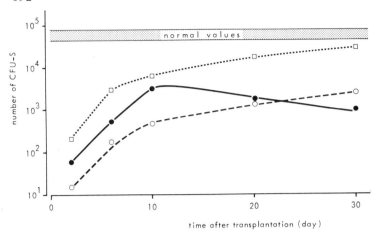

Fig. 9. Reappearance of CFU-S in mouse femoral bone marrow (*dashed line, open circles*) and spleen (*solid line, closed circles*) after irradiation and transplantation. The *dotted line (open squares)* represents the sum of CFU-S in all bone marrow and spleen taking into account that one femur contains 6.5% of all bone marrow

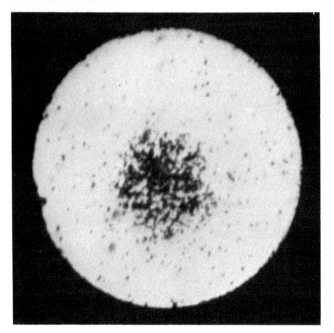

Fig. 10. Picture of an *in vitro* colony containing both granulocytes and monocytes, after fixation and May Grünwald Giemsa staining. The colony originates from an early myeloid progenitor cell which was deposited 7 days earlier in a 10 μl-well of a Terasaki tray in culture medium containing colony stimulating factor

has been described (Fig. 10) (Bradley and Metcalf, 1966; Metcalf and Foster, 1967; Van den Engh, 1974; Van den Engh and Bol, 1975; Iscove and Sieber, 1975; Nakeff and Daniels, 1976; Gregory, 1976; Metcalf et al., 1980; Wagemaker, 1980; Metcalf and Burgess, 1982). It has also become possible to achieve colonies from pluripotent stem cells *in vitro* (Fauser and Messner, 1978; 1979; Humphries et al., 1981; Neumann et al., 1981; Nakahata and Ogawa, 1982a; 1982b). The type of blood cell in the in vitro colonies depends strongly on the regulatory factors which are added to the growth media. Some of the regulatory factors have been purified to a high degree by conventional biochemical methods (Metcalf et al., 1980; Wagemaker, 1980). Analyses of the regulatory factors and of the composition of the colonies has revealed different steps in the formation of blood cells. It can be demonstrated that the earliest steps in haematopoietic differentiation require the presence of a growth stimulus which is specific for the line of differentiation but not for the stage (Fig. 11). Erythropoietin, for instance, is needed for each cell division which results in the formation of red blood cells (Gregory, 1976). A factor called burst promoting activity (BPA) is required in addition to erythropoietin in order to obtain red blood cell colonies from the earliest committed erythroid progenitor cells (Wagemaker, 1980). Another factor (stem cell activating factor; SAF) triggers the pluripotent stem cells to self renewal. SAF and BPA are different molecules (Wagemaker, 1980).

The time kinetics of colony formation *in vitro* is of importance for the identification of the progenitor stage which produces the colonies. An example of the time course of colony maturation is shown in Fig. 12 (Bol and Williams, 1980). Three progenitor cell types which represent consecutive stages in the myeloid differentiation lineage all produce colonies of mature macrophages. The earliest of these requires 14 days to produce the end cells, the other two require 11 and 8 days, respectively. In addition, the size of the colonies is dependent on the stage of differentiation of the progenitor cells. As to be expected, the later stages produce smaller colonies than do the earlier ones. The incidence of the three cell types in normal bone marrow is also in agreement with a consecutive order: per 10^5 cells plated, 30 to 50 colonies which were formed by the earliest of the three progenitor cell types, 80 to 120 colonies from the next and 100 to 150 from the third are observed.

In order to employ the haematopoietic system as dosimeter the properties of the committed progenitor cells must also be taken into account. Abramson et al. (1977) have demonstrated that self renewal of committed progenitor cells occurs in mice. They transplanted bone marrow cells which contained chromosome markers into

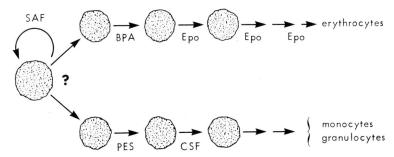

Fig. 11. Schematic representation of the requirement of regulatory factors in two differentiation pathways of haematopoiesis (*SAF* stem cell activating factor; *BPA* burst promoting activity; *Epo* erythropoietin; *PES* 18 h post endotoxin serum; *CSF* colony stimulating factor)

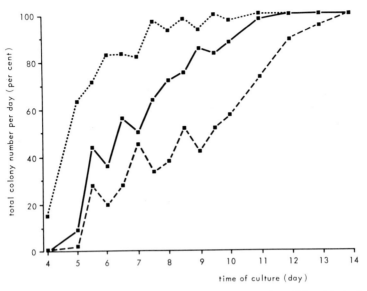

Fig. 12. Kinetics of the maturation of macrophage colonies which originate from cells with different stages of differentiation. *Dotted line:* colonies which originate from progenitor cells (CFU-C) with a density of 1.070g·cm^{-3}; *solid line:* colonies from CFU-C with a density of 1.075 g·cm^{-3}; *dashed line:* colonies from CFU-C with a density of 1.080 g·cm^{-3} (from Bol and Williams, 1980)

lethally irradiated mice. Almost one year later, some of the mice were found to contain lymphocytes and myelocytes with different markers within the same animal. This indicates that also progenitor cells other than the pluripotent stem cell are capable of both self renewal and the production of offspring. Up to now 3 years later, none of these mice have developed leukemias or other apparent malfunctions of the haematopoietic system or other organs in spite of the presence of radiation induced chromosome markers (R. Miller, personal communication). This indicates that the observation of the presence of chromosomal aberrations provides only limited possibilities to the use of the haematopoietic system as a dosimeter. Only in typical cases of chronic granulocytic leukemia in man a chromosome aberration, the so-called Philadelphia chromosome is, generally, present in the leukemic cells (Nowell and Hungerford, 1961).

2.4 Cell Separation

Several investigators realized 10 to 20 years ago that cell separation techniques could be used not only to isolate or enrich for certain cell types but also to determine cytophysical and cytochemical properties of the cells. Leif and Vinograd (1964) and Shortman (1968) developed equilibrium density centrifugation techniques to determine the buoyant density of cells. Miller and Phillips (1969) described velocity sedimentation methods for determining the relative sizes of spherical cells. The sedimentation rate could be shown to be related to the cell cycle stage for homogeneous cell populations (McDonald and Miller, 1970). Zeiller and co-workers (1972) developed the free flow cell electrophoresis for determining the net cellular electrical charge. All of these techniques could be performed without affecting the viability of the cells. Therefore, these methods were and still are of aid in further characterizing and identifying the cell types which play a role in the haematopoietic system which can be detected only by the *in vivo* and *in vitro* colony assays.

Bone marrow cells are separated into fractions which differ slightly according to the separation parameter: density, size or charge. Similar aliquots of each fraction are subsequently submitted to the colony assays to enumerate the progenitor cells. The distribution of the colonies over the fractions is then determined one week later. The progenitor cell can thus be characterized with respect to the parameter of the separation without purification of the cell type (Haskill and Moore, 1970; Haskill et al., 1970; Metcalf et al., 1971; Metcalf and McDonald, 1975; Williams and van den Engh, 1975; Bol et al., 1977; Byrne et al., 1977; Bol et al., 1979; Wagemaker and Visser, 1981; Francis et al., 1981). Monette et al. (1974) demonstrated by the velocity sedimentation method that the CFU-S are a heterogeneous cell population. Visser et al. (1977) established the sizes and densities of these CFU-S subpopulations by determination of the sedimentation rates and of the buoyant density. Most of the CFU-S were found to be in a quiescent state of the cell cycle (Vassort et al., 1973) and to have a diameter of 7.0 μm, which was calculated from the density and the sedimentation rate using Stokes' law concerning the sedimentation of spheres in viscous media. The density of these cells was 1.070 g.cm^{-3} at 4°C. The remaining CFU-S (20 to 40% in normal bone marrow) were proliferating and had a density of 1.075 g.cm^{-3}; the diameters of G1 and G2/M cells were 7.3 and 9.2 μm, respectively.

Density gradient separation is also useful in this separation and distinction of the three progenitor cell types which represent consecutive stages in the myeloid differentiation pathway. The three cell types have densities of 1.070, 1.075 and 1.080 g.cm^{-3}, respectively (Bol et al., 1977; 1979). The density increases with further differentiation. The sedimentation rates of these cells differ slightly due to the differences in density. The cell sizes therefore are similar (7.5 μm for cells in the G1 phase of the cell cycle).

3 Stem Cell Radiosensitivity and Bone Marrow Syndrome Mortality

The sensitivity of pluripotent haemopoietic stem cells in mice to ionizing radiation can be determined by enumeration of the stem cells by use of the CFU-S assay. Figure 13 shows an extrapolated curve of the CFU-S survival versus gamma (^{137}Cs; 1 Gy/min) irradiation dose. The D_0 of this curve is 0.80 Gy for n = 1. The figure illustrates that at 7.8 Gy, 0.005% of the CFU-S survive. A healthy adult mouse contains about 6 x 10^4 CFU-S, so that after 7.8 Gy there are 3 surviving CFU-S per mouse. Fifty per cent of the animals survive 7.8 Gy without treatment (LD$_{50}$ = 7.8 Gy gamma-irradiation). Experiments as shown in Fig. 6 indicate that a transplant of 10^4 cells containing 2.5 CFU-S protects 50% of supralethally irradiated animals. Van Bekkum and Schotman (1974) showed that the protection capacity of grafted stem cells is similar to that of endogenous stem cells which survive irradiation. The above calculations confirms this.

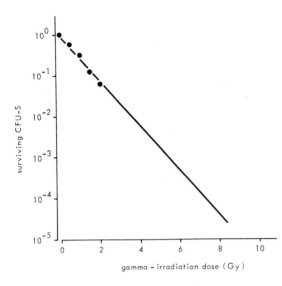

Fig. 13. Fraction surviving CFU-S versus gamma irradiation dose. BC3 mice were irradiated (^{137}Cs; 1 Gy · min^{-1}; total body), femoral bone marrow cells were taken shortly after irradiation and injected into lethally irradiated syngeneic recipients to enumerate the CFU-S content. The extrapolated curve represents a radiosensitivity with a D_0 of 0.8 Gy and $n = 1$

They indicate that the lethality of the radiation in this dose range can be fully explained by the sensitivity of the CFU-S or the pluripotent haemopoietic stem cells to ionizing radiation. It should be noted here that these results were obtained with specific pathogen-free laboratory animals. The presence of microorganisms such as Pseudomonas species is known to affect the mortality as well as the graft size needed for 50% protection (Wensinck et al., 1957).

The radioprotective effect of bone marrow grafts is due to both the self-renewal capacity of the stem cells and the extensive proliferation. The haematopoietic system is capable of producing sufficient numbers of mature blood cells of the types which are crucial for maintaining life (platelets and granulocytes) in the first few weeks after irradiation under laboratory conditions. Transfusions of these critical cell types have become general practice in the clinic during the first weeks after bone marrow transplantation in humans. The mortality resulting from ionizing radiation, however, can be expected to have a similar explanation in the human and other species as in mice, viz., the radiosensitivity of the pluripotent haematopoietic stem cell. The LD_{50} for the mouse, the rat, the rhesus monkey and the dog is different (7.0, 6.75, 5.25 and 3.7 Gy of X-rays, respectively). These differences cannot be explained by different radiosensitivities of the stem cells only (Vriesendorp and Van Bekkum, 1980); a too low D_0 of only 35 rad would be needed for the dog pluripotent haematopoietic stem cell to explain the dog LD_{50}. Another explanation for the differences in LD_{50} are differences in kinetic parameters of the haematopoietic system in the different species. The time required for a stem cell to finally produce an end cell would have to be longer for larger species, whereas the survival time of these end cells would have to be shorter. However, both the kinetics of disappearance of granulocytes and thrombocytes after total body irradiation and the regeneration of these cells after bone marrow transplantation are similar in all species. The differences in LD_{50} among species could be explained by the assumption that the number of stem cells per body unit differs among species, whereas the D_0 of the stem cells is between 0.5 and 0.75 Gy (X-rays) for all species (Vriesendorp and Van Bekkum, 1980). These calculations resulted in an estimated bone marrow rescue dose for an adult man of 4×10^9 autologous cells and a LD_{50} for total body irradiation of 3.15 Gy (X-rays). The LD_{50} value for man obtained from the analysis of radiation accidents is 3.0 Gy which is close to the above calculated value. In addition, it has recently been shown that the D_0 of the human pluripotent haematopoietic stem cell is 0.91 Gy gamma-irradiation (Neumann et al., 1981), which is close to the value reported for mouse CFU-S.

4 Sensitivity to Regulatory Factors

The adequate production of blood cells to meet body demands requires a set of specific messenger molecules. A number of possible candidates to play such a role *in vivo* are in the process of being purified by biochemical methods. The chemical structures of these substances are not known. Therefore, the haematopoietic system is employed to quantitate the purity of the regulatory factor. The best known example of such a factor is erythropoietin, a stimulator of red blood cell formation. This substance is produced by unidentified cells which reside primarily in the adult kidney and in the foetal liver. Low oxygen levels in the body trigger these cells to release erythropoietin by further unknown pathways. The amount of erythropoietin in human or animal plasma, renal extracts, fractions after purification steps, etc., are determined in posthypoxic polycythemic mice (Cotes and Bangham, 1961; Camiscoli et al., 1968). Briefly, mice are first made polycythemic by exposure to low air pressure for two weeks. Subsequently, the mice are kept at ambient air pressure. Due to the exposure to low oxygen pressures, they then contain more than enough red blood cells for at least one week. Therefore, these animals do not produce erythropoietin themselves and, consequently, they do not produce new red blood cells. Aliquots of fluids in which the erythropoietin is to be quantified are administered to such animals along with ^{59}Fe. The erythropoietin will stimulate red blood cell formation which requires iron uptake. The amount of newly formed red blood cells can be determined from the ^{59}Fe uptake. The assay is linear over a wide range. International reference preparations for calibrating the system are available.

The regulatory factor which stimulates the pluripotent stem cell to produce itself is a target in purification studies. The number of CFU-S in *in vitro* culture system decreases to 5–10% within three days if no stimulatory factors are added (e.g., Worton et al., 1968). Löwenberg and Dicke (1975; 1977) demonstrated that the number of CFU-S in suspension can be maintained for 3 days by addition of conditioned medium of embryonic fibroblasts. Conditioned media from lectin-stimulated mouse spleen cultures (Cerny, 1974) and from human leukocyte cultures (Wagemaker and Peters, 1978) · have similar properties. Dexter et al. (1977) demonstrated maintenance of CFU-S in cultures during several weeks. The conditioned media can be separated into fractions of proteins which differ in molecular weight, size and charge by biochemical separation methods. Each fraction can then be tested for its maintenance activity for CFU-S in suspension. CFU-S numbers are determined by the spleen colony assay described above in Sect. 2.2. The spleen colony assay is employed as a dosimeter for the relative concentration of so-called stem cell activating factor (SAF) in this way. In addition these methods provide knowledge concerning the chemical properties of SAF (Wagemaker, 1980).

5 Concluding Remarks

The use of the haematopoietic system as a biological dosimeter requires knowledge concerning the identity and proliferation characteristics of more than ten different possible target cells which play important roles in the blood cell formation. Conventional methods for analysing these cell types are *in vivo* and *in vitro* culture techniques in the presence of specific colony stimulating factors and cell separation techniques. The haematopoietic system is in a complex equilibrium with continuous self-renewal and differentiation of the pluripotent stem cells, an amplification compartment where the committed stem cells produce large numbers of specialized progeny and the mature end cell compartment where cells function and die. These compartments are located in different parts of the body, so that migration also plays a role. Therefore, in addition to knowledge concerning the target cells, the time kinetics of the production and loss of the various blood cell types and of the migration should be known in order to employ the haematopoietic system as a biological dosimeter.

References

Abramson S, Miller RG, Phillips RA (1977). The identification in adult bone marrow of pluripotent and restricted stem cells of the myeloid and lymphoid systems. J Exp Med 145:1567–1579

Becker AJ, McCulloch EA, Till JE (1963). Cytological demonstration of the clonal nature of spleen colonies derived from transplanted mouse marrow cells. Nature 197:452–454

Bekkum DW van (1977). The appearance of the multipotential hemopoietic stem cell. In: Baum SJ, Ledney GD (eds.) Experimental Hematology Today. Springer, Berlin Heidelberg New York, pp 3–10

Bekkum DW van, Schotman E (1974). Protection from haemopoietic death by shielding versus grafting of bone marrow. Int J Radiat Biol 25:361–372

Bekkum DW van, Engh GJ van den, Wagemaker G, Bol SJL, Viser JWM (1979). Structural identity of the pluripotent hemopoietic stem cell. Blood Cells 5:143–159

Bol S, Williams N (1980). The maturation state of three types of granulocyte/macrophage progenitor eells from mouse bone marrow. J Cell Physiol 102:233–243

Bol S, Visser J, Williams N, Engh GJ van den (1977). Physical characterization of haemopoietic progenitor cells by equilibrium density centriguation. In: Bioemendal H (ed.). Cell Separation Methods. Elsevier/North-Holland Biomedical Press, pp 39–52

Bol S, Visser J, Engh GJ van den (1979). The physical separation of three subpopulations of granulocyte/macrophage progenitor cells from mouse bone marrow. Exp Hemat 7:541–553

Bradley TR, Metcalf D (1966). The growth of mouse bone marrow cells *in vitro*. Aust J exp Biol Med 44:287–300

Burton DI, Ansell JD, Gray RA, Miclem HS (1982). A stem cell for stem cells in murine haematopoiesis. Nature 298:562–563

Byrne P, Heit W, Kubanek B (1977). The in vitro differentiation of density subpopulations of colony-forming cells under the influence of different types of colony-stimulating factor. Cell Tissue Kinet 10:341–351

Camiscoli JF, Weintraub AH, Gordon AS (1968). Comparative assay of erythropoietin standards. Am NY Acad Sci 149:40–45

Cerny J (1974). Stimulation of bone marrow hemopoietic stem cells by a factor from activated T cells. Nature 249:63–66

Chen MG, Schooley JC (1968). A study on the clonal nature of spleen colonies using chromosome markers. Transplantation 6:121–126

Cotes PM, Bangham DR (1961). Bio-assay of erythropoietin in mice made polycythaemic by exposure to air at a reduced pressure. Nature 191:1065–1067

Curry JL, Trentin JJ (1967). Hemopoietic spleen colony studies. I. Growth of differentiation. Devl Biol 15:395–413

Dexter TM, Moore MAS, Sheridan APC (1977). Maintenance of hemopoietic stem cells and production of differentiated progeny in allogeneic and semiallogeneic bone marrow chimeras *in vitro*. J Exp Med 145:1612–1616

Engh G van den, Bol S (1975). The Presence of a CSF enhancing activity in the serum of endotoxin-treated mice. Cell Tissue Kinet 8:579–587

Engh GJ van den (1974). Quantitative *in vitro* studies on stimulation of murine haemopoietic cells by colony stimulating factor. Cell Tissue Kinet 7:537–548

Fauser AA, Messner HA (1978). Granuloerythropoietic colonies in human bone marrow, peripheral blood and cord blood. Blood 52:1243–1248

Fauser AA, Messner HA (1979). Identification of megakaryocytes, macrophages and eosinophils in colonies of human bone marrow containing neutrophilic granulocytes and erythroblasts. Blood 53:1023–1027

Fowler JH, Wu AM, Till JE, McCulloch EA, Siminovitch L (1967). The cellular composition of hemopoietic spleen colonies. J Cell Physiol 69:65–72

Francis GE, Bol S, Berney JJ (1981). Proliferation capacity, sensitivity to colony stimulating activity and buoyant density: linked properties of granulocyte-macrophage progenitors from normal human bone marrow. Leukemia Res 5:243–250

Gregory CJ (1976). Erythropoietin sensitivity as a differentiation marker in the hemopoietic system: studies of three erythropoietic colony responses in culture. J Cell Physiol 89:289–302

Haskill JS, Moore MAS (1970). Two dimensional cell separation: comparison of embryonic and adult haemopoietic stem cells. Nature 226:853–854

Haskill JS, McNeill TA, Moore MAS (1970). Density distribution analysis of *in vivo* and *in vitro* colony forming cells in bone marrow. J Cell Physiol 75:167–180

Humphries RK, Eaves AC, Eaves CJ (1981). Selfrenewal of hemopoietic stem cells during mixed colony formation in vitro. Proc Natl Acad Sci USA 78:3629–3633

Iscove NN, Sieber F (1975). Erythroid progenitors in mouse bone marrow detected by macroscopic colony formation in culture. Exp Hematol 3:32–43

Jacobson LO, Marks EK, Robson MJ, Gaston E, Zirkle RE (1949). The effect of spleen protection on mortality following X-irradiation. J Lab Clin Med 34:1538–1593

Kadish JL, Basch RS (1976). Hematopoietic thymocyte precursors. I. Assay and kinetics of the appearance of progeny. J Exp Med 143:1082–1099

Lahiri SK, Putten LM van (1969). Distribution and multiplication of colony forming units from bone marrow and spleen after injection in irradiated mice. Cell Tissue Kinet 2:21–28

Lahiri SK, Keizer HJ, Putten LM van (1970). The efficiency of the assay for haemopoietic colony forming cells. Cell Tissue Kinet 3:355–362

Lajtha LG (1965). Response of bone marrow stem cells to ionizing radiations. In: Ebert M, Howard A (eds) Current Topics in Radiation Research I. North Holland, Amsterdam, pp 139–163

Leif RC, Vinograd J (1964). The distribution of buoyant density of human erythrocytes in bovine albumin solutions. Proc Natl Acad Sci US 51:520–528

Löwenberg B, Dicke K (1975). Studies on the *in vitro* proliferation of pluripotent haemopoietic stem cells. In: Leukemia and Aplastic Anemia. Il Pensiero Scientifico, Rome pp. 377–391

Löwenberg B, Dicke KA (1977). Induction of proliferation of hemopoietic stem cells in culture. Exp Hematol 5:319–331

Matioli G, Vogel H, Niewisch H (1969). The dilution factor of intravenously injected hemopoietic stem cells. J Cell Physiol 72:229–234

McCulloch EA, Till JE (1964). Proliferation of hemopoietic colony-forming cells transplanted into irradiated mice. Radiat Res 22:383–397

McDonald HR, Miller RG (1970). Synchronization of mouse L-cells by a velocity sedimentation technique. Biophys J 10:834–842

Metcalf D, Burgess AW (1982). Clonal analysis of progenitor cell commitment to granulocyte or macrophage production. J Cell Physiol 111:275–283

Metcalf D, Foster R (1967). Behaviour on transfer of serum stimulated bone marrow colonies. Proc Soc Exp Biol 126:758–762

Metcalf D, Johnson GR, Burgess AW (1980). Direct stimulation by purified GM-CSF of the proliferation of multipotential and erythroid precursor cells. Blood 55:138–147

Metcalf D, McDonald HR (1975). Heterogeneity of *in vitro* colony and cluster-forming cells in the mouse marrow. Segregation by velocity sedimentation. J Cell Physiol 85:643–654

Metcalf D, Moore MAS, Shortman K (1971). Adherence column and buoyant density separation of bone marrow stem cells and more differentiated cells. J Cell Physiol 78:441–450

Micklem HS, Ford CE, Evans EP, Gray J (1966). Interrelationships of myeloid and lymphoid cells studied with chromosome-marked cells transplanted into lethally irradiated mice. Proc R Soc Lond B Biol Sci 165:78–102

Miller RG, Philips RA (1969). Separation of cells by velocity sedimentation. J Cell Physiol 73:191–201

Monette FC, Gilio MJ, Chalifoux P (1974). Separation of proliferating CFU from G^0 cells of murine bone marrow. Cell Tissue Kinet 7:443–450

Nakahata T, Ogawa M (1982a). Clonal origin of murine hemopoietic colonies with apparent restriction to granulocyte-macrophage-megakaryocyte (GMM) differentiation. J Cell Physiol 111:239–246

Nakahata T, Ogawa M (1982b). Identification in culture of a class of hemopoietic colony-forming units with extensive capability to self-renew and generate multipotential hemopoietic colonies. Proc Natl Acad Sci USA 79:3843–3847

Nakeff A, Daniels-McQueen S (1976). In vitro colony assay for a new class of magakaryocyte precursor colony-forming unit megakaryocytic (CFU-M). Proc Soc Exp Biol Med 151:587–590

Neumann HA, Löhr GW, Fauser AA (1981). Radiation sensitivity of pluripotent hemopoietic progenitors (CFU-c-GEMM) derived from human bone marrow. Exp Hematol 9:742–744

Nowell PC, Hungerford DA (1961). Chromosome studies in leukemia. II. Chronic granulocytic leukemia. J Nat Cancer Inst 27:1013–1036

Shortman K (1968). The separation of different cell classes from lymphoid organs. II. The purification and analysis of lymphocyte populations by equilibrium density gradient centrifugation. Aust J Exp Biol Med Sci 46:375–396

Silini G, Pous S, Pozzi LV (1968). Quantitative histology of spleen colonies in irradiated mice. Brit J Haemat 14:489–500

Siminovitch L, McCulloch EA, Till JE (1967). The distribution of colony-forming cells among spleen colonies. J Cell Comp Phys 62:327–336

Till JE, McCulloch EA (1961). A direct measurement of radiation sensitivity of normal mouse bone marrow cells. Radiat Res 14:213–222

Vassort F, Winterholer M, Frindel E, Tubiana M (1973). Kinetic parameters of bone marrow stem cells using *in vivo* suicide by tritiated thymidine or by hydroxyurea. Blood 41:789–796

Visser J, Engh GJ van den, Williams N, Mulder D (1977). Physical separation of the cycling and noncycling compartments of murine hemopoietic stem cells. In: Baum SJ, Ledney GD (eds) Experimental Hematology Today. Springer, Berlin Heidelberg New York, pp 21–27

Vriesendorp HM, Bekkum DW van (1980). Role of total body irradiation in conditioning for bone marrow transplantation. In: Thierfelder S, Rodt H, Kolb HJ (eds) Immunobiology of Bone Marrow Transplantation. Springer, Berlin Heidelberg New York, pp 349–364

Wagemaker G, Peters MF (1978). Effects of human leukocyte conditioned medium on mouse hemopoietic progenitor cells. Cell Tissue Kinet 11:45–56

Wagemaker G (1980). Early erythropoietin-independent stage of in vitro erythropoiesis: relevance to stem cell differentiation. In: Baum S, Ledney GD, Bekkum DW van (eds) Experimental Hematology Today. Karger, Basel pp. 47–60

Wagemaker G, Visser TP (1981). Analysis of the cell cycle of late erythroid progenitor cells by sedimentation at unit gravity. Stem Cells 1:5–14

Wensinck F, Bekkum DW van, Renaud H (1957). The prevention of *Pseudomonas aeruginosa* infections in irradiated mice and rats. Radiat Res 7:491–499

Williams N, Engh GJ van den (1975). Separation of subpopulations of in vitro colony forming cells from mouse bone marrow by equilibrium density separation. J Cell Physiol 86:237–246

Worton RG, McCulloch EA, Till JE (1969). Physical separation of hemopoietic stem cells differing in their capacity for self-renewal. J Exp Med 130:91–103

Wu AM, Till JE, Siminovitch L, McCulloch EA (1967). A cytologic study of the capacity for differentiation of normal hematopoietic colony-forming cells. J Cell Physiol 69:117–184

Wu AM, Till JE, Siminovitch L, McCulloch EA (1968). Cytological evidence for a relationship between normal and hematopoietic colony-forming cells and cells of the lymphoid system. J Exp Med 127:455–464

Zeiller K, Schubert JCF, Walther F, Hanning K (1972). Free-flow electrophoretic separation of bone marrow cells. Electrophoretic distribution analysis of *in vivo colony* forming cells in mouse bone marrow. Zeitschrift f Phys Chemie 353:95–104

Assessment of Environmental Insults on Lymphoid Cells as Detected by Computer Assisted Morphometric Techniques

George B. Olson and Peter H. Bartels

Department of Microbiology, University of Arizona, Tucson, Arizona 85721, USA

1 Introduction

Analysis of the texture and distribution of chromatin in cell nuclei has resulted in the recognition and differentiation of: (i) different types of normal cells, (ii) cells in different stages of the cell cycle, (iii) neoplastic cells in various stages of differentiation, and (iv) cells altered by irradiation, chemicals and biological treatment [1–14]. The differentiation may be made by an analysis of chromatin stained in stoechiometric process with the Schiff stain using the Feulgen procedure. The information resides in the condensation, granularity and general spatial arrangement of the stained chromatin [15–16]. Small, gradual changes in the organization of the chromatin reflect changes in the normal physiology, maturity and differentiation of the cells [15–18].

In individual studies, peripheral blood lymphocytes and splenocytes obtained from mice subjected to various environmental insults showed gradual changes in the texture of chromatin. These changes allowed the cells to be classified as normal or altered [11–13, 16]. The insults included exposure of the mice to irradiation, treatment with cyclophosphamide and infection with murine leukemia virus. In each case, certain changes in chromatin texture were important as specific diagnostic features and could be correlated both to the magnitude of the response and the time after exposure.

Ability to detect and monitor specific environmental insults with great sensitivity prompted the following study — the use of splenocytes and PBL as indicator cells to detect, monitor and differentiate between physically, chemically and biologically induced environmental insults. To be tested in this study is the hypothesis that different types of chemical, physical and biologic insults cause distinctive changes in the nuclear chromatin of the exposed cells. Should this assumption be warranted, one could: (i) monitor water, soil, air and food samples for undesirable chemicals and biological vectors (viruses), (ii) monitor people living near nuclear energy sites and people in radiation-related professions, and (iii) monitor cell samples obtained from people exposed to viral infections.

This paper presents data from (i) a study in which PBL and splenocytes obtained from mice exposed to various doses of x-ray irradiation were examined for changes in biologic function and chromatin texture as a function of irradiation dose and time after irradiation, and (ii) a study in which lymphocytes were removed from mice exposed to chemical, biological, and physical environmental insults and analyzed to differentiate each insult from normal and from the other insults.

Biological Dosimetry. Edited by W. G. Eisert and M. L. Mendelsohn
© Springer-Verlag Berlin Heidelberg 1984

2 Materials and Methods

2.1 General Protocol

In the first experiment, 10-week-old CBA mice (Jackson Laboratory, Bar Harbor, Maine) were divided into four groups, housed in plastic boxes and exposed to 0, 5, 50 and 500 rads whole body irradiation. X-ray irradiation was generated by a General Electric x-ray machine using 250 Kv peak, 15 ma, HVL 2.3 mm copper with a Thoraeus filter. The absorbed dose was 99 rads per minute at a distance of 26 cm. Individual subgroups of mice were irradiated 1, 3, 7, 15 and 30 days prior to time of death.

Prior to death, each mouse was bled from the suborbital venous plexus for total leukocyte and differential counts. All mice were then killed by cervical dislocation and processed as follows: (1) body weight was determined; (2) peripheral blood was obtained by cardiac puncture as a source of peripheral blood lymphocytes (PBL); (3) spleen weights were determined; and (4) single cell suspensions from individual spleens were prepared. Cardiac blood samples from the mice of each group were combined, and the PBL were separated on a Hypaque-Ficoll gradient. Splenic lymphocytes were separated in the same manner. Analysis of these lymphocyte suspensions included in part: (i) determination of their relative numbers of thymus-derived (T) and bone-marrow-derived (B) cells by standard immunofluorescence, (ii) determination of their in vitro responsiveness to the lymphocyte mitogens phytohemagglutinin (PHA), concanavalin A (Con A) and lipopolysaccharide (LPS) and (iii) morphometric analysis of Feulgen-stained nuclei. A detailed analysis of all the biological features has been described elsewhere [19].

The single cell suspensions of PBL and splenocytes were placed upon poly-L-lysine coated microscope slides, allowed to adhere to the slides for 25 min in a humid atmosphere, then dipped 5 times in PBS and fixed for ten minutes in 90% methyl alcohol plus 10% glacial acetic acid [3]. The slides were saved for staining and analysis as described below.

In the second experiment, 8-10-week-old, male BALB/c mice were divided into four groups. One group served as control animals and the three other groups were subjected to the following environmental insults: (a) 500 rads whole body x-ray irradiation as described for the first experiment, (b) an IP injection of 0.1 ml suspension (2×10^3 splenic FFU) of Friend leukemia virus, and (c) an IP injection of 40 mg/Kg BW cyclophosphamide. In each case, the animals were killed at the time clinical and biological assessment indicated maximal change from normal. The times as determined by previous work are 3 days post irradiation [20], 30 days post virus infection [13], and 4 days post injection of the drug [16]. PBL and splenocytes were removed from each animal and prepared for staining and analysis as stated for Experiment 1.

2.2 Staining and Scanning of Cells

Lymphocytes fixed with 90% methyl alcohol plus 10% glacial acetic acid were hydrolyzed for 20 min at room temperature in 5N HCl and stained with Schiff reagent in standard fashion at a pH of 2.3. Digitized images of Feulgen-stained cells were recorded on a Leitz MPV II microscope photometer operated on line to a PDP 11/45 computer (Digital Equipment Corporation). A scanning spot size of 0.5 X 0.5 sq. μm and a wavelength of 540 μm were employed. Only morphologically intact cells were scanned, and a minimum of 100 cells from each preparation was selected in random fashion for analysis. All recorded cell images were edited immediately to eliminate extraneous data points from the scanned area.

The digitized optical density (OD) values for each cell file were subjected to five feature extraction programs to obtain the following features:

(i) CONDEN: determines 3 features for granularity and dispersion of condensed and non-condensed chromatin, i.e., CNSl, CNS2, CNS3.

(ii) CELSCN: determines 7 features concerned with heterogenicity of chromatin and run length test statistics concerned with dispersion of isostained chromatin, i.e., SCS1, SCS2, etc.

(iii) HIST10: determines the relative frequency of occurrence of OD values, plus total OD, relative nuclear area and average OD, i.e., BIN1. . . . 18, TOD, RNA and AVGN (21 features in total).

(iv) REDUCE: determines 24 features describing the transition probabilities of a Markovian dependence scheme, i.e., TPRB1, etc.

(v) POLAR: determines 17 features describing the radial OD profile along the radius of the cell, i.e., PO1, etc.

The resultant files were processed by an automated feature evaluation and selection program (FMERIT) to ascertain those features that best served to characterize the various environmental alterations. Linear stepwise discriminant analysis and supervised learning programs were employed to differentiate and classify normal cells from altered cells.

3 Results

3.1 Evaluation of Biologic and Clinical Changes in Mice Following Whole Body Irradiation

Figure 1 shows the changes in 4 different biologic features as a function of irradiation dose and time after irradiation. The features represent (A) percent change in spleen weight to body weight ratio between irradiated and normal mice, (B) percent change in the percent of cells bearing theta and IgM surface membrane markers in irradiated and normal mice as determined by immunofluorescence techniques, (C) percent change in PHA induced DNA synthesis of cells from irradiated mice compared to cells from non-irradiated mice, and (D) percent change in LPS induced DNA synthesis of cells from irradiated mice compared to cells from normal mice.

The time and dose dependent change from normal reflected by these data is clearly seen in Fig. 2. The mean values of the four features for each irradiation level were averaged to obtain an index to reveal the change from normal. Data obtained from mice three days after exposure to 500 rads show the greatest change.

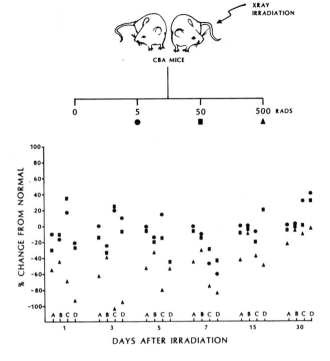

Fig. 1. Percent change from normal for various biologic features as a function of irradiation dose and time after irradiation. (A) percent change in spleen weight to body weight ratio between irradiated and normal mice (B) change in the proportion of cells bearing theta and IgM surface membrane markers in irradiated and normal mice as determined by immunofluorescence techniques (C) percent change in PHA induced DNA synthesis of cells from irradiated mice compared to cells from non-irradiated mice (D) percent change in LPS induced DNA synthesis of cells from irradiated mice compared to cells from normal mice

3.2 Evaluation of Chromatin Features from Irradiated and Non-Irradiated Mice

Evaluation of the biologic data showed the greatest change from normal to occur in mice 3 days after exposure to 500 rads. Therefore, feature files of the splenocyte images of this group were compared to feature files of normal splenocytes to determine which features have discriminatory value. Table 1 presents the selections made by FMERIT. Each feature has a value for the measure of detectability greater than 0.24. The features for the most part represent the relative frequency of occurrence of various OD values, the transitional probabilities of OD values and the distribution of OD values as a function of cell radius.

The cell images and chosen features for the 2 splenocyte populations were submitted to a supervised learning program to establish a decision rule to differentiate radiation altered lymphocytes from unperturbed cells. The resultant decision rule was then applied by another program, CLASFY, to classify splenocytes and PBL obtained at different times following irradiation. Figure 3 presents the results. On the third day

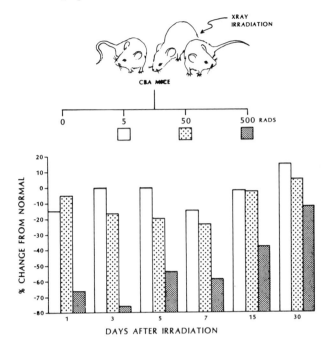

Fig. 2. Averaged percent change from normal for biologic features as a function of irradiation doses and time after irradiation

Table 1. Evaluation of features by FMERIT. Cells from control mice compared to cells from mice 3 days post exposure to 500 rads

Feature	d' values from ROC curve
BIN5	.357
TPRB4	.338
TPRB7	.328
BIN4	.320
PO7-2	.317
PO2-2	.257
CNS-1	.239
AVG-2	.239

after exposure, about 80% of the splenocytes from mice exposed to 500 rads were classified as abnormal. About 50% of the cells from the 50 rad exposure were morphologically abnormal. Splenocytes from mice exposed to 5 rads contained 20% or less altered cells at any time period. A similar trend, but of lesser magnitude, was observed in the

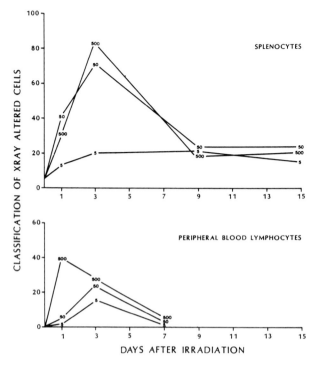

Fig. 3. Classification of x-ray altered splenocytes and PBL from mice as a function of irradiation dose and time after irradiation

classification of PBL. The only noted change was that a greater number of PBL was classified as abnormal form the 500 rad exposure group on day 1. Thus the dose-dependent, time-dependent sequence seen in the biologic features is reflected in the analysis of the nuclear chromatin.

Cells classified as abnormal may be placed into a separate file so the characteristics of the individual cells for each exposure and time period can be studied. In this study, the radiation-altered cells obtained from mice 1 and 3 days after exposure to 5, 50 and 500 rads were placed into 3 files. Figure 4 shows the overall separation of cells exposed to 50 and 500 rads from non-irradiated cells when compared by a stepwise linear discriminant analysis program (DISCRM). The separation from control cells was more defined for splenocytes exposed to 500 rads than for cells exposed to 50 rads, but the mean values of the discriminant function scores for the two sets of irradiated cells were not significantly different.

A detailed comparison of the features important in the recognition of the selected irradiated cells is shown in Fig. 5 and 6. Figure 5 presents the distribution of OD values of Feulgen stained nuclear material. Irradiation results in a dose dependent shift toward lower OD values. Figure 6 contrasts the changes in the radial OD profile, relative nuclear area and total OD values for three groups of cells. The total OD value was not significantly different among the three groups; however, irradiation caused a

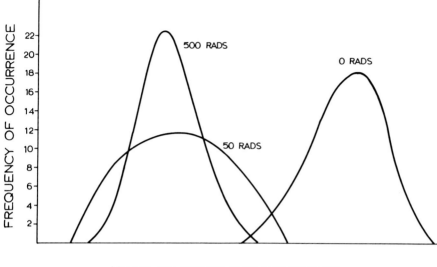

Fig. 4. Distribution of discriminant function scores for splenocytes from non-irradiated mice and from mice given different amounts of radiation

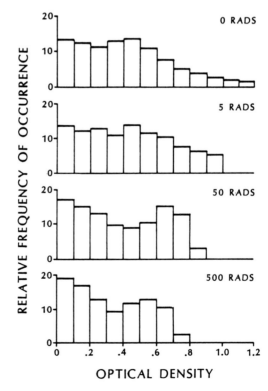

Fig. 5. Relative frequency of occurrence of optical density values in cells obtained from irradiated and non-irradiated mice

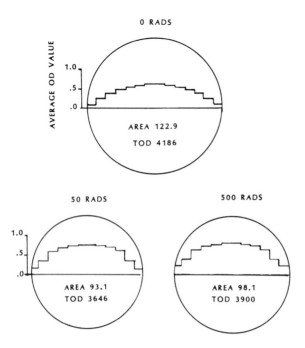

Fig. 6. Comparison of relative nuclear areas, total optical density and distribution of radial optical density profiles for cells from irradiated and non-irradiated mice

decrease in relative nuclear area and rearrangement of chromatin material as indicated by the increase in radial OD values near the center of the cell nucleus.

3.3 Evaluation of Chromatin Changes in Cells Exposed to Different Types of Environmental Insults

Figure 7 shows a composite of Feulgen stained lymphocytes obtained from control BALB/c mice and age matched BALB/c mice exposed to 500 rads x-ray irradiation, 40 mg/Kg body weight cyclophosphamide and 2000 FFU of Friend leukemia virus. The cells were removed from the experimental animals at the time the animals were experiencing maximum changes from normal. These times were 3 days post irradiation, 30 days post virus infection and 4 days post treatment with the chemical. In each case the cells were hydrolyzed and stained with Schiff stain according to the Feulgen procedure as if they were control cells. Inspection of the photographs does not provide to the unaided eye any detectable difference in morphology or staining characteristics. However, comparison of the features of the control group to each experimental group by FMERIT showed marked differences. Table 2 shows the selection of features important in these three comparisons.

The separation of control cells and cells of each experimental group can be accomplished using the cell populations as training sets, the selected features from the

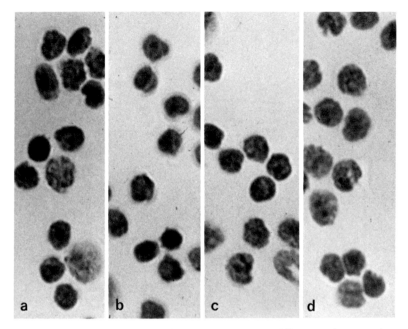

Fig. 7a–d. Photograph of splenocytes removed from mice subjected to different environmental insults. (**a**) control mice, (**b**) virus infected mice, (**c**) drug treated mice, (**d**) x-ray irradiated mice

Table 2. Features which differentiate cells from control animals and cells removed from animals exposed to different environmental exposures. Control compared to each experimental group by FMERIT

Virus		Drug		X-ray	
	ROC		ROC		ROC
PO6-2	.493	CNS1-1	.426	TPRB 14	.427
PO5-2	.486	PO1-2	.376	TPRB 13	.423
PO7-2	.467	PO5-2	.370	BIN7	.412
TEE-4	.459	TPRB 15	.347	BIN8	.362
TPRB 14	.455	PO2-2	.342	PO4-2	.322
TPRB 16	.444	TPRB 22	.292	BIN1	.285
BIN7	.423	TPRB 14	.236	PO3-2	.155

FMERIT evaluation and a stepwise linear discriminant program, DISCRM. Figure 8 presents the separation achieved in each case. The modes of control and experimental cells are projected onto the first discriminant function, and good separation is evident.

This represents an extremely important point. One can identify and separate virus infected cells from normal cells by a variety of immunologic techniques, but techni-

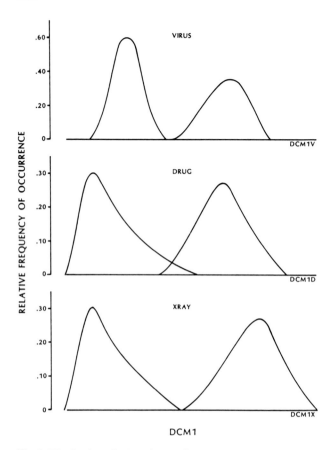

Fig. 8. Distribution of values for the first discriminant function for cells from control mice compared to cells from virus infected mice (DCM1V), from chemical treated mice (DCM1D) and from x-ray irradiated mice (DCM1X)

ques to separate irradiated or chemically altered cells from normal cells do not exist. When generating the discriminant function, the DISCRM programs list the features responsible for the discrimination and identify each cell in the experimental group that has been classified as a control or test cell. Therefore, training sets of drug treated cells, irradiated cells and virus altered cells can be obtained and analyzed to determine the differences in features and feature values important to the differentiation. Figure 9, for example, shows the relative frequency of occurrence for OD values in training sets. A comparison to control cells shows: (i) virus treated cells show an increase in the frequency of occurrence of OD values between 0.5 to 1.0 OD value, (ii) drug treated cells show both a decrease and an increase in OD values, and (iii) x-ray altered cells show a decrease toward lower OD values.

The training sets of cells representing the three different environmental insults and the control cells can be submitted simultaneously to the multiple stepwise discriminant analysis program to examine the overall separation between all groups and to obtain

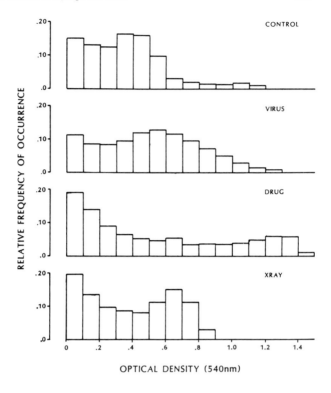

Fig. 9. Relative frequency of occurrence of optical density values in cells obtained from mice exposed to different environmental insults

an appreciation of their distribution in multiple dimensional feature space. The results may be viewed in several ways. Figure 10 presents a territorial map which shows the distribution of cells. The axes represent the first and second discriminant functions. There appear to be certain assigned areas for each group, but one has no impression of the statistical significance of the separation. Such statistical information can be presented by a bivariate plot of the first and second discriminant function values for each group as shown in Fig. 11. The bivariate plot shows the bivariate mean for each group, the 95% confidence region of the mean, 50% tolerance ellipse for each data set and the Bayesian decision boundaries. The bivariate plot also shows clearly the directionality in the separation of the data sets.

The specific space assigned to each data set and the number of cells from a given sample that would have to be detected in that space before the cell sample could be considered to be abnormal can be determined. The procedure is as follows: (i) determine the space for a particular test group by using the bivariate mean of the control group as a vertex of sector and determine the sector angle by projecting its two sides to touch tangentially the appropriate Bayesian decision boundaries. The sector reflects the area and direction of the space assigned to a given test group; (ii) establish the 75%, 90%, 99% tolerance ellipse for the control group. This will further divide the

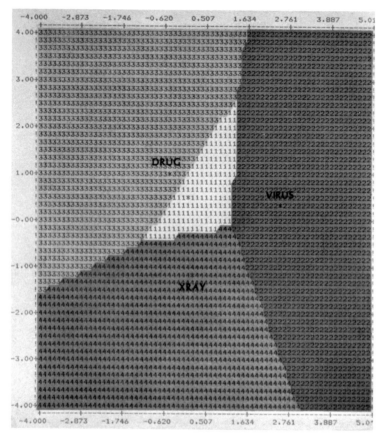

Fig. 10. Territorial map showing the spatial distribution of cells obtained from mice exposed to different environmental insults. Results expressed on basis of the first and second discriminant functions

space allocated to each test group into 0–50%, 51–75%, 76–90% and 91–99% tolerance region segments; (iii) determine the expected frequency of cells in each tolerance region segment of the control group, (each region should contain a minimum count of 5 cells); (iv) determine the number of cells observed within each tolerance region segment of the assigned space for the test group and (v) compare the expected and observed frequencies in a Chi Square test and test for significance.

Figure 12 presents the results of this study based upon 250 cells per sample. In the virus test sample, an observation of 13 cells (increase of 8 over controls) in the 91–99% confidence region is significant at an a value of 0.005. This represents an increase in cells of only 3% in the test sample. For irradiated cells, an increase of only 4% in these confidence regions is significant. For cyclophosphamide treated cells, an increase of 7.2% cells over controls was significant at the 0.005 level. Such an analysis shows the value of considering not only distance, but also directionality in the location of cells, in the decision space and how directionality can be used to indicate the specificity of an insult.

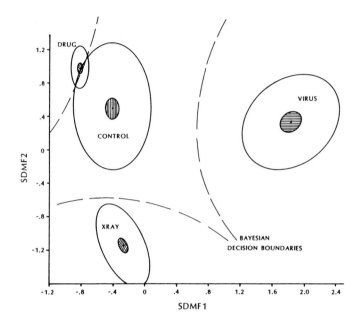

Fig. 11. Bivariate plot of first and second discriminant functions for cells exposed to different environmental insults. Smaller ellipses represent the 95% confidence regions for the estimates of the bivariate means. The larger ellipses represent the 50% tolerance regions for the groups

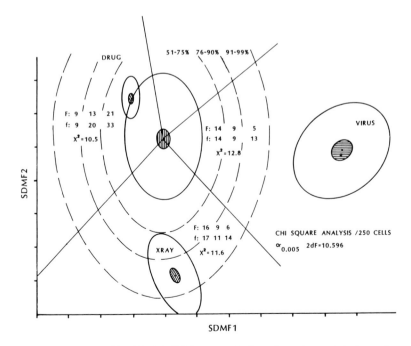

Fig. 12. Bivariate plot as shown in Fig. 12 with Chi Square analysis to show the statistical sampling requirements for significant detection of cells in the various sections. (*F*) expected number of cells, (*f*) number of cells required for significance

4 Discussion

Lymphocytes removed from mice subjected to whole body x-ray irradiation show dose-time dependent functional and morphological changes. Analysis of the arrangement and texture of the chromatin of the cells reveals that the relative frequency of occurrence of OD values in the chromatin shifts toward lower values indicating lighter staining chromatin as the concentration of irradiation increases. Features descriptive of the texture of chromatin can be employed in a decision rule to classify cells obtained from irradiated animals. The decision rule evaluates the feature values in any tested cell and classifies it as a normal cell or as an irradiation-altered cell.

Irradiation altered cells from animals given as little as 5 rads whole body irradiation can be detected as a moderate proportion of 20% or less of the cell sample. Application of the decision rule shows the number of cells classified as irradiated cells to be dependent upon the total body radiation and the time after exposure. This finding suggests that such diagnostic features and their values may be useful as a biologic dosimeter. Use of circulating lymphocytes as internal biologic dosimeters would have great diagnostic value over a stationary dosimeter which could detect radiation directed only to the part of the body containing the dosimeter.

The lymphocyte has long been recognized for its sensitivity toward irradiation [19]. Lymphocytes of different types show altered physiologic function, altered morphology and selective survival rates depending upon the dose of irradiation. The present study extends these observations and defines for the pathologist and biologist in a quantitative fashion the features to be associated with irradiation induced damage and the trend in feature values to be associated with increasing damage of the chromatin. Interpretation of such information may also provide clues for the employment of more defined analytic procedures and staining techniques to further advance the detection of irradiation induced damage.

These studies, furthermore, show that lymphoid cells can be used to differentiate between animals subjected to different types of environmental exposures. In the model, choice physical, chemical and biological insults were administered to mice and at the time maximum pathologic changes were known to have occurred, lymphoid cells were removed, stained as if they were normal cells, scanned and analyzed. The results reveal (i) changes occurred in the texture and organization of the chromatin of cells which proved to be diagnostic for the different exposures, (ii) decision rules based upon diagnostic features can be made and used to detect and quantitate different types of environmental exposures from normal and from each other, (iii) multivariate analysis of cell populations from various exposures revealed each population had directionality and spatial arrangement in a multi-dimensional decision space, and (iv) one can determine the number of cells which must be scanned from a given sample before differentiation at proper significance levels can be made.

All cells analyzed by this procedure are treated as if they came from a normal homogeneous cell population. This assumption is known to be merely a pragmatic approximation as different environmental insults produce unique changes to cell morphology and cell populations [21, 22]. Therefore, the changes in chromatin patterns, which are detected may reflect several different in vivo changes. In part, the changes may represent: (i) actual textural changes within the same cell, i.e., damage or alteration of the exposed cell, (ii) changes in the normal distribution of cells in the cell cycle, (iii) loss of a certain cell type, due to cell death or entrapment which is then replaced by a transient, opportunistic cell type, (iv) loss of a cell type, which is replaced by immature cells of the same type, and (v) any combination of the above. The actual change which occurs may not be important to the diagnostic process. The important point is that the decision rule for detecting a certain environmental insult is based upon unique morphological differences or changes, whatever they may be, between the normal and the test sample.

The technique offers a means of monitoring and quantitating environmental insults. It may complement existing methods, such as analysis of epidemiologic survey data, in vitro tests such as the Ames test and chromosomal aberration tests [23–25]. In fact, it may offer potential advantages such as improved specificity and high efficiency. Full automation, though, would require: (i) data acquisition systems capable of scanning several thousand cell images per minute [26], (ii) optimized software to allow for fast data reduction and data handling, (iii) the development of computer systems capable of real time processing of the massive data stream, and (iv) the improvement of classification and decision procedures for the thorough assessment of multimodal cell populations in their proportions, modality and modes descriptive statistics. Furthermore, additional research effort will be required to provide a better understanding of the biologic meaning of the differentiating morphologic changes detected in the chromatin. The information obtained from such a process may provide clues for further significant improvements in preparation methods and staining techniques; thus allowing for increased specificity and discrimination.

5 References

1. Olson GB, Anderson RE, Bartels PH (1974) Differentiation of murine thoracic duct lymphocytes into T and B subpopulations by computer cell scanning techniques. Cell Immunol 13: 347
2. Olson GB, Bartels PH (1979) Characterization of murine T and B cells by computerized microphotometric analysis. Cell Biophysics 1:229
3. Olson GB, Donovan RM, Bartels PH, Pressman NJ, Frost JK (1980) Microphotometric differentiation of human T and B cells tagged with monospecific immunoadsorbent beads. Anal Quant Cytology 2:144
4. Dormer P, Abmayr W (1979) Correlation between nuclear morphology and rate of deoxyribonucleic acid synthesis in a normal cell line. J Histochem and Cytochem 27:188
5. Gray JW (1974) Cell cycle analysis from computer synthesis of DNA histograms. J Histochem and Cytochem 22:642
6. Nicoloni C, Kendall F, Giaretti W (1977) Objective identification of cell cycle phases and subphases by automated image analysis. Biophysical J 19:63
7. Abmayr W, Giaretti W, Gais P, Dormer P (1982) Discrimination of G1, S and G2 cells using high resolution TV scanning and multivariate analysis methods. Cytometry 2:2
8. Wied GF, Bartels PH, Bibbo M, Chen M, Reale FR, Schreiber H, Sychra JJ (1979) Discriminant analysis on cells from developing squamous cancer of the respiratory tract. Cell Biophysics 1:39

9. Sychra JJ, Bartels PH, Bibbo M, Taylor J, Wied GL (1977) Computer recognition of abnormal ectocervical cells. Comparison of efficacy of contour and textural features. Acta Cytol 21:765
10. Koss LG, Bartels PH, Sychra JJ, Wied GL (1978) Computer discriminant analysis of atypical urothelial cells. Acta Cytol 22:382
11. Anderson RE, Olson GB, Howarth GL, Wied GL, Bartels PH (1975) Computer analysis of defined populations of lymphocytes mediated in vitro. Amer J Pathology 80:21
12. Olson GB, Anderson RE, Bartels PH (1979) Computer analysis of defined populations of lymphocytes irradiated in vitro. Human Path 10:179
13. Olson GB, Bartels PH (1981) Computer discrimination of splenocytes and peripheral blood lymphocytes from mice infected with Friend murino leukemia virus. Pattern Recogn 13:37
14. Nair KE, Bartels PH, Mahon DC, Olson GB, Oloffs PC (1980) Image analysis of hepatocyte nuclei from chlordane treated rats. Anal and Quant Cytology 2:285
15. Bartels PH, Wied GL (1977) Computer analysis and biomedical interpretation of microscopic images: Current problems and future directions. Proceedings of the IEEE 6:252
16. Anderson RE, Olson GB, Bartels PH (1982) Computer assisted quantification of injury to cells and tissues. In: Topics of Environmental Pathology, Vol 2, Published by VAREP
17. Sandritter W, Kiefer G, Schluter G, Moore W (1967) Eine cytophotometrische methode zur objektiorirang der morphologue von Zellkerulon. Histochemie 10:341
18. DeCampos Vidal B, Schluter G, Moore GW (1973) Cell nucleus pattern recognition: Influence of staining. Acta Cytol 17:510
19. Anderson RE, Olson GB, Autry JA, Howarth GL, Troup GM, Bartels PH (1977) Radiosensitivity of T and B lymphocytes. IV Effect of whole body irradiation upon various lymphoid tissues and numbers of recirculating lymphocytes. J Immunol 118:1191
20. Olson GB, Anderson RE, Bartels PH (1982) Computer assisted morphometric analysis of radiation injury in murine lymphocytes. Acta Cytol 4:181
21. Fusenig NE, Amer SM, Boukamp P, Lueder M, Wost P (1977) Methods for studying neoplastic transformation of epidermal cells in culture by chemical carcinogens. In: Methods for carcinogenesis tests at the cellular level and their evaluation for the assessment of occupational cancer hazards. By Fondazione C. Erba, pp 53–70
22. Ray VA (1981) Comparison of mutagenesis and in vitro transformation models in detecting potential carcinogens. In: Sugimura T, Kondo S, Takebe H (eds) Environmental Mutagens and Carcinogens Alan R. Liss, Inc., pp 197–208
23. Ames BN, McCann J, Yamasaki E (1975) Methods for detecting carcinogens and mutagens with the Salmonella/Mammalian Microsome Mutagenicity test. Mutat Res 31:347
24. McCann J, Ames BN (1976) Detection of carcinogens as mutagens in the Salmonella Microsome test: Assay of 300 chemicals. Proc Natl Acad Sci USA 73:950
25. Thilly WG, Deluca JG (1979) Human lymphoblasts: Versatile indicator cells for many forms of chemically induced genetic damage. In: Scheutzle D (ed) Monitoring Toxic Substances, pp 13–27
26. Shack R, Baker R, Buchroeder R, Hillman D, Shoemaker R, Bartels PH (1979) Ultrafast laser scanner microscope. J Histochem Cytochem 27:153

Dose Related Changes in Cell Cycle of Bone Marrow and Spleen Cells Monitored by DNA/RNA Flow Cytometry

W.G. Eisert,[1] H.U. Weier,[1] and G. Birk[2]

[1]Gesellschaft fuer Strahlen- und Umweltforschung (GSF), Herrenhäuserstr. 2,
D-3000 Hannover 21, FRG
[2]Universität Hannover, Institut für Biophysik, Herrenhäuserstr. 2, D-3000 Hannover 21, FRG

Summary

Based on cell cycle analysis by flow cytometry on isolated mononuclear cells from bone marrow and spleen, dose-dependent alterations after radiation as well as after chemical treatment have been observed in C3H mice. Cells from the hemopoetic system are easily available and are ideally suited as biological endpoints for a biological dose determination in men. Such cells are also suitable for measurement of the additive or synergistic effect of two or more treatments at the same time. Based on computer analysis of two-parameter DNA/RNA measurements, alterations in the cell cycle kinetics can be monitored at as low as 80 rads whole-body radiation in mice. Similar results were obtained after oral treatment with formaldehyde in a single bolus. The data indicate that the sensitivity of this system to radiation may easily be enhanced. Preliminary results on the simultaneous treatment with radiation and oral administration of formaldehyde indicate a more than additive effect of both.

1 Introduction

The application of genotoxic chemicals as well as ionizing radiation to mammalian cells can result in induced mutations, sister chromatid exchange (SCE), cell death or alterations in cell proliferation. Formaldehyde (FA) is known to act as an alkylating agent of DNA when applied to cells (Alderson 1961). First estimations of rad — equivalents have been performed by Chanet et al. (1976) for yeast and mammalian cells in culture. To investigate possible interactions between FA and γ-radiation as mentioned by Sobels (1956), germfree animals were treated with various doses of FA or Co(60) — radiation alone and in combination.

SCE has been proposed as a sensitive indicator of mutagenesis (Carrano et al. 1978; Perry et al. 1975). However, an automated cytometric assessment of SCE remains difficult. In contrast, high resolution flow-cytometry including fluorescence and extinction scanning (Eisert, 1981) has found many applications in detecting minute alterations in whole cells. This technique can be applied to monitor treatment and classify hematologic disorders (Andreeff et al. 1981; Hiddemann et al., 1982) by performing

Biological Dosimetry. Edited by W. G. Eisert and M. L. Mendelsohn
© Springer-Verlag Berlin Heidelberg 1984

cell kinetic analysis based on simultaneous measurements of intracellular DNA and RNA as well as on cell size. Transient changes in cell cycle kinetics after an insult may turn out to be as sensitive an indicator as SCE when measured within a certain time frame. Our experience in monitoring effects of chemotherapy in leukemic patients stimulated us to measure the alteration of cell cycle kinetics of mouse mononuclear cells in bone marrow and spleen after an insult in a time dependent manner by determining the relative number of cells in different stages of the cell cycle and by deriving a dose relationship from those data.

2 Materials and Methods

Germfree C3H mice (age: 31 +/− 3 days) were irradiated with a Co(60) source at a dose rate of approximately 152 rad/min. Whole-body radiation doses from 80 to 400 rad were applied in different experiments. At certain time intervals after the radiation insult, one or two treated and one control animal were sacrificed for every measurement. Cells were washed out from the isolated spleen and femur of each animal using phosphate-buffered saline (PBS). Mononuclear cells from both sources were isolated using Ficoll gradient centrifugation. Cell preparations were checked microscopically for purity prior to the staining of the cells. For fluorescent staining of intracellular DNA and RNA using acridine orange (AO) we followed the two-step protocol of Darzynkiewicz (1979). Repetitive samples were analyzed by multiparameter flow cytometry (Eisert et al., 1975; Eisert, 1981).

For each distribution, the green (DNA) and red (RNA) fluorescence of approximately 5×10^5 cells were recorded simultaneously. Additional sets of data were recorded for simultaneous detection of DNA content and cell size in order to discriminate between cell aggregates and polyploid cells.

The relative number of cells in different stages of the cell cycle as well as the number of cell aggregates or isolated nuclei produced as an artefact during cell preparation were easily determined by fitting gaussian distributions to the prominent peaks of the two-parameter distributions. An example is given in Fig. 1. It shows the DNA vs RNA distribution of cells isolated from the spleen of a male mouse of age 32 days. The peak corresponding to the cells in the G_0/G_1 phase is easily identified by the analysis program. The computer calculates gaussian distributions by a least-squares fit. The second peak due to the G_2+M population is treated similarily. The area between the two peaks is defined as cells in the S-phase of the cycle. Those objects with similar levels in green (DNA) fluorescence as cells in G_0/G_1-phase but with very little red (RNA) fluorescence are classified as isolated nuclei. These fragments, cell aggregates as well as other data-points that can not easily be identified are shown in Fig. 1 marked as "? ?".

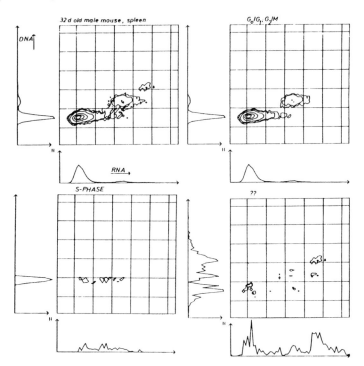

Fig. 1. The fraction of cells in the different stages of the cell cycle as well as aggregates/polyploidy and isolated nuclei can be determined by fitting normal distributions to the histogram. The absolute and relative numbers are calculated automatically and stored for later purposes. The example shows the original histogram (*upper left*), the calculated distributions for cells in G0/G1- and G2+M-phase (G0/G1,G2/M), the S-phase (S-PHASE), and nuclei, aggregates/polyploid and non-classified cells (? ? , *lower right*)

3 Results

The data on cell cycle kinetics obtained after various doses of γ-radiation show a non-linear relationship between the number of cells in different phases of the cell cycle and the applied dose. After 200 rad, the number of cells in S-phase isolated from the bone marrow show a significant decrease 12 h after the treatment (Fig. 2). A difference in the radiation-induced decrease of cells between male and female animals of the same age also was noted.

In contrast the number of cells in S-phase isolated from the spleen was found to be increased the second day after radiation (Fig. 3). However at 6 to 8 days after treatment with a dose of 200 rad no significant difference from control values was observed.

The data on cells isolated from the bone marrow after whole-body radiation with 80, 200 and 400 rads are summarized in Fig. 4. The number of cells in S-phase as well as those in G_2+M-phase is consistently decreased after radiation treatment.

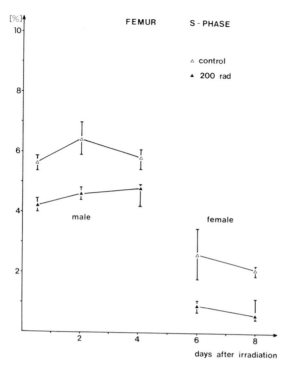

Fig. 2. The fraction of S-phase cells in irradiated bone marrow as a function of time after irradiation. A significant decrease can be observed 12h after treatment. The picture shows the time-dependence of the effect as well as the differences between male and female animals of the same age (32 days)

In contrast to the data from bone marrow, an increased fraction of cells in S-phase was noted in the population isolated from the spleen after radiation with 200 rad as shown in Fig. 5. In this case the fraction of cells in the G$_2$+M-phase decreased in a dose dependent manner. A block in the cell cycle was not detectable after the radiation treatment.

Additional data on the size of cell-to-cell aggregates were recorded but not used for the present analysis.

Cell cycle kinetics of cells from the bone marrow and spleen were analysed after the oral application of formaldehyde given as single bolus of 100 mg/kg BW and 200 mg/kg BW. A significant decrease of cells in the G$_2$+M-phase isolated from the spleen was noted after a dose of 200 mg/kg (Fig. 6). After a dose of 100 mg/kg however a transient increase of cells in the S-phase was detected (Fig. 6, left). At this dose level no significant change of the relative number of cells in G$_2$+M-phase was found.

In a series of preliminary experiments a bolus application of formaldehyde has been combined with a single dose of radiation (200 mg/kg BW + 80 rad). Within the limited number of experiments the cell cycle parameters were found to be changed to a significantly larger extent than after treatment with radiation or formaldehyde alone (Fig. 6, right). This effect appeared within one hour after treatment. Presently, investigations are concerned with this matter.

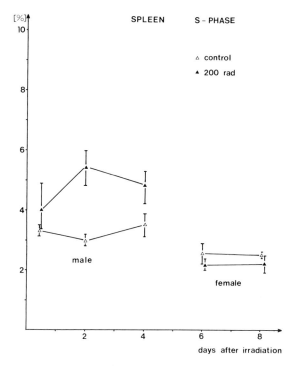

Fig. 3. The fraction of S-phase cells in irradiated spleen as a function of time after irradiation. In contrast to the relations shown in Fig. 2 the fraction of S-phase cells in spleen is significantly increased at the second day after irradiation. This could be verified in samples taken from male mice; samples derived from female animals at the 6th and 8th day showed no significant effect

4 Discussion

The number of cells in S- and in G_2/M-phase has been estimated by computer analysis of two-dimensional data based on simultaneous measurements of intracellular DNA and RNA. The model applied during the analysis assumes a gaussian distribution of cells in G_1- and in G_2/M-phases with respect to their DNA contents. Possible structural changes of the nuclear matrix after the insult are not expected to change the stoichiometry of the fluorescent dye bound to the DNA to a detectable extent. The limited data on the response of the hemopoietic system to a radiation or chemical insult reflect not only the alteration of the cell cycle kinetics but show also the compartmentation of the system. The interaction of different compartments of the hemopoietic system may also explain the different response of the number of cells in S-phase in bone marrow and in the spleen on day 1 to 4 after treatment.

The relative amount of cells in S-phase was consistently lower in female as compared to male mice. Presently we believe this may be due to an artefact in the cell prepara-

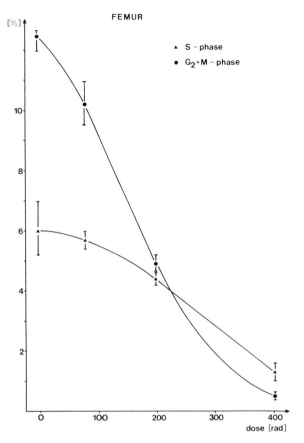

Fig. 4. The fraction of S- and G2/M-phase cells in bone marrow as a function of radiation dose. The fraction of cells in the S- and G2+M-phase decreases with increasing doses of γ-rays. At 0.8Gy the relative number of S-phase cells is not significantly altered

tion. Therefore only the ratios between treated and untreated animals have been analyzed. However this finding is presently under further investigation.

All data show consistently a dose-dependent decrease of the number of cells in G2/M-phase after radiation independent of the sample site. A dose-dependent decrease of the fraction in S-phase, however, was only found in the bone marrow (Fig. 4). In samples from the spleen the number of cells in the S-phase was increased after a dose of 200 rad during the first four days whereas after 400 rad the number dropped below the control. These results differ from data on the effect of radiation in cell culture as in our experiments no block within the cell cycle became obvious. Our results may have been affected by other regulatory effects within the hemopoetic system as well as by the release and transport of cells from different compartments. Such effects could have obscured a dose-induced block in the cell-cycle in these experiments.

The data on the effect of formaldehyde after oral administration of a bolus on the cell cycle of mononuclear cells isolated from the spleen seem to follow a similar trend as compared to the radiation treatment (Fig. 6). The number of cells in S-phase is

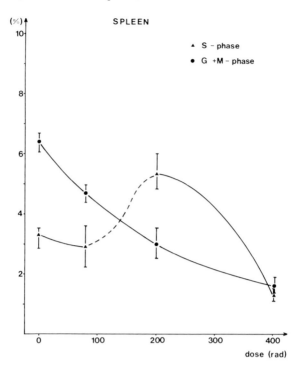

Fig. 5. The fraction of S- and G2/M-phase cells in spleen as a function of dose. Decreasing numbers of cells in the G2+M-stage were related to increasing doses. G2/M-phase cells decrease with dose, while an increased fraction of S-phase cells was found at 200 rad

increased over a period of 4 days after the application of 100 mg/kg BW. However, no change in the number of cells in the G2/M-phase after application of 100 mg/kg BW could be noted (Fig. 6, left).

In contrast, after a single dose of 200 mg/kg of formaldehyde given orally the number of cells in G2/M-phase is decreased for a period of 5 days, whereas the number of cells in S-phase undergo only minor changes (Fig. 6, middle).

The data on the simultaneous application of formaldehyde (bolus 200 mg/kg) and a single dose radiation (80 rad) so far are only preliminary (Fig. 6, right). However the noticable decrease of cells in the G2/M-phase is much larger than the decrease observed by one or the other treatment alone. These data are presently under investigation.

Monitoring alterations of the cell cycle of mononuclear cells taken as biological endpoint may prove an additive or synergistic effect of different treatments.

Additional informations based on cell — and cell-aggregate sizing needs further evaluation. This information is already available from the flow cytometric measurements and may provide additional clues to a possible formation of cell aggregates or the change of cell size after treatment.

The series of results after single dose radiation treatment showed that on the basis of the analysis of cells in the S-phase in the bone marrow, a single dose of 80 rad was detected with confidence. The slope of the dose-response curve indicates that doses

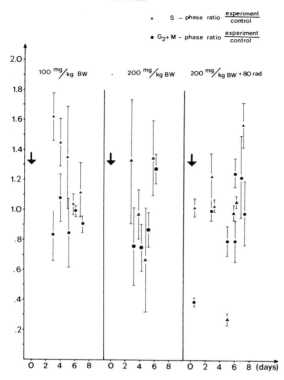

Fig. 6. The percentage of cells in the S- and G2+M-phase after treatment of the animals with two different doses of formaldehyde (FA) and a combination of FA with a single whole-body dose of 80rad. The figure shows the ratio (number of cells in samples from treated animals) / (number of cells in samples from control animals) as a function of the time. The day of treatment is indicated by an *arrow*

as low as 10 rad whole-body radiation may be analyzed with further refinement of the system. The simultaneous employment of immunological markers analysed simultaneously should further improve the sensitivity of a dosimetry system based on flow cytometric evaluation of cells that are easily available from victims of an accidential exposure. This underlines the potential of this approach to biological dosimetry in men.

5 References

Alderson Th (1961) Mechanism of Mutagenesis induced by Formaldehyde. Nature, Vol 191, No 4785

Andreeff M, Darzynkiewicz Z (1981) Multiparameter flow cytometry. Part II: application in hematology. Clin Bull 11(3), pp 120–130

Carrano AV, Thompson LH, Lindl PA, Minkler JA (1978) Sister Chromatid Exchange as an Indicator of Mutagenesis. Nature (London), Vol. 271, p 551–553

Chanet R, Magana-Schwencke N, Yoshikura H, Moustacchi E (1976) An Attempt to apply the Rad-Equivalence Notion to the Biological Effects of Formaldehyde. In: radiological protection. First European Symposium on Rad-Equivalence, Proceedings of the Seminar on Radiobiology-Radiation Protection – Orsay (France), 24–26 May 1976, p. 171–91

Darzynkiewicz Z (1979) Acridine Orange as a Molecular Probe in Studies of Nucleic Acids in Situ. In: Flow Cytometry and Sorting. J Wiley & Sons, New York, p 285–316

Eisert WG, Ostertag R, Niemann EG (1975) Simple flow – microphotometer for rapid cell population analysis. Rev Sci Instrum 46, pp 1021–1026

Eisert WG (1981) High resolution optics combined with high spatial reproducibility in flow. Cytometry 1(4), pp 254–259

Hiddemann W, Clarkson BD, Büchner T, Melamed MR, Andreeff M (1982) Bone marrow cell – count per cubic millimeter bone marrow: a new parameter for quantitating therapy induced cytoreduction in acute leukemia. Blood 59(2), pp. 216–225

Perry P, Evans HJ (1975) Cytological Detection of Mutagen-Carcinogen Exposure of Sister Chromatid Exchange. Nature (London), Vol 258, p 121–5

Sobels FH (1956) The effect of Formaldehyde on the Mutagenic Action of X-rays in Drosophila. Experientia, Vol XII/8, p 318–21

Hoechst 33342 Dye Uptake as a Probe of Membrane Permeability in Mammalian Cells

R.G. Miller,[1] M.E. Lalande,[1] and E.C. Keystone[2]

[1]Ontario Cancer Institute, 500 Sherbourne Street, Toronto, Ontario, Canada, M4X 1K9
[2]Rheumatic Disease Unit, Wellesley Hospital, Toronto, Ontario M4X 1K9, Canada

The compound Hoechst 33342 (H0342), first developed as a pharmaceutical agent (Loewe and Urbanietz, 1974), was later shown to bind quantitatively to DNA and to become fluorescent on doing so (Latt and Stetten, 1976). Unlike other fluorescent DNA stains, H0342 is readily taken up by living cells and is relatively non-toxic (Arndt-Jovin and Jovin, 1977). Thus it can be used for DNA analysis of living cells which can then be sorted on the basis of DNA content and subsequently analyzed functionally. In the course of such studies we ran into an "artefact" in that we found subpopulations of cells stained with H0342 which differed in fluorescence intensity per cell but did not differ in DNA content (Lalande and Miller 1979). Subsequent investigation, reviewed below, showed that when either short dye incubation times or low dye concentrations are used, the rate at which the DNA becomes labelled is limited by the rate at which the dye is transported across the cytoplasmic membrane of the cell. Our work suggests at least three useful applications of this phenomenon:
1. Detection of a certain type of drug resistant cell.
2. Distinction of cell classes (here B cells and T cells).
3. Detection of activation of T cells before they have started to become blasts (on the basis of size change) and before they have entered cell cycle (on the basis of DNA synthesis).

1 Studies on Drug-Resistant Cells (Lalande et al. 1981)

Ling and Thompson (1974) have isolated a series of Chinese hamster ovary (CHO) cell lines showing varying resistances to the drug colchicine and have shown that the extent of resistance correlates directly with a reduction in drug permeability. Subsequent studies (Carlsen et al. 1976) showed that the kinetics of colchicine uptake are characteristic of an unmediated diffusion process.

The mutants were established by a step-wise selection procedure. Thus wild-type CHO cells (Aux B1) were exposed to a dose of colchicine such that almost all cells were killed. A clone (CH^{R}A3) was established from the surviving cells and was 6-fold less sensitive to the drug. Cells from this clone were grown up and exposed to an even higher concentration of colchicine such that almost all cells were killed. Another clone (CH^{R}B3) was established from these surviving cells and showed 21-fold less sensitivity to the drug.

Biological Dosimetry. Edited by W.G. Eisert and M.L. Mendelsohn
© Springer-Verlag Berlin Heidelberg 1984

Similarly, an even more resistant clone, CH^RC5, was established (see Table 1). An additional clone, I10-1, was established from CH^RC5, and is a clone which appears to have reverted to a drug sensitivity almost that of the wild-type Aux B1 cells. These CHO mutant lines show extensive cross-resistance to a number of unrelated compounds. Although, when different CHO lines are compared, the numerical extent of cross-resistance may vary from compound to compound, the rank order of resistance amongst the different lines is invariant or nearly so (Bech-Hansen et al. 1976).

The rate at which these cell lines take up HO342 appears to be governed by the same process. Viable cells from each line (10^6/ml) were incubated in serum free tissue culture medium containing 10 μM HO342 for 30 min at 37°C and then stored on ice until analysis. Fluorescence spectra were then measured in a flow cytometer-cell-sorter designed and built at the Ontario Cancer Institute but basically similar to those in widespread use elsewhere (Miller et al. 1981). Fluorescence was excited using the combined 351- and 364 nm lines from a Spectra Physics 164-05 argon ion laser adjusted to emit 25–50 mw. Table 1 lists the relative modal intensity of the G_1 peak for each of the cell lines. The modal fluorescence intensity of this peak decreases in step with increasing resistance to colchicine. When the same cell lines were fixed before staining with HO342 and flow cytometric analysis, the modal fluorescence intensity of the G_1 peak was essentially the same in all cases, indicating that the cell lines all have the same G_1 DNA content. Therefore, the reduction in HO342 fluorescence staining intensity of the mutants compared to the wild-type cells does not reflect a difference in DNA content but may be associated with the mechanism of resistance to colchicine, i.e. reduced permeability.

The kinetics of uptake of HO342 dye by Aux B1 and mutant A3 cells were next measured. In both cases, the fluorescence intensity increased rapidly initially, slowly decreased, and then reached a stable plateau, identical for the two lines. Aux B1 however, reached the plateau much more quickly (in 30 rather than 90 min). The rate of increase for at least the first two minutes appeared to be linear.

The effect of concentration on the rate of uptake of HO342 was measured by incubating Aux B1 cells for 1 min in medium containing HO342 dye in various concentrations from 0.2 to 5 μM and then measuring the fluorescence intensity of the G_1 peak. The data, when analyzed on a Lineweaver-Burk plot, fit a straight line whose intercept was not significantly different from zero. Thus, no evidence of saturation was seen, consistent with an unmediated diffusion process. Similar results were seen for colchicine (Carlsen et al. 1976).

As another approach to determine whether the uptake of HO342 into viable cells was mediated by the same mechanism as that previously observed for colchicine, the effect of a metabolic inhibitor, KCN, on the kinetics of dye uptake was examined both in the presence and absence of glucose. Metabolic inhibitors such as KCN stimulate the colchicine uptake of CHO cells by increasing membrane permeability to the drug, an increase that can be prevented by the addition of glucose (See et al. 1976). These effects are more marked for the membrane mutant lines, indicating that resistance to colchicine in CHO cells involves an energy-dependent barrier. The uptake of HO342 was affected in precisely the same way, further confirmation that the rate of colchicine uptake and the rate at which DNA stains with HO342 are governed by the same process.

Table 1. Comparison of colchicine resistance and HO342 fluorescence intensity in wild-type and drug resistant cell lines

Cell line	Relative colchicine[a] resistance	Relative decrease in HO342[b] fluorescence
Aux B1	1.0	1.0
CH^{R}A3	6	2.1
CH^{R}B3	21	4.6
CH^{R}C5	184	7.4
I10-1	3	1.3

[a]Taken from Ling (1975)

[b]Ratio of modal fluorescence intensity of the G_1 peak for Aux B1 to the modal intensity of the cell line being examined

2 Studies on Resting Mouse Lymphocytes

When cells from either mouse spleen or LN were stained with HO342 and analyzed by flow cytometry as in the previous section, two peaks of fluorescence intensity were seen, hereafter called LI (low intensity) and HI (high intensity) (Lalande and Miller, 1979, Lalande et al. 1980). See Fig. 1. These peaks do not reflect differences in DNA content as when the same cell suspensions were fixed before staining, only one dominant peak was seen, corresponding to the DNA content of cells in the G_0/G_1 phase of the cell cycle. This was as expected as almost all cells in mouse lymph node and spleen are resting, interphase small lymphocytes. The LI-HI intensity difference may, however be a result of a difference in HO342 dye permeability in two different lymphocyte subpopulations.

When kinetics of dye uptake were measured for the LI and HI peaks in spleen cells exposed to 2.5 μM HO342 dye, both peaks initially increased rapidly in fluorescence intensity and ultimately plateaued at the same fluorescence intensity, taking 90 and 60 min respectively to do so. This behaviour is similar to that seen for CHO cells. Next, fluorescence intensity after 2 min of uptake was measured as a fraction of dye concentration. The data fitted a straight line passing through the origin when plotted on a Lineweaver-Burk plot, precisely as seen for CHO cells. To investigate further this similarity, murine spleen cells were incubated in medium containing 5 μM HO342 and radioactive colchicine. Cells of the LI and HI peaks were then sorted from each other using our flow cytometer-cell sorter and radioactive counts per cell from each of the sorted fractions determined by standard methods. Cells of the HI peak incorporated 2.5 times more radioactive colchicine during the incubation than did cells from the LI peak, whereas the ratio of the HI to LI fluorescence intensities was about 1.9. It was concluded that murine lymphocytes take up HO342 dye by the same unmediated diffusion process as CHO cells and that this process is the same as is used for colchicine uptake (Lalande et al. 1981).

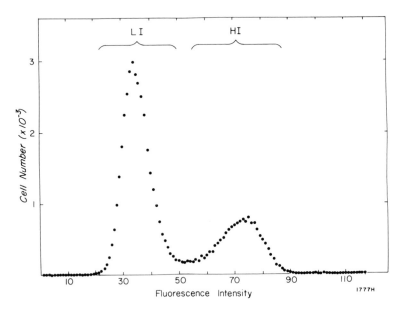

Fig. 1. Fluorescence analysis of HO342 — stained murine LN cells

What, then, do the HI and LI peaks represent? When the cells in the two peaks were characterized on the basis of various surface markers and on the basis of differential adhesion to nylon wool, it was found that the LI cells were all T cells and that the HI cells were nearly all B cells (Lalande et al. 1980, Loken 1980). Thus, there is an intrinsic diffference in the unmediated diffusion rate for resting B and T lymphocytes. Preliminary results suggest that other cell subpopulations may be recognizable on this basis. Thus, for example, as many as five subpopulations of cells can be recognized in murine bone marrow.

3 Studies on Mouse Lymphocytes Following Exposure to ConA (Lalande and Miller, 1979)

A number of plant lectins are specifically mitogenic for T cells. In the mouse system, on the addition of either PHA (phytohemagglutinin) or ConA (ConcanavalinA) all or nearly all T cells become blasts and start to proliferate. Blasts can first be detected about 16 h after addition of mitogen and cells can be detected entering S phase of the cell cycle about 4 h later. When these cultures are labelled with HO342 dye, there is a definite shift of cells from the LI to the HI peak, first detectable around 3 h after culture initiation and extremely pronounced by 12 h after culture initiation. It appears that T cells are shifting from the LI to the HI peak as an early step in their activation process and that an increase in the unmediated diffusion process is an early step in T cell activation.

4 Studies on Mouse Lymphocytes Responding in an MLR (Mixed Lymphocyte Reaction) (Lalande et al. 1980)

The MLR has been widely used as an *in vitro* model system for studying T cell activation. In the mouse, lymphocytes (responders) from one inbred strain are cultured with inactivated (e.g., by irradiation) stimulator cells from another strain which is different in one or more regions of its major histocompatibility (H-2) complex. After 4 to 6 days in culture, the cells from the responder strain will produce cytotoxic T lymphocytes (CTL) which specifically lyse targets bearing H-2 surface products identical to those of the original stimulating strain. The CTL arise as the result of the activation, differentiation and proliferation of a subpopulation of T lymphocytes, the cytotoxic T lymphocyte precursor (CLP) population.

Upon stimulation with a specific H-2 alloantigen, it has been estimated that from 2 to 10% of the responder cells are activated to undergo DNA synthesis. The frequency of specifically activated CLP is, however, only 0.1% to 0.2%. The CLP, therefore, account for a maximum of 10% of the observed DNA synthesis. The bulk of the proliferative pool remain poorly characterized as to the number, function and specificity requirements of its different cell populations, although T helper cells, macrophages and T suppressor cells have been implicated.

A limitation to the analysis of the events involved in CLP activation was that no method existed for isolating specifically activated CLP at early times after allogeneic stimulation. This problem has been solved by the demonstration that such CLP show increased HO342 uptake which enables them to be detected and separated from other CLP and from stimulator cells within 12 h of culture initiation. These CLP are in a state of partial activation and can go on to form CTL in the absence of stimulator cells provided appropriate supernatant factors are provided.

5 Studies on Human Peripheral Blood Lymphocytes

Work from this lab has shown that mouse T and B lymphocytes can be distinguished from each other on the basis of HO342 uptake rate. Williams et al. (1982) have shown that the same holds true for rat. We therefore asked whether the same approach would be valid for human lymphocytes and, in addition, whether one could also distinguish activated human T lymphocytes from resting T lymphocytes as was found to be the case for mouse (Keystone et al. 1983).

Peripheral blood lymphocytes (PBL) were obtained from healthy volunteers using standard procedures and stained with HO342. Spectra showing LI and HI peaks essentially identical to those of Fig. 1 were found once the appropriate staining conditions were determined. The only important difference from the mouse protocol appeared to be that human lymphocytes take up HO342 at a much faster rate overall so that 10-fold lower dye concentrations had to be used to obtain good resolution of the LI and HI peaks. When PBL were incubated with PHA or ConA, there was a pronounced shift of cells from the LI to HI peak, precisely as seen for the mouse.

Preliminary studies have been made on PBL from patients with rheumatoid arthritis and systemic lupus erythematosus. Both of these are long term, chronic diseases throught to be autoimmune in nature. Among other things, they are characterized by high levels of activated T cells. Fluorescence analysis of PBL stained with HO342 reveals that most of the cells, even those not activated by other criteria, are in the HI peak, suggesting that the majority of the T cells in these patients are in a state of partial activation. The possible significance of this observation is being currently investigated.

References

Arndt-Jovin DJ, Jovin TM (1977) Analysis and sorting of living cells according to deoxyribonucleic acid content. J Histochem Cytochem 25:585–591

Bech-Hansen NT, Till JE, Ling V (1976) Pleiotropic phenotype of colchicine-resistant CHO cells: Cross-resistance and collateral sensitivity. J Cell Physiol 88:23–31

Carlsen SA, Till JE, Ling V (1976) Modulation of membrane drug permeability in Chinese hamster ovary cells. Biochim Biophys Acta 455:900–912

Keystone EC, Albert S, Ehman D, Miller RG (1983) Manuscript in preparation

Lalande ME, Miller RG (1979) Fluorescence flow analysis of lymphocyte activation using Hoechst 33342 dye. J Histochem Cytochem 27:394–397

Lalande ME, McCutcheon MJ, Miller RG (1980) Quantitative studies on the precursors of cytotoxic lymphocytes. VI Second signal requirements of specifically activated precursors isolated 12 hr after stimulation. J Exp Med 151:12–19

Lalande ME, Ling V, Miller RG (1981) Hoechst 3342 dye uptake as a probe of membrane permeability changes in mammalian cells. Proc Natl Acad Sci USA 78:363–367

Latt SA, Stetten G (1976) Spectral studies on 33258 Hoechst and related bis benzimidazole dyes useful for fluorescent detection of deoxyribonucleic acid synthesis. J Histochem Cytochem 24:24–28

Ling V (1975) Drug resistance and membrane alteration in mutants of mammalian cells. Can J Genet Cytol 17:503–515

Ling V, Thompson LH (1974) Reduced permeability in CHO cells as a mechanism of resistance to colchicine. J Cell Physiol 83:103–116

Loewe H, Urbanietz J (1974) Basisch substituierte 2,6-bis-benzimidazolderivate, eine neue chemotherapeutisch aktive Körperklasse. Arzneim Forsch 24:1927–1933

Loken MR (1980) Separation of viable T and B lymphocytes using a cytochemical stain, Hoechst 33342. J Histochem Cytochem 28:36–39

Miller RG, Lalande ME, McCutcheon MJ, Stewart SS, Price GB (1981) Usage of the flow cytometer-cell sorter. J Immunol Meth 47:13–24

See YP, Carlsen SA, Till JE, Ling V (1974) Increased drug permeability in Chinese hamster ovary cells in the presence of cyanide. Biochim Biophys Acta 373:242–254

Williams JM, Shapiro HM, Milford EL, Strom TB (1982) Multiparameter flow cytometric analysis of lymphocyte subpopulation activation in lectin-stimulated cultures. J Immunol 128:2676–2681

Probing Macromolecular Structures by Flow Cytometric Fluorescence Polarization Measurements

W. Beisker and W.G. Eisert

Arbeitsgruppe Cytometrie, Gesellschaft für Strahlen- und Umweltforschung (GSF),
Herrenhäuserstr. 2, D-3000 Hannover, FRG

Introduction

After an insult to biological tissue structural changes are believed to take place prior
to or concomittend to large extend to changes in the amount of any component of
the individual cell. Although methodes measuring the amount of cellular constituents
in a single living cell have been able to detect changes as little as the amount of a
single chromosome, however, at this level of sensitivity alterations occuring at low or
moderate levels of exposure will most likely not be noticed. If on the other hand a
typically well balanced equilibrium between cells in different states of growth or different stages of the cell cycle get altered by any insult, a measurement of the entire
distribution within the population will very sensitively reflect the impact of the insult.
This information, however, is obtained by the analysis of the entire population only
and no conclusion about the individual cell can be drawn [6]. Methodes, that detect
structural alterations within each individual cell seem to be more suited to monitore
the effect of low level exposure. Immunological methodes detect various structural
alterations on the molecular level using antibodies against newly formed molecular
structures [1].

Immunological methodes may be developed to very high levels of sensitivity and
stecitivity, however, in order to raise the necessary antibodies, the newly formed molecular structure must be known before hand.

Alternatively, a markermolecule that can be attached to a structure of interest within a cell and which can be monitored for its mobility continuously will allow to follow
the time course of alterations of that structure [4, 8, 16]. This is of advantage for the
observation of reversible alterations, that are believed to take place after an insult.

Measuring the rotational mobility of a markermolecule by the degree of depolarization of the fluorescent light emitted by the markermolecule after excitation with
highly polarized light is one way of detecting structural changes on the macromolecular level. The methodes for measuring the degree of polarization (or depolarization)
of fluorescent light on a single cell level have been refined within the last years to
become a sensitive tool for tracing structural changes and rearrangements [5, 11, 17,
18]. The application of this technique however is limited to the availability of specific
markermolecules, that enter a specific cellular structure and do not cause rearrangements on binding by is o᾿ n.

Biological Dosimetry. Edited by W. G. Eisert and M. L. Mendelsohn
© Springer-Verlag Berlin Heidelberg 1984

The fluorescence depolarization measurements so far have been sucessfully applied to research on the cellular membrane [7, 9, 10, 14, 15].

A variety of markermolecules have been used, that enter the lipid bilayer of the outer cellular membrane and arrange itself either parallel or perpendicular to the surface of the membrane. Under heat treatment the altered arrangement of the fatty-acid chains allow a markermolecule a higher degree of rotational freedom, causing the emitted fluorescence to be depolarized to a larger extent [7, 8, 9, 16].

In lipid vesicles fluorescence polarization measurements have been compared with melting profiles [7, 12, 13].

Most data on fluorescence depolarization measurements so far have been obtained from experiments using suspensions of cells or vesicles in a cuvette. This, however, reduces the available information from the individual cell to an average value over the entire cell population. Therefore as data become available on a single cell level, which can then be correlated with other data on the same cell such as DNA — or protein-content, fluorescence polarization measurements have a great potential for uncovering minute alterations after an insult to the tissue. The following will describe a new double focus arrangement to measure the fluorescence polarization of the individual cell and a data aquisition system that determins the corrected value of the polarization (p) on each individual cell before the next cell enters the laser beams.

Materials and Methodes

Various optical arrangements have been published in the past to employ fluorescence polarization measurements in a flow cytometric system [2, 3, 5, 10, 11, 17]. Sofar all systems did use either standard particles or aqueous solutions for system calibration prior and after the actual measurement. Since a highly stable adjustment of the two optical and electronical pathways to equal sensitivity and linearity is crucial to the precision of the measurement, minor drifts of the system response will cause large variations in the results due to the algorithm determining the degree of polarization

$$p = \frac{I_1 - I_2}{I_1 + I_2} \ .$$

To overcome most of these problems in data aquisition the authors described in an earlier paper a flow system, which uses epi-illumination for fluorescence excitation by a highly polarized laser beam [5]. The sample was split into two parts which were measured consecutively in a single experiment. During the first part of the measurement the polarization of the exciting beam was kept parallel to the plane of polarization of one fluorescence detector whereas during the second part of the experiment the polarization of the excitation beam was switched by 90 degrees and was now parallel to the plane of polarization monitored by the other fluorescence detector. This scheme made the rotation of the polarization of the excitation beam necessary in the middle of the experiment.

Assuming that the degree of polarization is not different in the two subgroups of the sample both sets of data should show the same value for p. Any deviation from this symmetry then is believed to be due to a mismatch of the two detector channels. The deviations from symmetry are the taken to correct the two sets of data. The computer in its final computation takes also asymmetries into account due to unmatched optical components or transitions between different refractive indices in the light path. These constant asymmetries however are to be determined before hand and stored as correction function in the computer.

Allowing a single cell to pass through two laser beams in a sequence both being polarized, however the plane of polarization being perpendicular to each other, the same principle idea applies to the measurement of the fluorescence polarization of a single cell. Both excitation laser beams are generated from one beam by using a bifringent beam splitter. This beam splitter will seperate the two orthogonal components of the polarization into two spacially seperated parallel beams. The separation as well as the orientation of the planes of polarization are properties of the beam splitter and are namely due to the cut of the crystal being used.

Figure 1 shows the optical arrangement for the two excitation light beams. The unpolarized incoming light beam is split into two components, the plane of polarization being perpendicular to each other. The dichroic mirror directs the two beams into the microscope objective being used to form two foci in the flow chamber. In our system the distance between the two beams typically is 40 microns. The flow system forces a cell or a particle to pass through the two beams in sequence. At both focus positions the excitation is measured along the major cell axis by a photodiode. Simultaneously as the cell pass a focus the fluorescent light is directed through the dichroic mirror into an other polarizing beam splitter (Fig. 2). The fluorescent light is thus split into two polarized components. The plains of polarization being orthogonal to each other and one being parallel to the polarization of the excitation beam. Therefore one detector monitores fluorescence being parallely polarized to the excitation, the other monitores the perpendicular component. As the same particle or cell enters the second excitation beam after a given time interval due to flow velocity the same pair of photomultipiers will measure the two polarization components of the fluorescent light. However, since the plane of polarization of the excitation beam is orthogonal with respect to the plane of the first beam the measurement results in the inverted situation. The same detector that measured the parallel component in the position of the first beam now measures the perpendicular component and vice versa. Again the two sets of data obtained from the same particle at the two foci should lead to an identical value for the degree of polarization. Any asymmetry leading to differences in the results however can now be corrected for the individual measurement.

An online data – aquisition system uses a Z-80 microprocessor for control and data storage (Fig. 3). Impulses detected simultaneously by the two photomultipliers are correlated with the extinction pulse as detected by either one of the two photodiodes. This correlates the fluorescence detection with the information on the position of the particle with respect to the two excitation beams. As the particle or cell passes the two laser beams, two pairs of data are stored and digitized (X1, Y1:= first beam, X2, Y2 ;= second beam). Further analysis is only proceed if a number of feasibility criteria are met. One being that the second set of data must be detected within a certain time

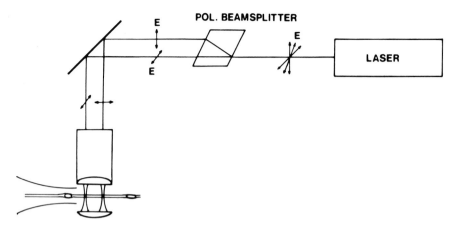

Fig. 1. Optical arrangement of the excitation beams. A single unpolarized laser beam is split into two polarized components by means of a polarizing beam splitter. The two polarized beams exit the beam splitter parallel to each other. A dichroic mirror reflects both beams into the focussing to allow epi – illumination fluorescence excitation. Two foci are formed at the position of the cells in the flow channel after hydrodynamic focussing of the cells

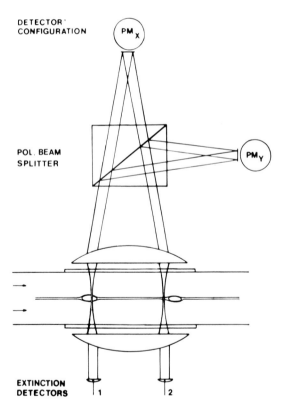

Fig. 2. Optical arrangement of the fluorescence and excitation detectors. A photodiode measures the extinction as particles or cells pass each laser focus. After passing through the dichroic mirror (not shown) the fluorescent light is split into two orthogonally oriented polarization components by a polarizing beam splitter. Light of each component is collected on a photomultiplier (PM X and PM Y). As the cell passes the second photomultiplier the measurement is repeated and the same photomultipliers again measure the two polarization components however this time with inverted parallel and orthogonal correlation

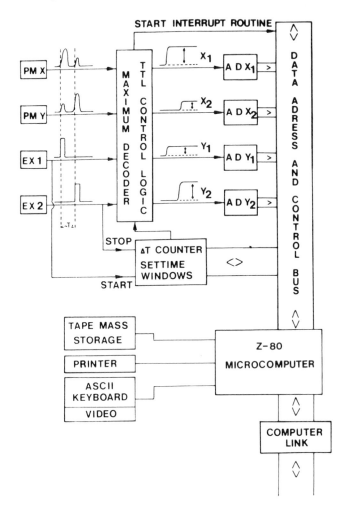

Fig. 3. Data aquisition. The pulses from the two photomultipliers PM X and PM Y are checked for coincidence with the pulses from either one of the two extinction photodetectors (EX 1 and EX 2). According to the time sequence the four fluorescence measurements (PM X_1; PM Y_1 and PM X_2 and PM Y_2) are simultaneously digitized by four A/D – converters (AD X_1; AD X_2; AD Y_1; and AD Y_2). From the pulse sequences of the extinction channels the flow velocity is calculated and a time window for the acceptance of the correlated pair of fluorescence data is set. Data are then transfered to a Z-80 microprocessor for calculation of p-values and storage of data either in the memory of the microcomputer or transfer to a host computer

window, which is related to the velocity of the particle stream in the flow chamber. The velocity however, is continuously measured by the time a single particle needs to travers from one focus to the other, the distance between the two being well known. If these criteria are met the degree of polarization is calculated for each individual cell, taking the numerical correction for asymmetries into account. The corrected value for p is stored into memory. The time necessary for processing the data of an individual cell is in the order of 150 microseconds. This still allows a very fast data collection in flow. During data aquisition the microprocessor generates a one or two dimensional histogram of the data being stored so far on a video screen. The histogram is updated rather frequently resulting in a "life-display" of the data aquisition. This enables an immediate control of the measurement.

Furthermore additional parameters such as cell length, multiangle light scattering and multiwavelength total fluorescence intensity are probed simultaneously and may be stored correlated to the polarization measurements.

Results and Discussion

The installation of the system has been completed not too long ago. Therefore only preliminary data on biological tissue is available at this time. An example on the rotational mobility of the dye propidium iodate (PI), which intercalates the cellular DNA is given in Fig. 4. L 1210 – cells have been stained and analyzed (PI 5 mg/100 ml). The resulting histogram of polarization values of this population centers around p = 0.16 (A). However if the same population of cells are treated with 1 molar MgCl solution after staining the histogram is shifted towards higher p-values. The mean polarization after cell treatment now being p = 0.24 (B). This shift may be interpreted as condensation of the DNA – matrix by MgCl – treatment restricting the rotational mobility of the intercalated dye.

The inverse has been observed, when cells were exposed of heat. Figure 5 gives an example of this behavior. Density specified human mononuclear blood cells were stained with propidium iodate (5 mg/100 ml). The histogram of the fluorescence polarization showed a mean value of p = 0.22 (A). However, if the cells were treated with heat (95°C for 10 min) and then stained with PI the mean value of the polarization dropped to p= 0.14 (B). Although the variation in the degree of polarization is considerably larger in the human blood cells as compared to L 1210 mouse tumor cells the drop of the mean after heat treatment is quite marked.

Since the system is capable of measuring other cellular parameters at the same time, it allows a correlation of structural informations with parameters related to cell function or cell growth. The latter being of importance for the application of this technique to biological dosimetry. The data obtained so far have confirmed our expectations. Due to the improved stability and sensitivity one promising field of application which is currently under investigation in this laboratory is the detection of reversible and irreversible structural alterations after an insult to the tissue which may lead to a deeper understanding of the underlaying mechanism of action and subsequently also open a new way for evaluating dose in human material.

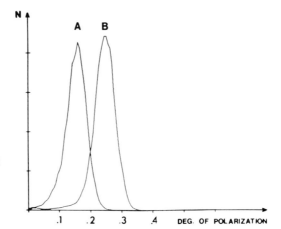

Fig. 4. Fluorescence polarization histograms of L 1210 – cells stained with propidium iodate. *A* untreated sample, *B* after treatment with 1 molar MgCl. Mean p increases from p = 0.16 (A) to p = 0.24 (B)

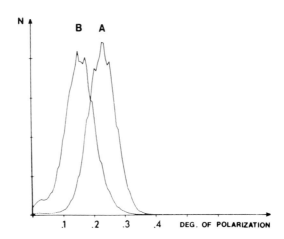

Fig. 5. Fluorescence polarization histograms of mononuclear cells isolated from human peripheral blood stained with propidium iodate. *A* untreated sample, *B* stained after heat – treatment (95°C, 10 min). Mean p-value decreases from p = 0.22 (A) to p = 0.14 (B)

References

1. Adamkewicz et al. (1984) This book
2. Arndt-Jovin DJ, Jovin TM (1976) Cell separation using fluorescence emission anisotropy; Membranes and Neoplasia; New Approaches and strategies. A R Liss Inc, New York, pp 123–136
3. Axelrod D (1976) Carbocyanine dye orientation in red cell membrane studied by microscopic fluorescence polarization. Biophys J 26:557–573
4. Cercek L, Cercek B (1979) Involvement of mitochondria in changes of fluorescence excitation and emission polarization spectra in living cells. Biophys J 28:403–412
5. Eisert WG, Beisker W (1980) Epi-illumination optical design for fluorescence polarization measurements in flow systems. Biophys J 31:97–112
6. Eisert WG, Weier H-U, Birk G (1984) Dose related changes in cell cycle of bone marrow and spleen cells monitored by DNA/RNA flow cytometry. This volume
7. Galla HJ, Hartmann W, Sackmann E (1978) Lipid-protein-interaction in modell membranes: Binding of melittin to lecithin bilayer vesicles. Ber Bunsenges Phys Chem 82:918–922

8. van Hoeven RP, van Blitterswijk WJ, Emmelot P (1979) Fluorescence polarization measurements on normal and tumour cells and their corresponding plasma membranes. Biochem Biophys Acta 551:44–54

9. Jähnig F (1979) Structural order of lipids and proteins in membranes: Evaluation of fluorescence anisotropy data. Proc Natl Acad Sci USA 76:6361–6365

10. Jovin TM (1979) Fluorescence polarization and energy transfer: Theory and application. In: Melamed MR, Mendelsohn ML, Mullaney PF (ed) Flow cytometry and sorting. John Wiley & Sons New York, pp 137–165

11. Lindmo T, Steen HB (1977) Flow cytometric measurements of the polarization of fluorescence from intracellular fluorescein in mammalian cells. Biophys J 18:173–187

12. Luken DW, Esfahany M, Devlin TM (1980) Effect of sterols on diphenylhexatriene fluorescence in lecithin vesicles. FEBS Letters 114:48–50

13. Pike CSP, Berry JA, Raison JK (1979) Fluorescence polarization studies of membrane phospholipid phase separations in warm and cool climate plants. CIW/DPB Publ No 660:305–318

14. Radda GK, Vanderkooi J (1972) Can fluorescent probes tell us anything about membranes? Biochim Biophys Acta 265:509–549

15. Shinitzky M, Dianoux A-C, Gitler C, Weber G (1971) Microviscosity and order in the hydrocarbon region of micelles and membranes determined with fluorescent probes. I. Synthetic micelles. Biochemistry 10:2106–2113

16. Täljedal I-B (1979) Polarization of chlorotetracycline fluorescence in pancreatic islet cells and its response to calcium ions and D-glucose. Biochem J 178:187–193

17. Udkoff R, Norman A (1979) Polarization of fluorescein fluorescence in single cells. J Histochem Cytochem 27:49–55

18. Udkoff R, Chan S, Norman A (1981) Identification of mitogen responding lymphocytes by fluorescence polarization. Cytometry 1:265–271

Physiological and Preparatory Variation of Nuclear Chromatin as a Limiting Condition for Biological Dosimetry by means of High Resolution Image Analysis

W. Giaretti[1*], W. Abmayr[2], G. Burger[2], and P. Dörmer[1]

[1]Abteilung für Experimentelle Hämatologie, Gesellschaft für Strahlen- und Umweltforschung, Landwehrstraße 61, D-8000 München, FRG

[2]Institut für Strahlenschutz, Gesellschaft für Strahlen- und Umweltforschung, Ingolstädter Landstraße 1, D-8042 Neuherberg, FRG

Summary

A human cell system suitable for serving as a biological dosimeter should show high specificity that is ideally only responding in a quantitative and reproducible manner to the dose received. If high resolution image analysis is chosen as the method forr detecting such changes, functional or preparatory influences on the image of the target cells should be well known to control them suitably. In the present study the Feulgen staining procedure was chosen to study such possible influences on the nuclear chromatin structure using mouse L fibroblasts as a model cell population. In order to investigate the variability primarily due to functional heterogeneity, the Feulgen hydrolysis behaviour of GO, G1, S and G2 cells was studied. It could be shown that depolymerization and depurination slightly depend on the cellular phase in the cycle, indicating a different acid stability of the chromatin matrix. It could further be shown that other preparatory conditions as, for example, the way of spreading the cells or the conventional trypsinization are critically influencing the features extracted from the nuclear images. The results obtained so far will be used to develop a strategy for an optimum selection and preparation of human cells for dosimetric purposes using high resolution image analysis.

Introduction

One of the serious hopes in biological dosimetry is that some type of easily accessible human cell might be found to be a suitable monitor for radiation or chemical damage to the organism. Although almost every cell system is able to repair sublethal or potentially lethal damage, the kinetics of these mechanisms seem to be slow enough in some cases to keep the cells reasonably long in an altered state and possibly detectable by means of quantitative morphological methods.

*Present address: Laboratorio di Biofisica Istituto Nazionale per la Ricerca sul Cancro
Viale Benetetto XV n.10, 16132 Genova, Italy

Biological Dosimetry. Edited by W. G. Eisert and M. L. Mendelsohn
© Springer-Verlag Berlin Heidelberg 1984

It has been suggested that the interaction of DNA with the protein matrix may be used to detect differences in chromatin stability (expressing itself in structural features of the chromatin or in its resistance to acid treatments) between cells differing in gene activity [1, 2]. The same mechanism may also be useful for studying alterations or damages to the DNA target due to radiation and chemicals. In the present study we have not yet investigated the expected effects following radiation or chemical damage. We have rather focused our interest on the prerequisites of dosimetry, namely the investigation of the dependence of chromatin structure of cells on their physiological status and on preparatory conditions. For this purpose experiments have been performed with cells in the GO-G1-S-G2 phases of the cell cycle and the effects due to the preparatory procedures of acid hydrolysis and trypsinization have been evaluated.

Material and Methods

Cell Preparation

The experiments were performed with mouse L fibroblasts. The cells were grown for 3 days, then the culture was deprived of serum and a medium change was made every 3rd day. On day 10 the cells were subcultivated at a density of 10^4 cells/cm^2 and stimulated to proliferate by adding serum. The kinetics of the cells after stimulation studied by flow cytometry [3] showed that 90% of the cells remained in the GO-G1 phase for about 15 h (lag-phase) before the onset of DNA synthesis. Unless stated otherwise, the cells were grown directly on the microscopic slides in order to avoid any further manipulation such as trypsin treatment. Cells subcultivated for 5 h in the absence of serum were used as controls for the GO unstimulated fraction, whereas samples grown in the presence of 10% fetal calf serum served as the G1 stimulated fraction. Samples obtained 40 h after stimulation were incubated with ^{3}H-thymidine in order to provide fractions of G1, S and G2 cells. The autoradiographic protocol is described in detail by Dörmer [4]. Cells were fixed in absolute methanol for 48 h. The acid hydrolysis procedure for Feulgen staining was performed at 60°C using 1N HCl. In order to assess the depurination process hydrolysis times of 6, 8 and 10 min were found to be suitable whereas for studying the depolimerization pattern hydrolysis lasted 12, 14, 15 and 20 min.

Lag-phase cells were used in one experiment to analyze the effect of the trypsin treatment (0.25% trypsin at 37°C for 1 min) on detaching the cells from the glass. The control cells were grown on the same Petri dish and detached from the glass mechanically with the use of a rubber policeman. Both samples were then smeared identically onto microscopic slides.

Measurements and Analysis

An AXIOMAT microscope (C. Zeiss, Oberkochen, West Germany) was used with an apochromat 50/1.32 oil immersion objective and a 40/0.6 substage condenser. Illumination was provided by a 100 W halogen lamp equipped with a 550 nm filter with a band-width of 20 nm.

A Zeiss scanning stage with a maximum speed of 10 KHz allowed cells to be relocated with great accuracy for the purpose of performing Feulgen absorption measurements and autoradiographic classification on the same cells.

Scanning was performed with a high-resolution TV-plumbicon camera (R. Bosch, West Germany). The pixel size was $(.5 \,\mu m)^2$. The camera was interfaced with an AP 120 B array-processor (Floating Point System Inc, Portland, USA). Shading correction of the sampled image was obtained by software (pointwise multiplication) in the AP. The scan-ned data were transferred from the AP via a Siemens PR 330 16bit minicomputer to a disk memory (Fig. 1) and to magnetic tapes for further evaluation. The raw cell data were analyzed in a Siemens 7760 computer with BS 2000 operating system and the BIP software package [5]. The statistical analysis was performed with the use of the BMDP-software package [6]. After segmentation of the Feulgen nuclei morphological photometrical and textural features were derived as described in detail elsewhere [3]. Useful information was gained when the frequency of occurrence of low and high opti-cal density values was evaluated. The histogram analysis performed for different hydro-lysis conditions allows to obtain a differential pattern of the decondensed and conden-sed regions of the Feulgen chromatin and might in principle be useful in detecting chan-ges due to specific alterations of the condensation-decondensation pattern of the chro-matin.

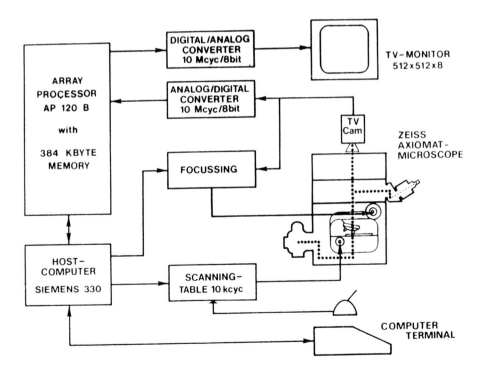

Fig. 1. System overview with the Axiomat microscope, the TV-camera and the Array-processor

Results

Total mean extinction values in arbitrary units and standard deviations are shown in
Fig. 2 for increasing values of hydrolysis time namely 6, 8, 10, 15 and 20 min for G1,
S or G2 cells. It is evident that the curves of depurination during the 6, 8, 10 min of
hydrolysis as well as of depolimerization in the 15 and 20 min samples show a similar
behaviour for the cells in the G1, S and G2 phases of the cell cycle. However, the ratios
of the total extinction values of G2 to G1 cells have the constant value of 2, as expected, only for 8, 10 and 12 min hydrolysis. In the case of 6 and 20 min hydrolysis the
G2/G1 ratio is 2.25 and 1.8 respectively, that is a relative variation of 12 and 10%.
The hydrolysis profiles of parameters other than DNA content and supervised multivariate analysis using different combinations of textural and chromatin features showed
that the correct classification (80 to 90%) of G1, S and G2 cells [7] is only slightly dependent on the hydrolysis times (Abmayr et al., in preparation). The Feulgen DNA pattern of cells stimulated to proliferate (G1 cells) as a function of hydrolysis is compared
to that of unstimulated cells (GO cells) in Fig. 3. There is clear evidence that the stainability of cells stimulated to proliferate by the addition of 10% serum after 5 hrs is higher
than for the unstimulated control in the depurination (6 min) and depolimerization
(14 min) regions of hydrolysis. In this case, where the DNA-content should be constant,
the ratios of the total absorption values are 1.3 at 6 min and 1.15 at 14 min hydrolysis.
Comparing these ratios to the plateau region at 8, 10, and 12 min, which has an average
ratio of 1.06, there is a maximum relative variation of 22%. The data reported in Fig. 4
show the relative frequencies of occurrence of optical density values for G1, S or G2
cells for a fixed hydrolysis time. The frequency of lower OD values is slightly higher for
G1 cells at every time of hydrolysis. On the other hand, the frequency of the highest
OD values is clearly systematically lower for the G1 cells compared to S and G2. At
20 min hydrolysis (not shown in the Figure) the three curves overlap to a great extent
and are shifted to lower OD values. In Fig. 5 the nuclear area of resting cells grown to
confluency has been plotted against DNA content. These data indicate that the use
of trypsin for detaching the cells from the glass influences the geometry of the Feulgen
nuclei. The use of mechanical means (rubber policeman) serves as the control. No differences can be seen in the Feulgen DNA content.

Discussion

The results reported in Fig. 2 suggest that during the processes of depurination and depolymerization [8–10] no differential effects of hydrolysis seem to exist for G1, S
and G2 cells. The total extinction values obtained separately for these cell cycle phases
indicate that the Feulgen stainability is nearly proportional to the DNA content for
all hydrolysis times tested. When the same type of analysis was performed on cells which
had received a serum stimulus to proliferate (G1 cells) and compared to unstimulated
cells (GO cells) there was some slight evidence that GO and G1 cells had a different de-

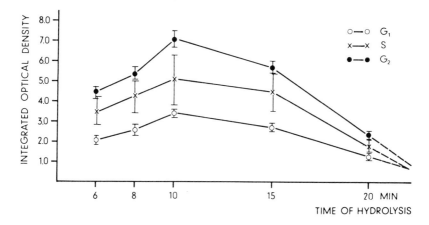

Fig. 2. Integrated optical density in arbitrary units (mean ± SD) versus hydrolysis times (min). Acid hydrolysis was in 1 N HCl at 60°C-. The sample size for each individual point was around 400 cells. The *lower curve* refers to G1 cells, the *mid-curve* to S and the *upper curve* to G2 cells

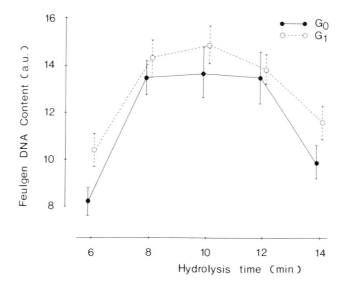

Fig. 3. The Feulgen DNA content (a.u.) (mean ± SD) as a function of hydrolysis time (min) for cells stimulated with 10% fetal calf serum for 5 h (*upper broken curve*) and for unstimulated cells (*lower continuous curve*). Acid hydrolysis was in 1N HCl at 60°C. The sample size for each individual point was around 400 cells

purination and depolymerization behaviour (Fig. 3). Similar results have been obtained in cells having different composition of the nucleoproteins [2] and on films containing condensed or swollen chicken erythrocyte nuclei [11].

One possible explanation may be that the interaction of DNA with proteins in the chromatin matrix is different for G0 and G1 cells. This would lead one to expect that

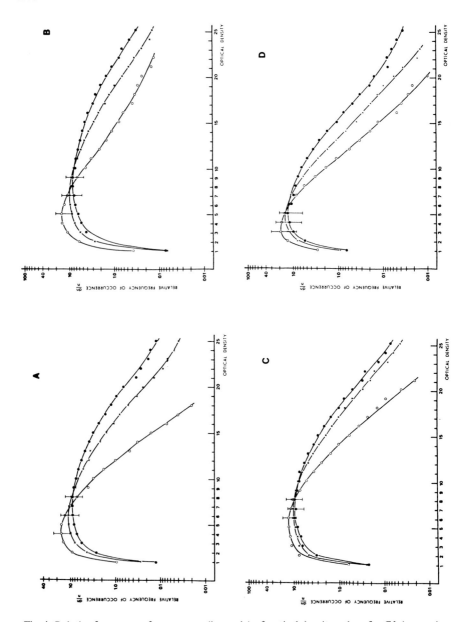

Fig. 4. Relative frequency of occurrance (log scale) of optical density values for G1 (o- - - -o), S (x- - - -x) and G2 cells (●- - - -●). For a fixed time of hydrolysis: A = 6 min, B = 8 min, C = 10 min, D = 15 min. In the region of high density values the *lower curve* refers to G1 cells, the *mid-curve* to S and the *upper curve* to G2 cells

structural alterations or damages to the DNA target may be more easily recognizable when hydrolysis profiles are evaluated and the specific alterations in the acid stability of the chromatin would most likely be detected in the depurination and depolymeriza-

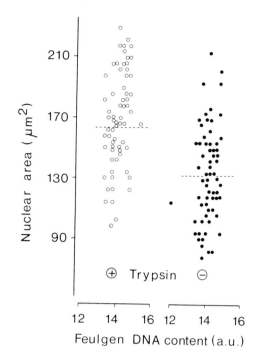

Fig. 5. Nuclear area (μm^2) of resting cells grown to confluency against DNA content (a.u.). Conventional trypsinisation for detaching the cells from the glass (*left side*). The use of a rubber policeman served as a control (*right side*). Acid hydrolysis was performed in 5N HCl at 28°C for 20 min

tion regions of hydrolysis. Apparently, in the optimum region of hydrolysis between 8 and 12 min structural alterations in the chromatin or different interactions of DNA with chromosomal proteins level off. The separate analysis of optical density distributions for G1, S and G2 cells (Fig. 4) could be interpreted similarly. These results might indicate that the G1 cells have a higher probability of being in a decondensed state when compared to S and G2 cells reflecting the fact that the interaction of DNA with proteins in the chromatin matrix can be different for G1, S and G2 cells and that the G2 cells have the highest level of condensation [12, 13]. It should be noted, however, that geometric conditions are another important factor which can alterate the profile of the frequency distribution of OD values. The coefficient of correlation of about 0.6 between DNA content and nuclear area of G1, S and G2 cells indicates that a G2 nucleus is larger but not twice as large as a G1 nucleus and that this implies a geometric effect on the OD values measured. As far as the results reported in Fig. 5 are concerned, the increase in nuclear area for the cells treated with trypsin most likely reflects a loss of nuclear as well as cytoplasmic proteins [14]. Changes in the chromatin structure detected by circular dichroism and viscosity measurements along with loss of nuclear and cytoplasmic proteins due to trypsin have been reported [14].

The present methodological approach has revealed considerable variability in results depending on the procedure applied and on the functional status of the cells. On the one hand it calls for rigorous reproducibility in handling and selecting the cells on the other it has revealed, in view of the purpose of this study, that some procedures may be more advantageous for detecting minute changes in the chromatin structure than others. Previous approaches have been undertaken using the same technique to monitor effects of radiation *in vitro* on easily accessible cells such as lymphocytes [15, 16]. However, the ideal conditions which would allow detection of small changes in the DNA target due to radiation or chemicals have not yet been devised. There is no doubt that lymphocytes represent a most suitable cell type for monitoring environmental damage, even though they are comprised of such a heterogenous subset of cells, such as B, T-helper, T-suppressor, natural killer and null cells. An abundant panel of highly specific antibodies capable of defining the different properties of the various lymphocyte subsets offer in fact promising means of overcoming their functional heterogeneity. In addition the availability of machines capable of sorting cells according to well-defined functional properties enables us to hope that reproducible conditions for a cellular dosimetry at the lower level range will be worked out.

Acknowledgments. Mrs. E. Dietel and Mr. P. Knoblauch are thanked for technical assistance. Thanks are also due to Mrs. K. Hiller for revising the manuscript and to Mrs. K. Küffer for expert typing

References

1. Anderson GKA, Kjellstrand PTT (1972) Influence of acid concentration and temperature on fixed chromatin during Feulgen hydrolysis. Histochemie 30:108–114
2. Kjellstrand PTT, Anderson GKA (1975) Histochemical properties of spermatozoa and somatic cells. I. Relations between the Feulgen hydrolysis pattern and the composition of the nucleoproteins. II. Differences in the Feulgen hydrolysis pattern induced through alteration of the nucleoprotein complex. Histochem J 7:563–583
3. Giaretti WA, Abmayr W, Dörmer P (1983) Multiparameter nuclear morphology of the G0-G1 transition. Anal Quant. Cytol 5:90–98
4. Dörmer P (1981) Quantitative carbon-14 autoradiography at the cellular level: principles and application for cell kinetic studies. Histochem J 13:161–171
5. Mannweiler E, Rappl W, Abmayr W (1982) Software for interactive biomedical image processing – BIP. Proceedings 6th International Conference on Pattern Recognition, IEEE Computer Society Press, p 1213
6. Dixon WJ, Brown MB (1977) BMDP Biomedical Computer Programs P-series 1977. University of California Press
7. Abmayr W, Giaretti WA, Gais P, Dörmer P (1982) Discrimination of G1, S and G2 cells using high-resolution TV-scanning and multivariate analysis methods. Cytometry 2:316–326
8. Kjellstrand P (1980) Mechanisms of the Feulgen acid hydrolysis. J Microscopy 119:391–396
9. Decosse JJ, Aiello N (1966) Feulgen hydrolysis: Effect of acid and temperature. J Histochem Cytochem 14:601–604
10. Böhm N (1968) Einfluß der Fixierung und der Säurekonzentration auf die Feulgen-Hydrolyse bei 28°C. Histochem. 14:201–211
11. Duijndam WAL, van Duijn P (1975) The influence of chromatin compactness on the stoichiometry of the Feulgen-Schiff procedure studied in model films I. Theoretical kinetics and experiments with films containing isolated deoxyribonucleic acid. II. Investigations on films containing condensed or swollen chicken erythrocyte nuclei. J Histochem Cytochem 23:882–900
12. Bradbury EM, Inglis RJ, Matthews HR, Langan TA (1974) Molecular basis of control of mitotic cell division in eukaryotes. Nature 249:553–556
13. Giaretti WA, Gais P, Jütting U, Rodenacker K, Dörmer P (1983) Correlation between morphology as derived by digital image analysis and autoradiographic labeling pattern. Anal. Quant Cytol 5:79–89

14. Maizel A, Nicolini C, Baserga R (1975) Effect of cell trypsinization on nuclear proteins of WI-38 fibroblasts in culture. J Cell Physiol 86:71–82
15. Anderson RE, Olson GB, Shonk C, Howarth JL, Wied GL, Bartels PH (1975) Computer analysis of defined populations of lymphocytes irradiated in vitro: I. Evaluation of murine thoracic duct lymphocytes. Acta Cytol 19:126–133
16. Olson GB, Anderson RE, Bartels PH (1979) Computer analysis of defined populations of lymphocytes irradiated in vitro. III Evaluation of human T and B cells of peripheral blood origin Hum Pathol 10:179–190

Morphologic, Kinetic, and Metabolic Effects

Automated Autoradiographic Analysis of Tumor Cell Colonies in vitro

Robert F. Kallman

Department of Radiology, Stanford University, Stanford, CA 94305, USA

1 Kinetics-Based Cancer Therapy

The studies reported below are basic to the design of therapeutic strategies against cancer and to the understanding of therapeutic effects. It is unlikely that clinical therapy could employ the same methodology, however, so this approach could not be incorporated into "kinetics-directed" therapy. As defined and implemented successfully by Barranco et al. (1982a,b), therapy is *kinetics-directed* when the kinetic properties of cell populations under treatment are actually determined and used to schedule the timing of multiple treatments. Thus, the substance of this paper is of potential importance to *kinetics-based* cancer therapy, i.e. using historical or empirical data to formulate models.

The changing sensitivity of actively proliferating cells (P cells) as they progress through the various phases of the cell cycle, indeed the timing of the cycle as well, have been the subjects of almost all kinetic approaches to cancer therapy. But what about non-cycling, or non-proliferating tumor cells? Although it is well known that tumors contain many non-proliferating cells (Q cells), the question is still not resolved whether such cells fall within a single tumor cell population with, say, a log-normal distribution of cell proliferating and in the absence of effective perturbation will never return to the proliferating condition. (Based upon our determinations of labeling index after prolonged exposure of tumors to ^{3}HTdR, we believe the latter to be the case.) Most commonly, the Q compartment is thought to be comprised of cells in an abnormally prolonged G_1 phase – the phase that is frequently termed G_0. Largely through the findings of Darzynkiewicz and coworkers (1980), there is evidence suggesting that quiescence can start at any of several points in the cycle – during G_2 as well as G_1, and even during S. But that evidence is drawn primarily from work on a special class of tumor cells; and it is not yet known whether the cells of typical solid tumors are regulated the same way.

2 Experimental Approach

We have worked with solid tumors in mice and have adopted the approach that was introduced by Barendsen et al. (1973). A unique feature of this approach is the use of a tumor that can be grown in the solid state *in vivo*, can be readily dispersed with mild enzymatic digestion, and whose cells can grow into colonies with very high efficiency *in vitro*. Without this feature, we would be unable to examine the properties of clono-

Biological Dosimetry. Edited by W. G. Eisert and M. L. Mendelsohn
© Springer-Verlag Berlin Heidelberg 1984

genic cells. We would be in the same state as most workers who extrapolate from labeling index data and ascribe to labeled cells the property of unlimited future cycling — and therefore malignancy. In our experiments (Kallman et al. 1980), EMT6 tumor cells (Rockwell et al. 1972) are identified as cycling by their ability to incorporate radioactively labeled DNA precursor, ^{3}HTdR, when it is presented to the tumor cells for 24 h, which is considerably longer than their mean cell cycle time. We infer from statistics obtained from classical percent labeled mitosis (PLM) methodology that less than 1% of the cycling cells of this tumor will fail to become labeled when the label is presented for 24 h; so these unlabeled tumor cells may be regarded as Q cells.

After tumors are exposed to label *in vivo* in syngeneic BALB/c mice, they are converted to a single cell suspension for plating *in vitro,* so that we may test their clonogenic capability. When these cells are cultured in Waymouth's medium containing 15% fetal bovine serum, typical macrocolonies develop from the clonogenic precursor cells in about 12–14 days. There is very extensive label dilution in the individual cells, as the amount of label is halved each of the 10–20 times cells divide to form a macrocolony. For this reason, it is impossible reliably to detect the label in macrocolonies, for the amount of tritium in the DNA of the individual cells would be only barely greater than background. Consequently, our analyses have been based upon the number of silver grains counted in autoradiographs of the cells of microcolonies that develop in 3 days of incubation and when the mean colony size is approximately 7 cells. The autoradiography is done by the dipping method, using Kodak NTB-2 emulsion and a 5-week exposure period.

We have carried out experiments of 2 different but closely related designs. In one, treatment (primarily with radiation) is administered when the tumors have grown to an appropriate size, then time is allowed to elapse, and then the 24-h period of thymidine administration is begun. Tumors are excised for cell suspension 1 h after the cessation of thymidine administration. In the other kind of experiment, treatments are administered approximately 1 h after completion of the 24-h thymidine course. All these experiments present a biological dosimetric problem of major magnitude: the need to determine accurately the labeling indices of single tumor cells and of their progeny through several generations *in vitro.* The number of grains contained in the progeny that constitute a given colony (the sum of the grain counts of the individual cells of the colony) can be related to the grain counts of undivided cells, i.e. before incubation is begun.

This project was started several years ago and the grain counts were performed by technicians and, increasingly, by students in my laboratory. The difficulties associated with the manual grain counts (subjective errors, tedium, boredom, etc.) plus the conviction that this is an important line of investigation led us to develop an instrument to aid in performing these simple observations and analyses.

The culture vessel used in these experiments is a commercially obtainable slide chamber (Lab-Tek). The plastic wall of this chamber is easily removed when the period of incubation is completed, and the cells can then be fixed for autoradiography, dipped in emulsion, etc. To obtain grain counts manually, the observer makes a raster scan at low magnification. He identifies colonies on the basis of cell clustering. He then examines each cell of a cluster (or colony) under high magnification, and counts the number of grains in each. Like most biologists confronted with the job of determining labeling indices, we did this using transmitted light and oil immersion optics. This is an exceedingly tedious and slow procedure because of the necessity of changing magnifications very frequently, recording data by hand, and having to do this with perhaps 1,000 cells per slide. The time required per slide was at least about 12 h and usually quite a bit longer. Major effort was devoted to develop the instrument described below, which performs most of these tasks automatically (Kallman et al. 1980; Kemper and Kallman, 1983).

3 Automated Grain-Counting

In SACCAS (the Stanford Automated Cell Colony Autoradiographic Scanner), a video image of cells visualized with a Zeiss Universal microscope is transmitted via a Sierra Scientific camera and video-graphic display hardware (Digital Graphics System). This hardware allows the viewing of any combination of the direct video image, a thresholded video signal (an analog quantization which can be selected over a range of 64 levels), and the display memory containing 32K words. In order to do a labeling determination of a given preparation, the observer first scans the entire slide under low magnification (8X primary). When cells appear on the monitor screen, he stops the scan and then maps the position of each cell in a given colony using a crosshair cursor. Thus, each cell is assigned to a colony or is designated as an individual cell.

Once the scan is complete the operator turns the nosepiece of the microscope to engage a 43X epi-darkfield objective which allows the specimen to be viewed under reflected darkfield conditions. The computer is then made to drive the stage to every cell successively of every colony mapped. Transmitted light is used so that the nucleus can be distinguished from the cytoplasm, a rectangular mask is drawn with a joystick to exclude most of the non-nuclear area of the cell, and the substage light is then blocked off so that the cell is illuminated only by incident light. Tritium grains viewed this way under incident darkfield stand out as bright spots against a relatively dark background. A thresholding circuit is used to separate the grains from the rest of the image within the rectangular mask, and this thresholded image is digitized into 16 gray levels and automatically counted by a simple peak detection algorithm. The ability of the machine to count tritium grains accurately and reproducibly depends upon proper thresholding of the grains from the background.

When the grain counting consistency of this machine was measured against the manual counting consistency of 2 experienced operators by plotting the mean of 4 replicate manual counts as a function of the mean of 4 automatic counts, a straight line was obtained with a slope of 0.95 and a correlation coefficient of 0.98. Thus, the consistency and reliability of this instrument are easily comparable to manual counts. The instrument provides greater objectivity in counting grains per nucleus, and it does so much more rapidly than is possible otherwise.

Though this paper is intended to present a biological dosimetric problem and a discussion of methods being employed to attack this problem, it is useful to summarize some of the potentially important findings that have already emerged. (Ironically, the data to be discussed were not obtained with the instrumentation that has just been described; rather, these are manually obtained data which are currently being duplicated in appropriate experiments analyzed with the aid of SACCAS.)

4 Q cell Recruitment and/or Cycle-Dependent Radiosensitivity

Experiments in which the 24-h labeling period followed X-irradiation at increasing times (Kallman et al. 1980) yielded some suprising results. If the only significant effect of irradiation was to cause recruitment of Q cells into P, there should be more cycling cells at the start of thymidine infusion and this should be easily seen by an elevation in labeling indices (LI) — both immediately upon plating (before replication of the cells) and after the cells have formed microcolonies. However, instead we found the LI to decrease progressively. This kind of evidence does not force us to discard the notion that Q cell recruitment can take place under these conditions, but obviously it is being over-shadowed by other changes. The most likely of these is the death of P cells owing to their presumably greater radiosensitivity. We have started to investigate the differential radiosensitivity of P and Q cells *in vivo* by first labeling the P cells by 24-h ^{3}HTdR infusion, then irradiating, and then converting the solid tumors into a single cell suspension. These can be cultured for autoradiography and cell survival can be analyzed by counting colonies. We might expect that the LI of individual irradiated cells determined immediately after they are plated and before they have had an opportunity to replicate would not differ from the LI of unirradiated cells. On the other hand, the colony LI might be lower than normal because P cells would have preferentially been rendered sterile so that colonies would tend to develop more from surviving Q cells.

What is known about the relative radiosensitivity of P and Q cells? The generally held notion that radiation is a cycle specific cytotoxic agent derives largely from the extensive literature on "age-response" functions of cells growing exponentially *in vitro* (Terasima and Tolmach 1963, Elkind and Whitmore 1967). Such age-response data suggest that mid-G$_1$ is a relatively resistant phase, and since Q cells tend mostly to be in the G$_0$ state and this resembles mid-G$_1$ we might infer that Q cells will be relatively radioresistant. The only data we have at the moment are derived from a single experiment and a poorly stained set of slides, so they must be regarded as tentative. These data suggest that the surviving fraction of irradiated Q cells as a whole may fall on a curve with a very broad shoulder followed by a steeply dropping exponential portion. Thus, survival is high at low doses then falls rapidly as dose increases past the shoulder of the curve. If this shape of Q cell survival curve can be confirmed, it must still be determined whether all Q cells have the same radiosensitivity or whether, like P cells, they exhibit heterogeneity — with perhaps the younger Q cells being radioresistant and the older and more senescent Q cells being considerably more sensitive.

It is entirely possible that the data obtained in the recruitment-type experiments summarized above are seriously flawed, for they are subject to a major systematic artifact. This is based upon the knowledge that radiation does not kill cells instantly, but rather that reproductive death may not occur until one-to-several generations later

(Hurwitz and Tolmach 1969; Thompson and Suit 1969). Therefore, every small colony that may have developed by, say, 3 days after tumor cells have been plated *in vitro* may not necessarily mature into a macrocolony a week or more later. So it is essential that microcolony LIs be corrected by quantitative colony abortion probability factors, and that these be obtained for any cell line used in this kind of investigation.

5 Multiple Field Time-Lapse Cinemicrography

We have very recently initiated experiments with newly designed equipment to provide such quantitative correction factors. We desired to make a large number of time-lapse movies from which we could derive colony abortion probabilities and/or cell generation times as a function of microcolony size or age. This must be done for different treatments and doses administered to colony-forming cells. In order to provide this kind of dosimetric data, each time-lapse movie must trace the development of a single cell into a typical macrocolony (containing more than 50 cells).

The procedure we have developed is as follows (Fig. 1): Cells are plated in Lab-Tek chamber slides in densities such that potentially clonogenic cells are a few millimeters apart. We designed and constructed a mini-incubator that is just large enough to contain 2 such culture slides. The slides rest on ledges slightly above the floor of the incubator so that the ambient warm ($37\pm0.1^{\circ}C$) humidified air (containing 5% CO_2) bathes them above and below. The mini-incubator is fixed to a specially machined clamp on the scanning stage of an Olympus inverted microscope, and positioned by X and Y stepping motors each of which has a 10 μm step size. The X, Y positions are programmed; and although we currently use a dedicated controller for this it will soon be replaced by a more versatile microcomputer as shown in Fig. 1. Because we wish to record images over the course of the evolution of a typical macrocolony from a single colony-forming cell, photography is performed at very low primary magnification, namely, 4X. Because it is exceedingly difficult to detect living unstained cells at this magnification using standard brightfield optics, we have used the Hoffman modulation contrast system (Hoffman 1977). Photographic exposure control is provided by commercially available automatic exposure equipment (Olympus), which controls the microscope lamp, a variable shutter, and the rate of film advance. The film advance is coupled to a Bolex 16 mm cine camera that is equipped with a 400-ft film magazine.

To do a typical experiment, 2 Lab-Tek slide chambers containing appropriate numbers of cells are sealed into the mini-incubator on the stage of the microscope. At least 6 hours later, an observer sits at the microscope, scans the cultures using the 4X objective lens, and centers within each of up to 99 fields a single presumptive clonogenic cell. The coordinates of each such field position are programmed in the controller, which can regulate the incubator movement from field to field at any desired frequency. After the field coordinates have been programmed, photography is begun. The master controller that we have used for this system is a Z-80 microcomputer (Northstar Horizon). Starting with the first field, the controller turns on the microscope lamp, the correct exposure is determined by the exposure meter, and a variable speed shutter

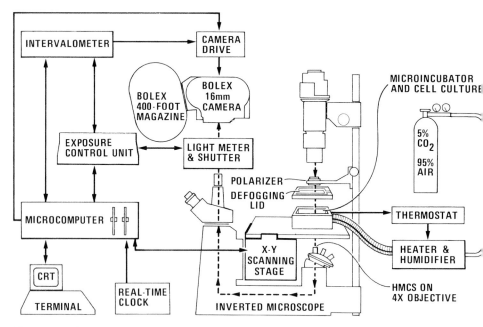

Fig. 1. Setup for photographing multiple microscope fields containing single mouse tumor cells as they grow *in vitro* to become macroscopic colonies. The Hoffman Modulation Contrast System (HMCS) consists of a modulator on the 4X objective lens, a polarizer between the object and the condenser, and a slit combined with another polarizer at the front focal plane of the condenser

(Olympus) is opened and closed. A signal is sent to the controller which then causes the film transport motor to advance the film in the Bolex camera to the next unexposed frame. Simultaneously, the controller directs the incubator to be advanced on the stage to the next preselected and programmed field. Completion of this step is confirmed to the controller which also receives a signal from the camera that the next unexposed frame is ready. The microscope lamp is turned on, the light intensity metered, and the shutter opened and closed proportionately. These steps are repeated as many times as there are fields. As each photographic exposure usually requires less than 1 s, the cells are allowed to grow in the dark for the vast majority of the time. The photography of 99 fields (plus a blank "field" that is used as a marker frame in the original film record) requires about 5–7 min. The controller then allows a predetermined time to elapse before starting the entire procedure again with the first field.

We have made several highly satisfactory films following this procedure and have used intervals of up to 30 min between successive exposures of the same field. The apparatus has run in a given experiment for as long as 8 days, utilizing about 32,000 successive frames (800 ft) of 16 mm film.

Although these films are technically satisfactory, the images are not in the sequence that allows them to be viewed as time-lapse movies. Our primary film consists of a series of images of different fields, then another series of images in the same order of the same fields, then another, etc. For this reason, it was necessary to devise an apparatus that would permit the images to be rearranged into time-lapse sequences. This

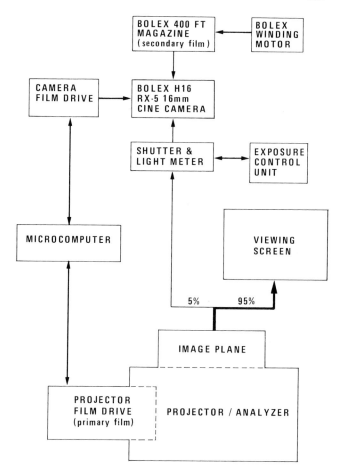

Fig. 2. Essential steps in the translation of primary 16 mm cinefilms obtained as in Fig. 1 into secondary time-lapse movies

was done by coupling a high-quality film projector with our microcomputer (Fig. 2). The projector will accept up to 1200 ft of 16 mm film, and has been adapted to accept our Bolex cinecamera with its 400-ft magazine. The primary developed film is loaded onto the projector and advanced to the first frame of the first field. The frame counter is then zeroed. Unexposed secondary film in the camera is also moved to its first frame. We have used the same Olympus shutter and exposure meter here for the secondary image as we use to record the initial primary image through the microscope. The shutter is made to open and close so that the first unexposed frame is exposed. While the secondary film is advanced to the next unexposed frame, primary film is advanced to the next frame that shows the same field. If, for example, 99 fields had been programmed and photographed (and the last field was followed by a blank frame as a kind of marker), the projector must advance the primary film 100 frames. In order that this be done within a reasonable length of time (hours to rephotograph a single experiment), the projector must be able to advance film significantly faster than most standard cineprojectors. The one we have used (made by Tagarno A/S, Horsens, Denmark) can advance film at up to 130 frames per second. Both the projector and the camera signal the computer when they have reached the desired frame, the computer then causes the next exposure to be made, and this is repeated over and over until a complete set of images of a given field has been rephotographed. Then the primary film is rewound under computer control to frame 2 of the first series to start the re-imaging of the next field. The procedure is repeated until all 99 fields have been rephotographed in the manner described above.

Development of this apparatus is still undergoing considerable refinement, so it is too early to attempt to report statistics on the growth of these cells. Ultimately, we propose to generate abortion probability factors that can be applied to the colony labeling indices produced with the aid of SACCAS. Then we will be able to offer a less flawed answer to the dosimetric problem that we set out to study at the very beginning.

Acknowledgements. This investigation was supported by Grants CAO3353 and CA10372, awarded by the National Cancer Institute, DHHS. Sincere thanks are due to Roger Furlong, Stavros Prionas and Nikolas Blevins for their valuable help in designing, constructing, interfacing and programming of the time-lapse equipment.

References

1. Barendsen GW, Roelse H, Hermens AF, Madhuizen HT, von Peperzeel HA, Rutgers DH (1973) Clonogenic capacity of proliferating and non-proliferating cells of a transplantable rat rhabdomyosarcoma in relation to its radiosensitivity. J Natl Cancer Inst 51:1521–1527
2. Barranco SC, Townsend CM, Costanzi JJ, May JT, Baltz R, O'Quinn AG, Leipsig B, Hokanson KM, Guseman LF, Boerwinkle WR (1982a) Use of 1,2:5,6-dianhydrogalactitol in studies on cell kinetics-directed chemotherapy schedules in human tumors *in vivo.* Cancer Res 42:2894–2898
3. Barranco SC, Townsend CM, Costanzi JJ, May JT, Baltz R, O'Quinn AG, Leipsig B, Hokanson KM, Guseman LF, Boerwinkle WR (1982a) Use of 1,2:5,6-dianhydrogalactol in studies on cell kinetics-directed chemotherapy schedules in human tumors *in vivo.* Cancer Res 42:2899–2905
4. Darzynkiewicz Z, Traganos F, Melamed MR (1980) New cell cycle compartments identified by multiparameter flow cytometry. Cytometry 1:98–108
5. Elkind MM, Whitmore GF (1967) The radiobiology of cultured mammalian cells. Gordon and Breach, New York
6. Hoffman R (1977) The modulation contrast microscope: Principles and performance. J Microsc 110, Part 3:205–222
7. Hurwitz C, Tolmach LJ (1969) Time lapse cinemicrographic studies of X-irradiated Hela S3 cells. I. Cell progression and cell disintegration. Biophys J 9:607–633
8. Kemper HL, Kallman RF (1983) SACCAS: A semi-automated counting machine with an accommodating threshold selector. Anal Quant Cytol 5:138–142
9. Kallman RF, Combs CA, Franko AJ, Furlong BM, Kelley SD, Kemper HL, Miller RF, Rapacchietta D, Schoenfeld D, Takahashi M (1980) Evidence for the recruitment of noncycling clonogenic tumor cells. In: Meyn RE and Withers HR (eds) Radiation Biology in Cancer Research, Raven Press, New York, pp 397–414
10. Rockwell SC, Kallman RF, and Fajardo LF (1972) Characteristics of a serially transplanted mouse mammary tumor and its tissue-culture-adapted derivative. J Natl Cancer Inst 49:735–749

11. Terasima T, Tolmach LJ (1963) Variations in several responses of HeLa cells to X-irradiaton during the division cycle. Biophys J 3:11−33
12. Thompson LH, Suit HD (1969) Proliferation kinetics of X-irradiated mouse L cells studied with time-lapse photography. II. Int J Radiat Biol 15:347−362

Flow Cytometric Quantification of Radiation Responses of Murine Peritoneal Cells*

N. Tokita and M.R. Raju

Life Sciences Division, Los Alamos National Laboratory, University of California, Los Alamos, NM 87545, USA

Summary

Methods have been developed to distinguish subpopulations of murine peritoneal cells, and these were applied to the measurement of early changes in peritoneal cells after irradiation. The ratio of the two major subpopulations in the peritoneal fluid, lymphocytes and macrophages, was measured rapidly by means of cell volume distribution analysis as well as by hypotonic propidium iodide (PI) staining. After irradiation, dose and time dependent changes were noted in the cell volume distributions: a rapid loss of peritoneal lymphocytes, and an increase in the mean cell volume of macrophages. The hypotonic PI staining characteristics of the peritoneal cells showed two or three distinctive G_1 peaks. The ratio of the areas of these peaks was also found to be dependent on the radiation dose and the time after irradiation. These results demonstrate that these two parameters may be used to monitor changes induced by irradiation (biological dosimetry), and to sort different peritoneal subpopulations.

Introduction

Quantification of radiation response is the basis for biological dosimetry. The conventional dosimeters used for low-LET irradiation have been peripheral lymphocyte counting and chromosome aberration enumeration. More recently, other dosimetric methods have also been reported, e.g., dosimetry based on spermatogenesis (Hacker et al. 1980) and on peripheral reticulocyte counting (Chaudhuri et al. 1979). The study of chromosome aberrations in peripheral lymphocytes provides an average dose estimate within a few days and has been used clinically (Purrot et al. 1972; Doloy et al. 1977).

The development of flow cytometry has made it possible to measure biological parameters rapidly and with a high degree of statistical accuracy. This technology has been applied to the field of biological dosimetry (Carrano et al. 1978; Hacker et al. 1980). In an attempt to shorten the assay period, i.e., obtaining results within 24 h after expo-

---

*This work was supported by Grant CA 22585 from the National Cancer Institute, U.S. DHEW and by the U.S. Department of Energy

sure, and to make the assay independent of circulating lymphocytes, we have measured the radiation responses of mouse peritoneal cells using flow cytometry. Two major sub-populations, lymphocytes and macrophages, exist in the peritoneal cavity of mice (Felix and Dalton 1955; Goodman 1964; Balner 1963). Several studies have shown that the number of peritoneal lymphocytes was reduced after irradiation, while the macrophage population remained relatively unchanged (Balner 1963; Bercovici and Graham 1964; Kornfeld and Greenman 1966). Because of cell size differences between lymphocytes and macrophages (Felix and Dalton 1955; Kornfeld and Greenman 1966) and our pre-liminary data indicating different propidium iodide staining characteristics of peritoneal subpopulations, the radiation response of murine peritoneal cells was measured in terms of cell volume distributions and DNA fluorescence staining intensity using flow cyto-metry with a view to develop a rapid biological dosimetry system.

Materials and Methods

Mice (strain CD-1, female, 8–12 weeks old, Charles River Laboratories) were used for this study. Six to ten mice were used per dose point. Whole body irradiation of mice was carried out with x rays (GE Maxitron, 250 KeV, 30 mA, 2 mm Cu HVL, a dose rate of 2 Gy/min) through a single portal. At various times after irradiation, 8 ml of alpha MEM medium (Grand Island Biological Company) containing 5 USP units per ml of sodium heparin and 190 mM morpholinopropanesulfonic acid (Sigma Chemical Company) at 37°C was injected intraperitoneally. Approximately 15 min after i.p. in-jection, the mice were sacrificed and the fluid removed from the peritonial cavity. The average number of cells obtained from unirradiated mice was $6-8 \times 10^6$ cells, while the total number of cells obtained from the peritoneal cavity of irradiated mice decrea-sed depending on the radiation dose and time after irradiation. After pipetting repeated-ly, aliquots of the cell suspension were used for cell counting using a hemocytometer or an electronic cell counter (Coulter Electronics). Cell volume distribution analyses were performed with the addition of a pulse height analyzer to the Coulter counter. The remaining cells were centrifuged at 300 g for 5 min. Part of the cells were used for histological staining; the remaining cells were fixed in 70% ethanol for 30 min at 4°C. The ethanol-fixed cells were stained with a hypotonic propidium iodide solution containing 50 μg/ml propidium iodide (Sigma) and 1 mg/ml sodium citrate in distilled water. This is a modification of the staining procedure of Crissman and Steinkamp (1973) and of Krishan (1975). The measurement of PI fluorescence as well as cell sor-ting were performed on flow cytometers available at Los Alamos using an argon-ion laser operating at 488 nm and the appropriate filters and detectors (Holm and Cram 1973; Steinkamp et al. 1973).

Results

The cell volume distributions of peritoneal cells measured 12 h after 0 to 11 Gy of x rays are shown in Fig. 1. Cells from each peak in the control were sorted onto slides and histological observations demonstrated that the small and large volume populations corresponded to lymphocytes and macrophages respectively. As the radiation dose was increased, the fraction of the cells in the first peak decreased progressively as a result of a rapid loss of peritoneal lymphocytes. The number of macrophages remained unchanged within 24 hs after irradiation.

Figure 2 shows the cell volume distributions of peritoneal cells measured 24 hours after 0 to 8 Gy of x rays. In addition to the changes in the relative areas of the two peaks, there is a shift of the second peak towards larger cell volumes at 5 and 8 Gy. A cytological examination revealed that the macrophages obtained from mice receiving 8 Gy appeared 20–30% larger than those from unirradiated mice. There were also an increased number of granulocytes in the peritoneal cavity of irradiated mice.

In order to quantitate the changes in peritoneal cell subpopulations after irradiation, the fraction of cells in the larger volume peak was measured by dividing the two populations at their inflection point and calculating the fraction of cells above the inflection point as a percent of the total population. Figure 3 shows the percent of the population contained in the large volume peak plotted as a function of radiation dose. The relative increase in the fraction of cells in the second peak was similar at 12 and 24 h after irradiation. However, the relative increase in second peak fraction was enhanced at 48 h.

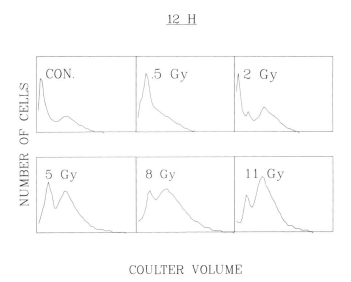

Fig. 1. Cell volume distributions of murine peritoneal cells obtained 12 h after irradiation. The cells comprizing the small and large volumes peaks in the control population correspond to lymphocytes and macrophages, respectively

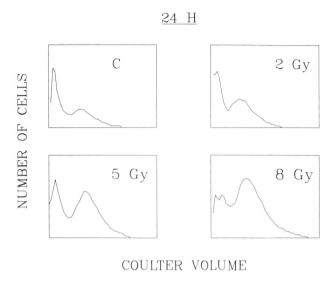

Fig. 2. Cell volume distributions of peritoneal cells obtained 24 h after 0 to 8 Gy x-irradiation. The cell volume distributions of cells taken after single doses of 5 and 8 Gy show an increase in the mean cell volume of macrophages

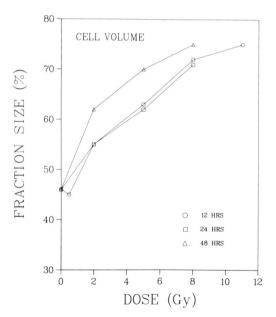

Fig. 3. The fraction of cells in the second peak of each cell volume distribution expressed as a percent of the total population plotted as a function of radiation dose

When ethanol-fixed peritoneal cells were stained with hypotonic PI solution without RNase, two or three peaks were observed in the G_1 fluorescence intensity region. Figure 4 shows the fluorescence intensity of the PI-stained peritoneal cells 24 h after irradiation. Both of the two prominent peaks in the control correspond to G_1 fluorescence intensity peaks. Cell sorting of the two populations onto slides followed by mic-

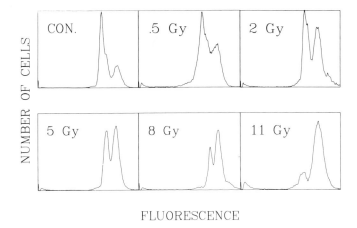

Fig. 4. Propidium iodide staining characteristics of G_1 peritoneal cells taken 24 h after irradiation

roscopic examination indicated that the cells in the lower fluorescence intensity peak appeared to be mostly lymphocytes along with some granulocytes, whereas the cells comprizing the higher fluorescence intensity peak were macrophages with some large lymphocytes. As the radiation dose was increased, the first peak area also decreased. Figure 5 shows the G_1 fluorescence intensity distributions of PI-stained cells collected 0 to 8 days after a single dose of 5 Gy. The relative second-peak area increased with longer times after irradiation.

The fraction of cells in the higher G_1 fluorescence intensity peak was estimated by the same method described for volume distributions and plotted as a function of radiation dose. Figure 6 shows the dose-response patterns measured at various times after irradiation. There is no significant change noted in the fraction of cells in the lower fluorescence intensity peak when the cells were sampled between 12 to 48 h after irradiation.

Discussion

There were two types of changes noted in the cell volume distributions of peritoneal cells after irradiation: (1) a shift in the ratio of small to large cells due to a rapid loss of lymphocytes, and (2) an increase in the mean cell volume of macrophages seen 24 h after irraditation. The rapid loss of peritoneal lymphocytes after irradiation has been well documented (Balner 1963; Bercovici and Graham 1963; Kornfeld and Greenman 1966). The increase in cell volume seen only after higher doses (Fig. 2) is significant. Geiger and Gallily (1973) observed numerous invaginations or "holes" on the outer surfaces of irradiated macrophages. It is not known whether the increase in size observed in our study is an artifact caused by suspending the cells in medium. Metcalf et al. (1977) reported that mouse bone marrow progenitors of granulocytes and macro-

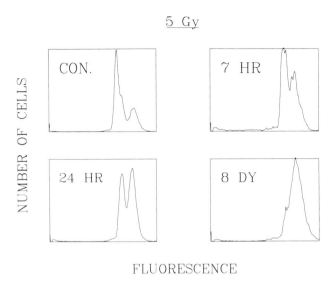

Fig. 5. Propidium iodide staining characteristics of peritoneal cells taken 0 to 8 days after a single dose of 5 Gy

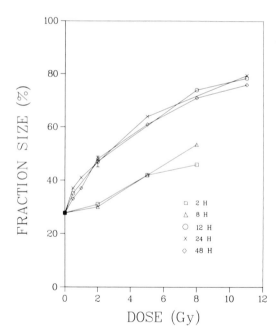

Fig. 6. The fraction of cells in the high fluorescence intensity peak as a percent of the total population plotted as a function of dose

phages increased in size 2 days after 2.5 Gy whole body irradiation and that all subpopulations were affected. In our study, however, the size increase appeared limited to the macrophage population.

The multimodal G_1 fluorescence intensity observed after PI staining of peritoneal cells indicates that staining characteristics differ for various types of cells even though the cells had been fixed in ethanol prior to staining. When the cells obtained at 0–8 days after irradiation (0 to 11 Gy) were stained with the method described by Krishan (1975), in which RNase was also added, no multimodal G_1 fluorescence intensity peaks were observed. The CV value for the G_1 peak was rather high (4–5%), however, and no cells were found with S-phase DNA contents (data not shown). Since the PI solution used in this study stains double-stranded RNA in addition to DNA (Crissman and Steinkamp 1973), the differences in double-stranded RNA content between the two subpopulations might have contributed to the separation of the control G_1 cells into distinct fluorescence intensity peaks. The changes in the relative size of the multimodal G_1 peaks of irradiated peritoneal cells appear to have been caused primarily by alterations in the cell populations. Geiger and Gallily (1974) reported that irradiation of macrophage donors caused activation of several macrophage functions, including a six-fold increase in RNA synthesis. It is yet to be determined if such an increase in RNA synthesis in irradiated macrophages has also affected the multimodal G_1 fluorescence intensity pattern seen in this study.

The results demonstrate that cell volume distributions as well as hypotonic PI staining may be used to monitor changes induced by irradiation (biological dosimetry), and to sort different peritoneal subpopulations.

Acknowledgements. The authors thank Mark Wilder for his assistance in cell sorting studies and Elvira Bain for computer graphics. We are also indebted to James P. Freyer, James H. Jett and Anita Stevenson for their comments and advice.

References

1. Balner H (1963) Identification of peritoneal macrophages in mouse radiation chimera. Transplantation 1:217–223
2. Bercovici B, Graham RM (1964) The immediate effects of local radiation on the cells in the peritoneal cavity of mice. Radiat Res 7:129–138
3. Crissman HA, Steinkamp JA (1973) Rapid, simultaneous measurement of DNA, protein and cell volume in single cells for large mammalian cell populations. J Cell Biol 59:766–771
4. Chaudhuri JP, Metzger E, Messerschmit O (1979) Peripheral reticulocyte count as biologic dosimetry of ionizing radiation. Acta Radiol Oncol 18:155–160
5. Doloy MT, LE GO R, Ducatez G, Lepetit J, Bourguignon M (1977) Utilisation des analyses chromosomiques pour l'estimation d'une d'irradiation accidentalle chez l'homme. Proc 4th Int Congr IRPA, Paris, 4:1199–1204
6. Felix MD, Dalton AJ (1955) A phase-contrast microscope study of free cells native to the peritoneal fluid of DBA/2 mice. J Nat Cancer Inst 16:415–428
7. Geiger B, Gallily R (1974) Effect of x-irradiation on various functions of murine macrophages. Clin exp Immunol 16:643–655
8. Geiger B, Gallily R (1974) Surface morphology of irradiated macrophages. J Reticuloendothel Soc 15:274–281
9. Goodman JW (1964) On the origin of peritoneal fluid cells. Blood 23:18–26

10. Hacker U, Schumann J, Göhde W (1980) Effects of acute gamma-irradiation on spermatogenesis as revealed by flow cytometry. Acta Radiol Oncol 19:361–368
11. Holm DM, Cram LS (1973) An improved flow microfluorometer for rapid measurement of cell fluorescence. Exp Cell Res 80:105–109
12. Kornfeld L, Greenman V (1966) Effects of total-body X-irradition on peritoenal cells of mice. Radiat Res 29:433–444
13. Krishan A (1975) Rapid flow cytofluorometric analysis of mammalian cell cycle by propidium iodide staining. J Cell Biol 66:188–193
14. Purrot RJ, Lloyd DC (1972) The study of chromosome aberration yield in human lymphocytes as an indicator of radiation dose. I. Techniques, Rep. NRPB-R2
15. Steinkamp JA, Fulwyler MJ, Coulter JR, Hiebert RD, Horney JL, Mullaney PF (1973) A new multiparameter separator for microscopic particles and biological cells. Rev Sci Instr 44:1301–1310

Cell Cycle Kinetics of Synchronized EAT-Cells After Irradiation in Various Phases of the Cell Cycle Studied by DNA Distribution Analysis in Combination With Two-Parameter Flow Cytometry Using Acridine Orange

Michael Nüsse

Gesellschaft für Strahlen- und Umweltforschung, Abteilung für Biophysikalische Strahlenforschung, Paul-Ehrlich-Straße 20, D-6000 Frankfurt/Main, FRG

Summary

Synchronized Ehrlich ascites tumour cells were irradiated with x-rays in various phases of the cell cycle (G1, G1/S, S, G2 and M-phase) and the progression through the remainder of the cell cycle and the following cell cycle was studied by DNA distribution analysis. Radiation-induced delays were observed in S- and G2-phase during the cell cycle. After the first division of irradiated cells, DNA distributions showed a broadening of the G1-peak which complicated evaluation of DNA histograms. Synchronized cells irradiated in various phases of the cell cycle were used to study the dose dependent broadening of the G1-peak after division. A linear increase of the coefficient of variation was observed as function of radiation dose (1–9 Gy).

   Using two-parameter flow cytometry after acridine orange staining, the fraction of cells in mitosis after irradiation could also be measured. Combining this technique with DNA distribution analysis it could be shown that only a proportion of a population of synchronized cells irradiated in M-phase underwent division. A dose dependent fraction of these cells did not divide but entered a polyploid state. Cells irradiated in G1-, G1/S- or S-phase also showed a dose dependent polyploidization but to a lesser extent. It is concluded that the high radiation sensitivity in M-phase may result from radiation induced defects in the division mechanism, whereas the high radiation sensitivity in G1/S-phase is caused by fixation of radiation lesions in S-phase prior to repair.

1 Introduction

DNA distribution analysis with one-parameter flow cytometric techniques is a widely used method for studying the perturbation of cell cycle progrssion of mammalian cells growing *in vitro* or *in vivo* cause d by different chemical or physical agents. Fractions of cells in G1-, S- and G2+M-phase can be measured from DNA distributions using suitable evaluation methods (e.g. Fried et al. 1980). Recently a two-parameter technique has been developed which allowed us to measure the fraction of cells in mitosis,

Biological Dosimetry. Edited by W. G. Eisert and M. L. Mendelsohn
© Springer-Verlag Berlin Heidelberg 1984

thus making it possible to discriminate between cells in G2- and M-phase (Darzynkiewicz et al. 1977). A combination of DNA distribution analysis with this technique can therefore extend the information on cell cycle kinetics in perturbed mammalian cell populations.

Irradiation of mammalian cells with ionizing or non-ionizing radiation causes delays in various phases of the cell cycle (e.g. Lindmo and Pettersen 1979, Nüsse 1981, Iliakis and Nüsse 1982a). The study of these delays can be important in understanding how the survival of cells is altered after these treatments, and how cells can counteract the effects of irradiation, e.g. by repair of potentially lethal damage (Iliakis and Nüsse 1982 a–c).

The aim of this work was to analyse the perturbation of cell cycle progression after x-ray irradiation in synchronized Ehrlich ascites tumour cells growing in vitro. DNA distribution analysis in combination with two-parameter flow cytometry to measure the fraction of cells in mitosis was used. Problems occurring in the analysis of DNA distributions of irradiated cells will be discussed. These include the dose dependent broadening of the G1-peak and the occurrence of non-dividing cells after irradiation both of which could be important for biological dosimetry. The kinetic behaviour of cells irradiated in various phases of the cell cycle will be discussed in relation to the change in survival (radiation sensitivity).

2 Materials and Methods

A strain of Ehrlich ascites tumour cells (EAT-cells) growing in *vitro* in suspension culture (Karzel 1965) was used. Details of growth conditions and growth characteristics of these cells have been reported previously (Iliakis and Pohlit 1979, Nüsse 1982). The cells were kept in a quasi-continuous culture by daily reduction of the cell concentration from 8×10^5 cells/ml to 2×10^5 cells/ml in fresh nutrient medium (A2-medium supplemented with 20% horse serum).

The cells were synchronized in G1-, G1/S- or S-phase by a method combining physical selection of G1-cells from an asynchronous population and additional G1/S block using excess thymidine (2mM). Details of this method are described elsewhere (Nüsse 1982). Cells synchronized in M-phase were obtained using an additional short block in mitosis (3 h) of presynchronized G1/S-cells by the reversible microtubule inhibitor nocodazole (0.4 μg/ml, EGA-Chemie, Germany). Details of this method are also described elsewhere (Nüsse and Egner, 1982, submitted for publication).

DNA distributions were measured in an ICP 21 (Phywe, Germany, now Ortho Instruments, Westwood, MA) after staining with Hoechst 33258 (Nüsse 1981) or in a Cytofluorograf 30L (Ortho Instruments) after staining with ethidium bromide. In the latter case the cells were centrifuged out of the growth medium and staining solution containing 10 mg/l ethidium bromide, 10 mg/l RNAse, 1000 mg/l Na-citrate, 584 mg/l NaCl and 0.3 ml/l Nonided P40, was added to the cell pellet. The coefficient of variation, CV, of the G1-peak was better than 3% with both methods. Fractions of cells in G1-, S- and G2+M-phase were evaluated from the area under the histograms assuming Gaussian distributions under the G1- and the G2+M-peak and attributing the remaining cells to the S-phase. This simple method gave results which were both accurate and reproducible to within $\pm$ 5%.

To determine the proportions of mitotic cells and interphase cells separately, a modification of the method described by Darzynkiewicz et al. (1977) was used. This method is based on the different stainability of double versus single stranded DNA to acridine orange (AO) in cells depleted of RNA. AO binds in its monomeric form to double stranded DNA by intercalation and fluoresces green (530 nm), whereas stacking of AO on single stranded DNA results in red fluorescence (630 nm). When fixed cells are acid treated (pH 1.4) and are stained at pH 2.6 with AO the DNA of mitotic cells shows a lower green and a higher red fluorescence by comparison with interphase cells thus making it possible to discriminate between M-phase and interphase cells. Cells (ca. $2x$ 10^6 cells) were rinsed in PBS+Mg^{2+} buffer (phosphate buffer 10^{-3} mol/l pH 7.4, NaCl 0.145 mol/l, MgCl$_2$ $2x10^{-3}$ mol/l), incubated in 1 ml of this buffer and fixed in 10 ml of 80% ethanol. It was found that the fixed cells could be stored for several months at 4°C. After fixation, cells were centrifuged and incubated in 0.5$-$1 ml of 40% ethanol containing 20 mg/l RNAse (1 h at 37^{o}C). Afterwards RNAse was mixed with 0.5 ml KCL-HCl buffer (0.2 mol/l, pH 1.4). After 30 sec 2 ml AO solution was added to stain the cells (10 μg/l AO in Na$_2$ HPO$_4$ (0.2 mol/l)-citric acid (0.2 mol/l) buffer (pH 2.6)). Stained cells were run in the cytofluorograf and red ($\rangle$ 610 nm) and green (530 nm) fluorescence measured simultaneously at 90° to the direction of the laser beam (argon laser, 50 mW).

Figure 1 shows a dot-plot of red and green fluorescence of asynchronous cells which were blocked in mitosis by a nocodazole treatment of 6 hours duration. Mitotic (M) and interphase (G1, S, G2) cells can be distinguished clearly. By proper window setting it was possible to measure the fraction of cells in mitosis.

Cells were irradiated in suspension with 150 kV x-rays (0.8 mm A1) having an effective photon energy of about 20 keV with respect to the photon attenuation coefficient in water. Absorbed dose was measured with an energy independent ferrous sulphate dosemeter (Frankenberg 1969). The absorbed dose was 10 Gy/min.

3 Results

3.1 Cell Cycle Kinetics of Irradiated Synchronized Cells: Radiation Induced Broadening of G1-peak

The cell cycle kinetics of synchronized cells irradiated in late S-phase with various doses (D= 0, 3, 6, 9 Gy) is shown in Fig. 2. Fractions of cells in G1-, S- and G2-phase measured by DNA distribution analysis and in M-phase measured by two-parameter flow cytometry using AO (closed symbols) or microscopic observation (open symbols) are shown as a function of time. Irradiation was performed at t= 3 h when synchronized G1/S-cells (t= 0) were in the middle of S-phase. The effect of X-irradiation on cell cycle progression can be seen clearly: a dose dependent S-phase delay ($\triangle$tS) and the radiation induced G2-block ($\triangle$tG2). Both delays, which depend on the point in the

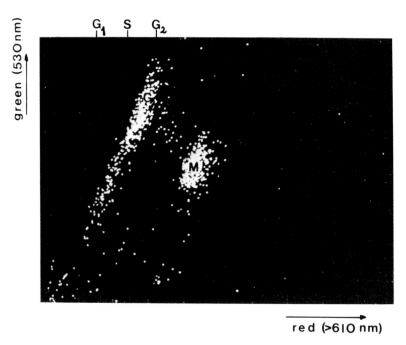

Fig. 1. Two parametric histogram showing interphase cells (G1, S, G2) and mitotic cells (M) after staining the cells with acridine orange. Exponentially growing cells were blocked in mitosis by a 6 h treatment with nocodazole (0.4 µg/ml)

cell cycle at which the cells were irradiated, have been described in detail elsewhere (Nüsse 1981) and will therefore not be discussed further. The fraction of cells in mitosis measured by microscopic observation and AO staining (lower panel of Fig. 2) shows the exit of cells from G2-block into the following M-phase. Results for unirradiated and irradiated cells are shown for parallel measurements made using the two methods. The results were similar showing the applicability of the flow cytometric technique for studying the entry of cells into, and the exit of the cells out of mitosis into the following G1-phase. The radiation-induced division delay can best be measured by the delayed increase of mitotic index but can also be measured by the delayed increase in G1-cells. Qualitatively similar results were obtained when cells had been irradiated in G1-, G1/S or G2-phase (data not shown, see Nüsse 1981). The study of cell cycle kinetics after irradiation by DNA distribution analysis can thus be extended using two-parameter flow cytometry with AO staining to measure the fraction of cells in mitosis. This was found especially useful when cells were irradiated in mitosis (see below).

The progression of cells through the FIRST cell cycle after irradiation can easily be measured by analysing DNA distributions. Problems occur however when DNA distributions after cell division (SECOND cell cycle) have to be evaluated. This is often the case when irradiated asynchronous cells (*in vitro* or *in vivo*) are analysed, because some cells divide after irradiation whereas other cells are still in the first cell cycle. Experiments with synchronized cells show that after the first cell division following irradiation,

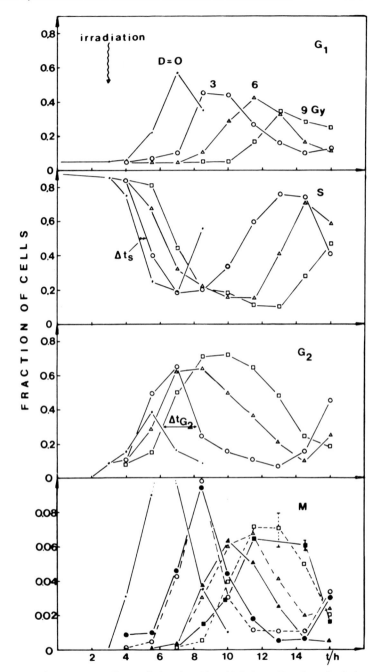

Fig. 2. Cell cycle kinetics of synchronized EAT-cells irradiated in mid S-phase. Fraction of cells in G1-, S- and G2-phase as function of time measured by DNA distribution analysis and in M-phase (*bottom panel*) measured by two-parameter analysis of mitotic cells using AO (*closed symbols*) or by microscopic observation *(open symbols)*. D = 0 (·), 6 Gy (△) and 9 Gy (□). Δt_S: progression delay in S-phase, Δt_{G2}: mean duration of G2-block including S-phase progression delay

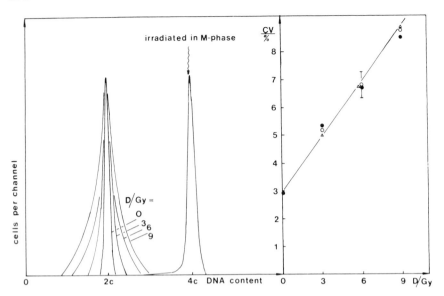

Fig. 3a,b. *Left panel (a):* Schematic diagram showing broadening of G1-peak of cells which have divided after irradiation in M-phase. The non-dividing cells are omitted from the diagram. *Right panel (b):* CV as function of dose of G1-peak of cells which have divided after irradiation in different phases of the cell cycle. (△) irradiated in G1/S-phase, (o) in mid S-phase and (●) in M-phase

the width of the G1-peak broadened. This is demonstrated in Fig. 3. Cells irradiated in M-phase with D=0, 3, 6, and 9 Gy (see below) begin to divide about 1 hour after irradiation independently of the applied dose. The G1-peak of divided cells shows a dose dependent broadening (Fig. 3a). The coefficient of variation (CV) of the G1-peak as function of dose is shown in Fig. 3b. A linear dependency of the CV as function of dose was observed. Cells irradiated in other phases of the cell cycle show the same broadening of G1-phase after division and the same dependency of CV on dose as cells irradiated in M-phase.

Thus an evaluation of DNA histograms to measure the fraction of cells in the cell cycle phases after irradiation should take into consideration this dose dependent broadening of the G1-peak of cells after division. This has been done in the present experiments with synchronized cells. Problems will always occur when asynchronous cells or cells *in vivo* have been irradiated because at later times after irradiation G1 cells which have already divided will overlap with G1 cells which have not yet divided. Unless this effect is corrected for the fraction of cells in G1-phase will probably be underestimated. A similar problem will occur with S-cells and G2-cells in the second cell cycle after irradiation.

3.2 Cell Cycle Kinetics of Irradiated Synchronized Cells: Consideration of Non-Dividing Cells

Besides the problem of broadening of the G1-peak after irradiation, we have observed a second effect in cells after irradiation (Nüsse 1981) namely that some cells progress through the first cycle but instead of dividing become polyploid. This effect occurs mainly in cells irradiated in M-phase. Figure 4 shows the cell cycle kinetics of cells irradiated in M-phase as measured by DNA distribution analysis using two-parameter flow cytometry and AO staining to show the fraction of cells in mitosis. For comparison are shown microscopic counts of mitotic cells and polyploid cells (cells with two or more nuclei). The irradiated cell population contained about 85% M-, 5% S- and 5% G2-cells. One hour after irradiation and incubation the cells began to divide. After 2 h most of the control cells had divided and were found in G1-phase. (Data on the fraction of M-phase cells from AO staining and microscopic observation agreed very well, but are not presented in Fig. 4 for the sake of clarity). Surprisingly, a dose-dependent fraction of irradiated M-cells did not divide but obviously entered a polyploid state in which the DNA content increased above that of M-phase cells. The fraction of these polyploid cells (PP-cells) has been measured by analysing DNA distributions, defining PP-cells as those with DNA content higher than the Gaussian distribution. The fraction of polyploid cells was simultaneously measured by microscopic observation. The non-dividing cells can also be registered in the two-parameter histograms after AO staining as cells with a higher red and green fluorescence than normal interphase cells but lying on the same axis as these and outside the area were mitotic cells are found. Figure 4, lower panel, shows the results. The first polyploid cells appeared three hours after irradiation, their number stayed constant during the following 10 h, decreasing at later times when cells having undergone division multiply again. The data from DNA distribution analysis and from microscopic observations again agreed very well within the experimental errors. Because of the large fraction of polyploid cells, especially after higher doses (6 and 9 Gy), the fraction of cells in G1- and S-phases after division was low (upper panels in Fig. 5). A small, dose dependent delay in G1-phase can be observed after irradiation in the preceding M-phase.

Figure 5 shows the fraction of nondividing polyploid cells as function of dose for cells irradiated in M-phase and also for cells irradiated in G1-, G1/S- and S-phase measured by the same techniques. The fraction of non-dividing cells arising after the first mitosis was much less in cells irradiated in the other phases of the cell cycle compared to that of cells irradiated in M-phase.

4 Discussion

Perturbation of cell cycle progression by irradiation not only changes the distribution of cells in the various cycle phases but leads to additional effects on the cells which change the general shape of the DNA distributions as they occur for unirradiated cells. The first effect is a broadening of G1-peak of irradiated cells after the first division.

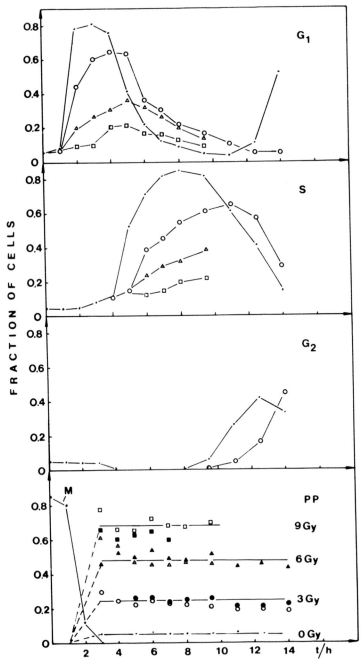

Fig. 4. Cell cycle kinetics of cells irradiated in M-phase. Fractions of cells in G1-, S- and G2-phase after irradiation measured by DNA distribution analysis and in M-phase measured by two-parameter analysis after AO staining or microscopic observation. (no differences between the methods were observed, therefore only the flow cytometric data are shown (·). The exit of cells after irradiation from mitosis was similar to the control data for all doses.) Fraction of polyploid (PP) cells measured by DNA distribution analysis counting all cells with DNA content higher than that of G2+M-phase (*open symbols*) and by microscopic observation *(closed symbols).* D=0 (·), 3 Gy (o), 6 Gy (△) and 9 Gy (□)

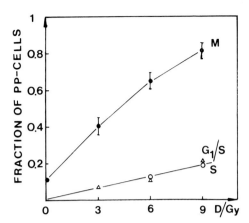

Fig. 5. Fraction of nondividing polyploid (PP) cells (number of PP cells during the cell cycle following the cell cycle the irradiation has been applied divided by the number of cells at time of irradiation) as function of dose applied in G1 or G1/S-phase (△), in mid S-phase (o) or in M-phase (●)

This broadening was found to be dose dependent and independent of the point in the cell cycle at which the cells have been irradiated (Fig. 3). A broadening of G1-peak after irradiation was observed earlier (Otto and Oldiges, 1980, Pinkel et al. 1980, Nüsse 1981). The reason for this increase in CV of the G1-peak could be a dose dependent variability of stain uptake or fluorescence properties or even a radiation-induced variability in DNA content. The first possibility seems not to be the case since staining of our cells with different dyes (Hoechst 33258 and ethidium bromide) or measuring DNA histograms with different instruments (in this case with ICP 21 and Cytofluorograf) led to qualitatively and quantitatively similar results (data not shown). The second possibility, a radiation induced variability of DNA content, was favoured by Otto and Oldiges (1980) and is supported by microscopic observations of irradiated cells after divison. A dose dependent number of micronuclei were observed in irradiated cells after division, originating from chromosome damage and unequal division of chromosomes (Midander and Revesz 1980, Molls et al. 1981). It can therefore be concluded that the reason for the increased CV in irradiated cells after division is the increased variability of DNA content originating from radiation induced chromosome damage (e.g. dicentrics and acentric fragments). With careful measurement of DNA distributions, especially with a low a CV of G1 peak as possible and in addition taking into consideration the cell proliferation it might be possible to perform biological dosimetry of irradiated cells *in vitro* or *in vivo*. The estimation of micronuclei in irradiated cells by flow cytometric determination of DNA histograms for a biological dosimetry will be a future goal of our investigations.

The second kinetic effect after irradiation is the occurence of non-dividing cells in the population. This results in a gradual increase in the cell fraction with a DNA content higher than that of G2+M-phase. These nondividing cells produced mainly after irradiation in M-phase and which later become polyploid cells result from damage to their division mechani· n, but most of these cells are able to progress through a subsequent cell cycle without undergoing division. Polyploid cells occur also after irradiation in the other phases of the cell cycle but to a much lesser extent (Fig. 5). This effect of irradiation on cell cycle kinetics could also be of importance for biological dosimetry with cells *in vitro* or *in vivo*. The fraction of cells with a DNA content higher than that of G2+M-cells, probably polyploid cells, could be an indicator of the dose received.

In this connection, the variation of radiation sensitivity during the cell cycle is of interest. EAT-cells, like most other mammalian cell strains, show a high radiation sensitivity at the G1/S border and in M-phase; G1-, S- and early G2-phase being much less radiation sensitive. Exponential survival curves can be obtained with G1/S- and M-cells (Do = 0.8 Gy) whereas survival curves of cells irradiated in other cell cycle phases show a marked 'shoulder'. Recently it has been shown that the occurence of a shoulder on the survival curve at the various stages of the cell cycle is caused by repair of potentially lethal damage (PLD) (Iliakis and Nüsse, 1982c, submitted for publication). From the results presented here using cell kinetic data it is shown that, although G1/S- and M-cells have the same exponential survival curves, the kinetic behaviour of cells irradiated in these sensitive phases is quite different. The reason for this high sensitivity of M-cells after irradiation is probably a defect caused in the division mechanism, whereas the high sensitivity of G1/S-border cells is probably connected with the onset of DNA synthesis by which radiation induced lesions are fixed before they can be repaired (Iliakis and Nüsse, 1982c, submitted for publication). A study of repair kinetics during the cell cycle in connection with cell cycle kinetic data is currently under way and may help to elucidate the differences between the types of damage in different phases of the cell cycle and their mode of repair.

5 References

Darzynkiewicz Z, Traganos F, Sharpless T, Melamed MR (1977) Recognition of cells in mitosis by flow cytometry. J Histochem Cytochem 25:875

Frankenberg D (1969) A ferrous sulphate dosimeter independent of photon energy in the range from 25 keV up to 50 MeV. Phys Med Biol 14:597

Fried J, Perez AG, Clarkson B (1980) quantitative analysis of cell cycle progression of synchronous cells by flow cytometry. Exptl Cell Res 126:63

Iliakis G, Pohlit W (1979) Quantitative aspects of repair of potentially lethal damage in mammalian cells. Int J Radiat Biol 26:321

Iliakis G, Nüsse M (1982a) Evidence that progression of cells into S-phase is not a prerequisite for recovery between split doses of U.V.-light in synchronized and plateau phase cultures of Ehrlich ascites tumour cells. Int J Radiat Biol 41:503

Iliakis G, Nüsse M (1982b) Conditions supporting repair of potentially lethal damage cause a significant reduction of ultraviolet light-induced division delay in synchronized and plateau phase Ehrlich ascites tumour cells. Radiat Res 91:483

Iliakis G, Nüsse M (1982c) Evidence that repair and expression of potentially lethal damage cause the variations in cell survival after x-irradiation observed through the cell cycle in Ehrlich ascites tumour cells. Int J Radiat Biol: in print

Karzel K (1965) Über einen in vitro in Suspension wachsenden permanenten Stamm von Ehrlich-Ascites-Tumorzellen. Med Pharm Exp 12:137

Lindmo T, Pettersen EO (1979) Delay of cell cycle progression after X-irradiation of synchronized populations of human cells (NHIK 3025) in culture. Cell Tissue Kinet 12:43

Midander J, Revecz L (1980) The frequency of micronuclei as a measure of cell survival in irradiated cell populations. Int J Radiat Biol 38:237

Molls M, Streffer C, Zamboglou N (1981) Micronucleus formation in preimplanted mouse embryos cultured in vitro after irradiation with x-rays and neutrons. Int J Radiat Biol 39:307

Nüsse M (1981) Cell cycle kinetics of irradiated synchronous and asynchronous tumor cells with DNA distribution analysis and BrdUrd-Hoechst 33258-technique. Cytometry 2:70

Nüsse M (1982) Synchronization of mammalian cells by selection and additional chemical block studied by DNA distribution analysis and BrdUrd-Hoechst 33258-technique. Cell Tissue Kinet 15:529

Nüsse M, Egner HJ (1982) Can nocodazole, an inhibitor of microtubule formation, be used to synchronize mammalian cells? Accumulation of cells in mitosis studied by two parametric flow cytometry using acridine orange and by DNA distribution analysis. Cell Tissue Kinet: submitted for publication

Otto FJ, Oldiges H (1980) Flow cytometric studies in chromosomes and whole cells for the detection of clastogenic effects. Cytometry 1:13

Pinkel D, Gledhill BL, Lake S, van Dilla MA, Wyrobek AJ (1980) Increased fluorescence variability in sperm from mice exposed to x-rays: is it due to induced DNA content variability? Bas Appl Histochem 24:350

Nuclear Morphology of Hepatocytes in Rats After Application of Polychlorinated Biphenyls

W. Abmayr[1], D. Oesterle[2], and E. Deml[2]

[1]Institut für Strahlenschutz
[2]Inst. f. Toxikol. u. Biochemie, Abteilung für Toxikologie, Gesellschaft für Strahlen- und
  Umweltforschung mbH München, D-8042 Neuherberg, FRG

Summary

The acute effect of Clophen A50, a mixture of polychlorinated biphenyls on nuclear morphology of hepatocytes in rats was studied. Adult female rats were treated p.o. with a single dose of 100 or 500 mg Clophen A 50/kg b.wt., and killed four days later. Sections of the liver were stained with the Feulgen reaction for DNA, scanned with a TV-camera and evaluated in a computer using multivariate analysis methods. Only tetraploid cells were investigated.

Significant changes in chromatin distribution pattern were found after the application of Clophen A50. The entropy of the nuclear optical density increased dependent on the dose, indicating an increase in chromatin clumping. Nuclear area was not affected.

Introduction

Polychlorinated biphenyls are ubiquitous and persistent environmental pollutants. The acute and chronic toxicity has been studied extensively. A weak carcinogenic as well as a potent promoting activity in rodent hepatocarcinogenesis has been demonstrated [1].

In a previous study using multivariate analysis we investigated the alterations in nuclear morphology of unaltered compared to preneoplastic cells in rat liver 40 weeks after application of diethylnitrosamine when Clophen A50 was applied as a promoting agent [2]. We demonstrated that TV-scanning and multivariate analysis, as developed for the analysis of single cells, can be applied successfully to the investigation of nuclear morphology in tissue sections [3, 4].

In the present experiment we studied the acute effect of two different doses of Clophen A 50 on nuclear morphology of hepatocytes in rats, in order to determine single features and feature combinations which show dose-dependent alterations, and to discriminate between nuclei in hepatocytes of untreated and treated animals. The method used may be helpful to quantify small changes in nuclear morphology which are not detectable by visual observation.

Material and Methods

Female Sprague-Dawley rats 6 weeks of age were given a single dose of 100 or 500 mg Clophen A 50/kg b.wt., dissolved in olive oil. Control rats received olive oil only. Four days later the rats were killed.

Pieces of the livers were fixed with buffered formaldehyde, dehydrated with ethanol and embedded in hydroxy-ethylmethacrylate. Two micron thick sections were stained with the Feulgen-reaction (60°C, 10 min hydrolysis 1 n HCl) for the demonstration of DNA.

For the analysis of nuclear differences two sections each were taken from three control rats, 3 rats treated with 100 mg Clophen A 50/kg b.wt. and 3 rats treated with 500 mg Clophen A 50/kg b.wt. From each section, 50 cells were measured randomly. Only intact tetraploid cells with a sharply delineated nuclear membrane and a clearly visible nucleolus were selected.

The section images of the liver cells were scanned in an Axiomat microscope (C..Zeiss, Oberkochen, West Germany) with a plumbicon camera (Robert Bosch, West Germany) interfaced to an Array Processor AP 120 B [2] (Floating Point Systems Inc., Portland, USA). The pixel distance was 0.25 micrometer (apochromat 100/1.32, oil immersion objective, 40/0.6 substage condensor). The pixel size was 0.25 micrometer. The measurements were performed at 550 nm wavelength with a bandwidth of 20 nm. Shading correction and automatic focusing were used for data acquisition. The Feulgen stained cells were automatically segmented using contrast enhancement by filtering and followed by thresholding applied to the contrast enhanced image. For feature extraction the BIP-program, a general purpose image processing system with a procedure interpreter was used [5].

For correction of differences in section thickness the mean optical density per unit area was measured, and image pixels (extinction values) have been shifted by adding or subtracting a constant value so that the mean optical density and the integrated optical density were standardized. Since the integrated optical density was measured in sections, the data do not represent total content of nuclear DNA.

41 features were investigated to detect differences of nuclear morphology between untreated rats and the rats treated with 100 and 500 mg/kg b.wt. Clophen A 50 (Table 1). The cell data were stored on magnetic tape. The statistical analysis of the cell features was performed with the BMDP software package [6].

Results

The variation in section thickness amounts up to 15% resulting in differences of the mean O.D. The mean-value distribution of the mean optical density in different sections is shown in Fig. 1a. The biased selection of the nuclei showed that their area remains constant independent from section thickness. Therefore the amount of DNA is constant after correction (Fig. 1b).

Table 1. Investigated features for nuclear morphology in rat hepatocytes

Index	Name	
1	MHISP 1	Relative frequency of occurence of
2	MHISP2	nuclear values <u>DN</u> in an image
3	MHISP3	N
.	.	Derived from the difference of the original image and
.	.	the median smoothed image
.	.	.
.	.	.
.	.	
18	MHISP 18	
19	MEDM1	Mean value of this image
20	MEDM2	Variance
2	MEDM3	Skewness
22	MEDM4	Excess
23	MEDMIN	Minimum value
24	MEDMAX	Maximum value
25	KEM1	Mean nuclear O.D.
26	KEM2	Variance
27	KEM3	Skewness
28	KEM4	Excess
29	KEFL	Nuclear area
30	KEMIN	Minimum value
31	KEMAX	Maximum value
32	TEE-4	Integrated optical density
33	KESPW	Difference KEMAX-KEMIN
34	KEVAR	C.V. of the nucleus
35	SEM	Standard error of the mean
36	ENTR	Entropy of the nuclear image
37	MEDSPW	Difference MEDMAX-MEDMIN
38	MEDVAR	C.V. ov the orig-med-image
39	MEDTS-4	Total sum
40	MEDSEM	Standard error of the mean
41	MEDENTR	Entropy of orig-med-image

In the two-group analysis cells from controls and the pooled cells from treated animals (100 and 500 mg/kg b.wt. Clophen A 50) were classified using the linear discriminant analysis program BMDP [6]. It ends up with 81.3% correct classification using 9 features and prior probability weighted according the number of cases of each group. The classification matrix, summary table for feature extraction and the histogram of the cannonical variable for the two-group case are shown in Fig. 2.

The most important feature is the entropy of the nuclear image (ENTR). This is an information measure which has the value of 0 if all image pixels are in 1 grey-level channel and the value of 8 if covered channels of the grey-level histogram are spread homogeneously over the whole grey-level range. ENTR is not influenced by thickness correction. The ENTR increases with the dose of Clophen A 50 indicating an increase of chromatin clumping (Fig. 3).

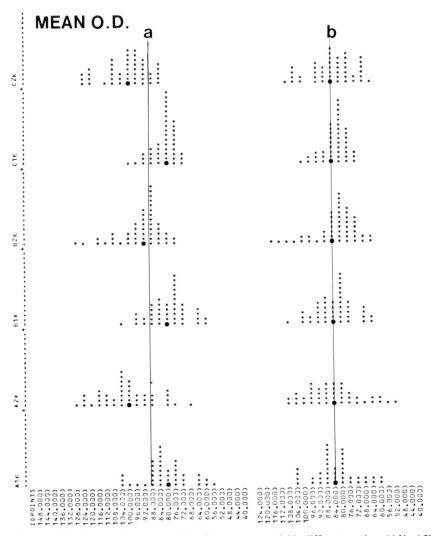

Fig. 1a,b. Distribution of the mean O.D. of hepatocyte nuclei in different sections (A1k, A2k, B1k, B2k, C1k, C2k). Two consecutive sections are from the same animal. **a** uncorrected section; **b** after correction for section thickness

MEDM3 represents the skewness of the O.D. of an image, derived from the difference of the median smoothed image and the original image (describing the chromatin structure) and increases with the dose of Clophen A 50 (Table 1). This indicates that the nuclear chromatin is more decondensed in cells after treatment with Clophen A 50 (Fig. 4).

In the three group analysis the following cells were used: from control animals, from animals treated with 100 mg/kg b.wt. Clophen A 50, and with 500 mg/kg b.wt. Clophen A 50. The rate of percent correct classification vs the number of features, included in the linear discriminant function ends up with about 72% correct classification using

A) CLASSIFICATION MATRIX

GROUP	PERCENT CORRECT	NUMBER OF CASES CLASSIFIED INTO GROUP —	
		CONTROL	TREATED
CONTROL	78.3	137	38
TREATED	83.2	46	228
TOTAL	81.3	183	266

B) SUMMARY TABLE FOR FEATURE SELECTION

STEP NUMBER	VARIABLE ENTERED	REMOVED	F VALUE TO ENTER OR REMOVE
1	36 ENTR		98.2506
2	21 MEDM3		58.0311
3	26 KEM2		34.6769
4	25 KEM1		26.7765
5	12 MHISP12		7.9460
6	11 MHISP11		4.3232
7	6 MHISP6		4.7125
8	13 MHISP13		4.2826
9	27 KEM3		6.4763

C) HISTOGRAM OF CANONICAL VARIABLE

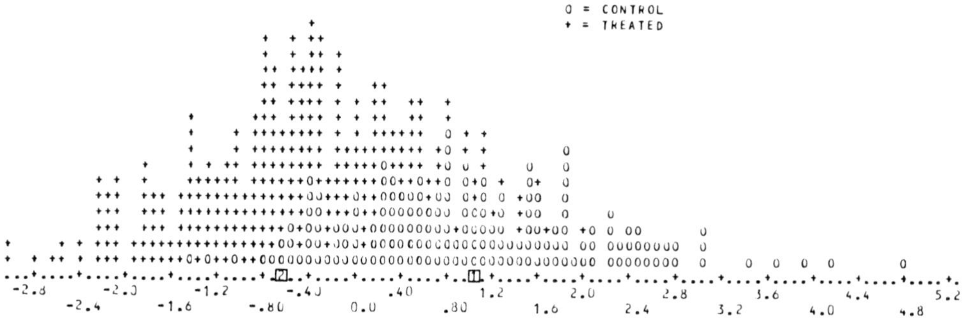

Fig. 2. Classification matrix, summary table for feature selection and cannonical variable for classification of cells from untreated (o) and Clophen A 50 treated (+) rats

12 features (Fig. 5). The features, included in the steps 5 to 12 do not improve the classification results significantly. The 5 most important features for discrimination in these three groups are the entropy of the nucleus (ENTR), the skewness of the O.D. of the nucleus (KEM3), the coefficient of variation of the nucleus (KEVAR), skewness of the O.D. of the difference image (MEDM3) and the maximal nuclear O.D. value (KEMAX). The nuclear area and the integrated optical density are not included in the discriminant function.

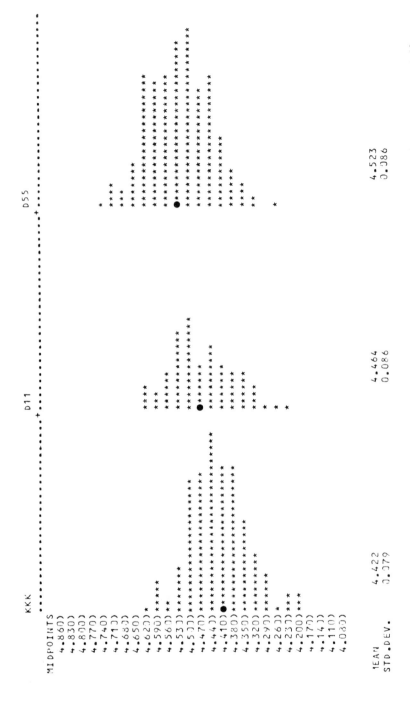

Fig. 3. Histogram of the entropy of the nuclear density (ENTR) for 3 groups: KKK = control, D11 = 100 mg/kg b.wt. Clophen A 50, D55 = 500 mg/kg b.wt. Clophen A 50. An increase of ENTR indicates an increase of chromatin clumping

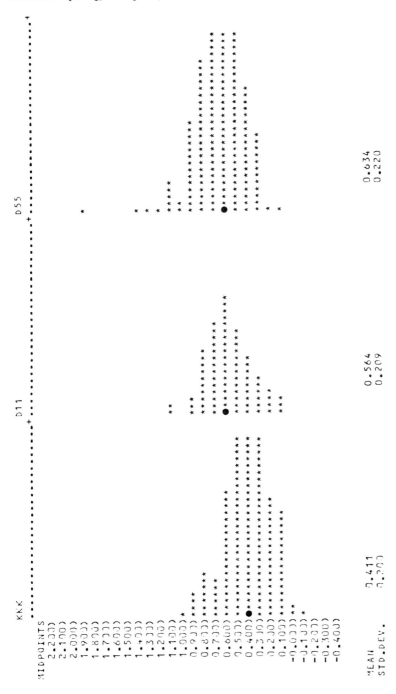

Fig. 4. Histogram of the skewness of the O.D. of the difference image (original minus median – smoothed image, MEDM3) for 3 groups: KKK = control, D11 = 100 mg/kg b.wt. Clophen A 50, D55 = 500 mg/kg b.wt. Clophen A 50. An increase of MED3 indicates an increase of chromatin decondensation

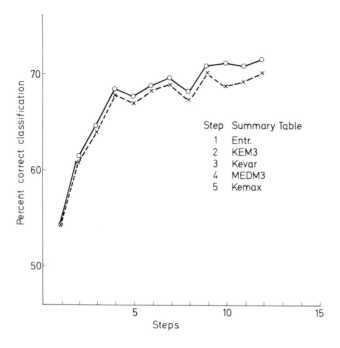

Fig. 5. Rate of percent correct classification vs the number of features included in the multivariate analysis. Three groups of cells were used: control, 100 mg/kg b.wt. Clophen A 50, 500 mg/kg b.wt. Clophen A 50. The steps 6 to 12 did not improve the results significantly. (o) indicates classification; (x) indicates cross-validation

Discussion

The preliminary data on the effect of PCBs (Clophen A 50) on nuclear morphology of hepatocytes show that cells with unaltered DNA content undergo statistically significant changes in their chromatin distribution pattern. The result of 81.3% correct classification into cells of untreated and treated animals could only be achieved using multivariate analysis methods. The entropy of the nucleus (ENTR) and the skewness of the O.D. of a difference image, describing the amount of condensed and decondensed chromatin (MEDM3) are the most important features. The nuclear area and the integrated optical density are not included in the discriminant function. Similar results are reported by Nair et al. in hepatocytes from Chlordane-treated rats [7].

Some textural features as ENTR and MEDM3 show a dose-dependent characteristic and indicate an increase in chromatin clumping and a decrease in the amount of condensated chromatin.

Two main problems arise from this type of section analysis:

1. The random selection of nuclei in hepatocytes is biased by the selection of intact tetraploid cells only, cells with a sharply delineated nuclear membrane and a clearly

visible nucleolus. This bias is necessary to avoid problems of section analysis dealing with the fact that all cells are cut at different levels (Holmes effect [8]) and by this cause an increase of the coefficient of variation of the nuclear area.
2. The analysis showed that it is extremely difficult to compare data of different sections because the variation of the section thickness influences photometric features as well as textural features, with the exception of ENTR, which is the most important feature. The correction for different thickness improves the accuracy of the three group analysis from 58% to about 72% correct classification.

Hobik and Grundmann [9] showed that the normal DNA-content of rat hepatocytes (tetraploid cells) did not change up to 15 days after application of diethylnitrosamine. Under the assumption that the DNA content is not altered after the application of Clophen A 50 the proposed type of thickness correction can be performed. However if no constant DNA content can be assumed, interferometric measurement of section thickness is necessary to correct DNA content for each individual section.

The used techniques of TV-scanning and multivariate analysis of hepatocyte nuclei of treated and untreated animals allow the detection of toxic effects not apparent by visual inspection. Significant changes in the chromatin condensation and distribution pattern can be quantified. Computer evaluation of microscopic images is highly sensitive and may be useful in the detection of small changes of chromatin pattern induced by cytotoxic agents.

Acknowledgement. The technical assistance of Ms E. Dietl and Ms Ch. Trautwein is highly appreciated.

References

1. McConnell EE (1980) Acute and chronic toxicity, carcinogenesis, reproduction, teratogenesis and mutagenesis in animals. In: Kimbrough RD (ed) Halogenated biphenyls, terphenyls, napthalenes, dibenzodioxins and related products. Elsevier, North-Holland: 109–150
2. Abmayr W, Deml E, Oesterle D, Gössner W (1983) Nuclear morphology in preneoplastic lesions of rat liver. Analytical and Quantitative Cytology 5:235–244
3. Abmayr W, Giaretti W, Gais P, Dörmer P (1982) Discrimination of G1, S and G2 cells using high-resolution TV-scanning and multivariate analysis methods. Cytometry 2:316–326
4. Abmayr W, Burger G, Soost HJ (1979) TUDAB project for automated cancer cell detection. Histochem and Cytochem 27:604–612
5. Mannweiler E, Rappl W, Abmayr W (1982) Software for interactive biomedical image processing – BIP. Proc. 6th International Conference on Pattern Recognition, IEEE Computer Society Press 1213
6. Dixon WJ, Brown MB (1977) BMDP biomedical computer programs P-series 1977. University of California Press
7. Nair KK, Bartels PH, Mahom DC, Olson GB, Oloffs PC (1980) Image analysis of hepatocyte nuclei from Chlordane-treated rats. Analytical and Quantitative Cytology 2:285–289
8. Williams MA (1977) In: Glauer AM (ed) Quantitative methods in biology. Practical methods in electron microscopy. North Holland, Amsterdam

9. Hobik HP, Grundmann E (1962) Quantitative Veränderungen der DNS und der RNS in der Rattenleber während der Carcinogenese durch Diäthylnitrosamin. Beiträge zur pathologischen Anatomie und allgemeinen Pathologie 127:25−48

Computer Graphic Displays for Microscopists Assisting in Evaluating Radiation-Damaged Cells

Peter H. Bartels and G. B. Olson

Department of Microbiology, University of Arizona, Tucson, AZ 85721, USA

Computer analysis of digitized images of cells and tissues has developed into a highly refined methodology. In particular, the nuclear chromatin has numerous features that allow reliable recognition and classification of cells. This was hardly a surprise to cyto-pathologists, since chromatin texture and staining properties have always provided valuable clues for visual diagnosis. However, two aspects of computer assessment came as a surprise. First, there was the consistency with which chromatin distribution patterns express the state of the cell, its differentiation, [1] and the sensitivity with which a change in the cell's environment is reflected in the chromatin. Second, chromatin texture provides highly discriminating, objectively measurable features [2, 3] that are not perceived by human visual assessment.

Numerous studies have shown that computer assessment is capable of making clear-cut discriminations in cases where human cytodiagnostic evaluation is not effective. Examples are the differentiation of atypical urothelial cells, [4] the discrimination of B and T lymphocytes, [5] and the expression of "marker features" in ectocervical intermediate cells for the presence of dysplastic or malignant disease [6]. Once certain features of the chromatin pattern are identified as diagnostic clues by the computer procedure, then human observers can learn to appreciate the differences, in some instances. However, on many occasions these differences are too subtle to permit reliable visual detection. Therefore, reliable substantiation of very subtle changes is of prime interest, particularly in environmental pathology, in order to detect very early stages of developing pathology, to examine the effects of subtoxic doses of chemicals, [7] or to detect small changes attributable to low-level irradiation.

The greatly enhanced capability to detect and substantiate changes in cells is based on statistical/analytical procedures. Results are usually expressed in numerical form. For each cell one may, for example, compute a discriminant function score, a distance to a standard normal cell population, or an atypicality index [8]. All of these computations have extensive applications and have proved valuable. If these analytical procedures were coupled with certain "human factors" considerations, the analytical power offered by computer diagnostic procedures would have a much more direct impact.

As a rule, the results from the computer analysis are not immediately available. When they are, they are usually not directly associated with the microscopic images of given cells. Furthermore, cytologists and pathologists are by training, experience, and natural inclination, superbly capable of evaluating visual information. A numerical result may not provide the same intuitive understanding as the visual diagnostic clues, it may not convey the same diagnostic weight or persuasive value, nor may it provide the same comprehension of the diagnostic situation.

Biological Dosimetry. Edited by W. G. Eisert and M. L. Mendelsohn
© Springer-Verlag Berlin Heidelberg 1984

To bring about better comprehension and utilization of computed diagnostic information, computer graphics were found to offer an ideal solution [9]. The numerical results are encoded as a color hue for a frame surrounding the image of a given cell. The diagnostician thus sees the cells as usual in the microscope, and he also sees the computer assessment in easily perceived color coding.

A near real-time display of the computer assessment has, in our experience, led to significantly increased acceptance and appreciation of the diagnostic clues provided by quantitative methods. The procedure is demonstrated in Fig. 1, which shows a slide of irradiated lymphocytes that were immediately mixed with control cells after irradiation. In this graphic display, the cells with feature values corresponding to control cells were encoded in green hues, and cells showing irradiation effects appeared in shades of yellow to red. In the black and white reproduction the green frames appear in a lighter shade. The color-coded frames not only aid in the assessment of a given cell, but they provide a strong visual impact concerning the state and homogeneity of an entire cell sample.

The computer graphic displays may also be used to display a series of cell images simultaneously with their encoded frames. These cells can be arranged according to some discriminant function score or according to the value of a selected feature. Very subtle changes in a given feature are not appreciated by visual observation when they are encountered randomly in the cell population. However, when viewed as an ordered sequence of increasing expression of the change, the visually required clues can often be developed.

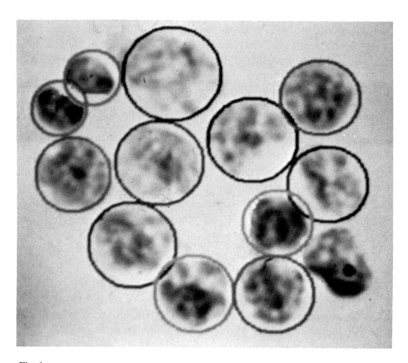

Fig. 1

It is important to distinguish between frame encoding described here for high-resolution, conventionally stained cell images, and computer representations of cell images on graphic displays. Computer representations, such as the addition of artificial color to accentuate condensed chromatin and its spatial dispersion, does not offer the conventional appearance of cells, nor does it preserve all the familiar visual diagnostic clues. In contrast, by incorporating the results from frame encoding, the diagnostician is able to fully utilize his experience, while complementing his human diagnostic acumen by objective, quantitative assessment. This may become a highly acceptable form of bringing analytical procedures into practical clinical cytologic assessment, and to the microscopic examination of cell materials studied in environmental pathologic experimentation.

References

1. Grundmann E, Stein P (1961) Untersuchungen ueber die Kernstruktur in normalen Geweben und im Carcinom. Beitr z Patholog Anat und zur allgem Pathologie 125:54−76
2. Bartels PH, Wied GL (1981) Automated image analysis in clinical pathology. Am I Clin Path 75: 489−493
3. Bartels PH, Olson GB, Lockart R, Wied GL (1980) Cytophotometric studies of cell populations. Cell Biophys 2:339−351
4. Koss LG, Bartels PH, Bibbo M, Freed S, Sychra J, Taylor J, Wied GL (1977) Computer analysis of atypical urothelial cells, I. Classification by supervised learning algorithms: Acta Cytol 21: 247−260
5. Durie BG, Chen YP, Olson BG, Salmon SE, Bartels PH (1978) Discrimination between human T and B lymphocytes and monocytes by computer data from scanning microphotometry. Blood 51:579−589
6. Wied GL, Bartels PH, Bibbo M, Sychra J (1980) Cytomorphometric markers for uterine cancer in intermediate cells. J Analyt Quant Cytol 2:257−263
7. Nair KK, Bartels PH, Mahon DC, Olson GB, Oloffs PC (1980) Image analysis of Feulgen stained rat hepatocytes: Effects of chlordane: Analyt Quant Cytol 2:258−289
8. Bartels PH, Bibbo M, Richards D, Sychra J, Wied GL (1978) Patient classification based on cytologic sample profiles, I. Basic measures for profile construction. Acta Cytol 22:253−260
9. Dytch HE, Bartels PH, Bibbo M, Pishotta FT, Wied GL (1982) Computer graphics in cytodiagnosis. J Analyt Quant Cytol 4:263−268

Flow Cytometric Measurement of Intracellular pH:
A Dosimetric Approach to Metabolic Heterogeneity

Oliver Alabaster,[1] Michael Andreeff,[2] Connie Spooner,[1] and Kathy Clagett[1]

[1]The George Washington University Medical Center, Washington, DC USA
[2]Memorial Sloan-Kettering Cancer Center, New York, NY 10021, USA

Introduction

Cell heterogeneity may be expressed in many ways such as variability in morphology, genetics, growth rate, cell cycle distribution, membrane antigen expression, clonogenicity, metabolic and biochemical activity, and therapeutic sensitivity. Virtually all forms of functional heterogeneity must ultimately be expressed as some difference in biochemical, metabolic or physiological patterns of activity. Not surprisingly, this may be clinically significant since growth rate, for example, can influence drug sensitivity [1]. However, quantifying metabolic heterogeneity in tumour cells, and assessing its significance in a dosimetric sense, requires the capability either to measure the relevant biochemical pathways in each cell or to find a suitable measurement from which differences may be inferred. Since the precise pattern of metabolic activity that determines sensitivity or resistance to cytotoxic therapy is largely unknown, and since biochemical techniques preclude measurements on individual cells, a flow cytometric approach is indicated.

Rapid measurement of the intracellular pH distribution by flow cytometry [2] provides an indirect measurement of the range of metabolic heterogeneity in a cell population, since the intracellular pH influences virtually all forms of biochemical activity. In particular, subtle changes in pH can influence protein structure and enzymatic activity. Thus, although the detailed molecular effects of differences in intracellular pH cannot be identified by flow cytometry, it is clear that a cell with a pH of 6 must be significantly different from one with a pH of 7.5.

Experiments were therefore designed to model the nutrition and gas gradients which are known to exist in solid tumours [3]. These gradients were shown to exist in a mouse mamary tumour, and were demonstrably responsible for influencing both the fraction of cells in DNA synthesis and the growth fraction. Using flow cytometric analysis of intracellular pH, it is now possible to determine the influence of nutritional dilution and restricted gas exchange on tumour cell metabolic heterogeneity, and to correlate these measurements with changes in the growth fraction; the latter being measured by the acid DNA denaturation technique [4].

Biological Dosimetry. Edited by W. G. Eisert and M. L. Mendelsohn
© Springer-Verlag Berlin Heidelberg 1984

Methods

The CA-46 human lymphoma derived cell line [5] was maintained routinely in RPMI 1640 growth medium containing 10% fetal calf serum, 1-glutamine and penicillin-streptomycin. Flasks were loosely-capped in an atmosphere of 5% CO_2 in air at 37°C, and were refed every 3—4 days by two thirds medium replacement.

Cells in log-phase of growth were subcultured in T25 flasks at a concentration of 0.25×10^6/ml of complete medium. Subgroups of T25 flasks were also established with complete culture medium diluted 1:1 and 1:2 with phosphate-buffered saline (pH 7.4) containing penicillin-streptomycin. Each flask contained 15mls of medium and was therefore maintained vertically. Half the flasks from each subgroup were then left loosely-capped while the other half were tightly closed, so limiting free gas exchange.

Measurements of intracellular pH were made at the time of subculturing (Day 0) and cells were also fixed in 70% ethanol, and stored at 4°C prior to subsequent analysis of the relative DNA content distribution and the relative acid denaturation of DNA. After seven days undisturbed in the culture medium, cells were harvested from triplicate flasks in each subgroup, counted on a Coulter counter, and tested for trypan blue exclusion to determine viability. Rapid measurement of intracellular pH was made on aliquots of cells within a few minutes of harvesting. The remaining cells were washed in PBS and then fixed in 70% ethanol as described above.

Measurement of Intracellular pH

Using 1,4-diacetoxy-2,3-dicyano-benzol (ADB) supplied by courtesy of Dr. G. Valet, the intracellular pH was measured by flow cytometry [2]. ADB is a nonfluorescent ester of the pH-dependent fluorescent dye, 2,3-dicyano-hydroquinone (DCH), and has the capacity to cross the cell membrane. Once in the cell, ADB is broken down by non-specific esterases to form the fluorescent compound DCH. This reaction takes place within a few minutes, and only occurs in living cells. About 90% of DCH is distributed in the cytoplasm, the remainder being distributed in cell organelles [2].

Freshly harvested cell samples were stained in 10—20 μg/ml of ADB at 37°C for 5—15 min before analysis in a Becton Dickinson FACS 4 flow cytometer equipped with a Spectra Physics 164-05 laser set at 350nm, with 100mw output. Using appropriate band pass filters, one photomultiplier collected fluorescence emission between 420—440nm (F1) and the second photomultiplier collected fluorescence emission between 500—580nm (F2). Using the FACS ratio circuitry, the F1/F2 ratio can be collected on each cell providing a ratio distribution for the population. This ratio distribution is proportional to the intracellular pH distribution. Calibration of the ratio distribution is achieved by establishing a series of Tris buffers at pH 5, 6, 7, 8 to which are added the same cells, 10 μg/ml ADB, and 20 μg of ammonium acetate. The latter collapses the pH gradient across the cell membrane, theoretically rendering the intracellular pH of all cells the same as the buffer. Using this technique, the coefficient of variation of log-phase CA-46 cells at pH 7 is usually between 4.9—6.0. Cells in different buffers are then analyzed, and the modal channels of the F1/F2 ratio distributions are then plotted against pH. In this way each channel of the original ratio distribution can be ascribed a pH value, so producing the intracellular pH distribution.

Measurement of G_0/G_1 Cells by Partial Denaturation of DNA in situ

In brief, the determination of the relative numbers of non-cycling G_0 and cycling G_1 cells depends upon the observation that, in chromatin, DNA is locally stabilized by positively charged macromolecules which provide counter ions for the phosphates of the DNA backbone. The extent of the interactions, as well as any cross-linking, may be estimated from the patterns of DNA denaturation which, in turn, provide information on chromatin structure in intact cells [4].

RNA'se-treated fixed cells are treated with acid at pH 1.4 for 30 s and subsequently stained with acridine orange (AO). The extent of DNA denaturation is then calculated as the ratio of red fluorescence (F 600; reflecting interactions of the dye with denatured DNA) to total cell fluorescence (F 600 + F530; representing interaction of AO with total DNA). A marked change in the sensitivity of DNA to both acid denaturation and heat has been observed during cell cycle progression and cycling cells can be distinguished from G_0 non-cycling cells [4].

Measurement of DNA Content Distribution

Cellular DNA was stained with propidium iodide containing RNA'se [6]. Cells were analyzed on the FACS 4 flow cytometer using laser excitation at 488nm. As in all other experiments data were collected, stored and analyzed by a DEC PDP 11/34 computer. All relative DNA content distribution were collected at the same amplifier gain settings and were standardized for the number of cells measured to facilitate visual interpretation. Calibration of the instrument was performed with standard particles (Coulter Electronics, Hialeah, FL, USA) which produced a C.V. of less than 1.5%.

Results

The relative DNA content distributions derived from cells in the 'open' flasks are shown in Fig. 1. It is evident that the cell cycle distribution changed with age in undiluted medium, and to an even greater extent when the medium was progressively diluted. The proportion of cells in S phase declined and the relative number of G_0/G_1 cells increased. These effects were not accompanied by significant changes in cell number, cell viability or extracellular pH. Corresponding intracellular pH distributions are shown in Fig. 2. Although the modal pH value did not change from day 0 to day 7 in cells grown in undiluted medium, there was a considerable increase in heterogeneity. As medium was diluted 1:1 and 1:2, there was a dosimetric reduction in the modal pH associated with comparable heterogeneity.

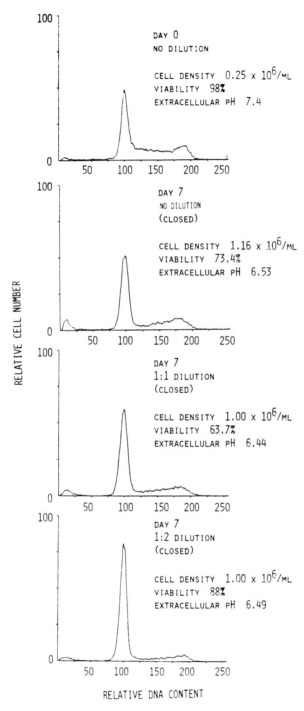

Fig. 1. DNA content distributions of CA-46 human lymphoma cells grown as suspension cultures in 'open' flasks. Day 0 cells were derived from well-fed stock cultures in log phase of growth, maintained in 5% CO_2 and air at 37°C, with flasks loosely capped. After 7 days in continuous culture, in unchanged medium at full strength and diluted 1:1 and 1:2, DNA content distributions, cell number, cell viability and extracellular pH were measured. Medium depletion and dilution produced a progressive reduction in S phase cells

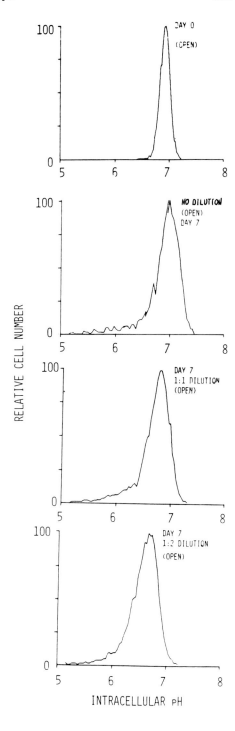

Fig. 2. Intracellular pH distributions of the same CA-46 human lymphoma cells that are shown in Fig. 1. Note the broadening of the intracellular pH distribution with depleted full strength medium, and progressive acidification of the cell population when medium was also diluted

RELATIVE DNA DENATURATION

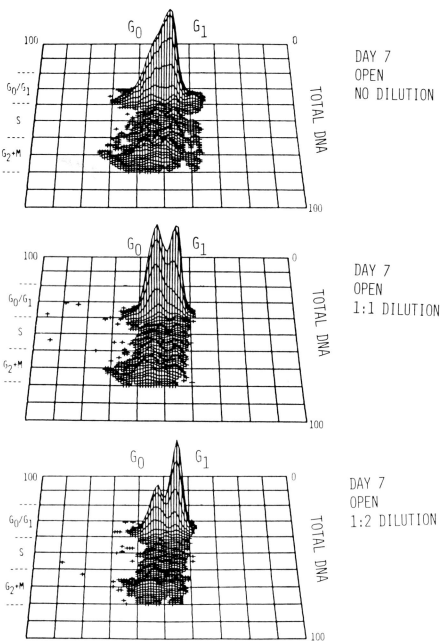

Fig. 3. Acid denaturation of cellular DNA from each of the three groups on day 7. Two cell populations can be seen extending from G_0/G_1 phase into S phase. Medium dilution (1:2) produced a reduction in G_0 cells

Analysis of the relative DNA denaturation suggested the presence of both non-cycling and cycling cell populations, with relatively fewer non-cycling cells when the medium was diluted 1:2 (Fig. 3). The presence of two populations is also evident in S phase, but it is not known whether one is definitely non-cycling.

Comparable analyses were performed on cells grown in closed flasks. The relative DNA content distributions shown in Fig. 4 indicate a similar trend toward a predominantly G_0/G_1 population, with a similar cell density. However, the extracellular pH was slightly lower, and trypan blue exclusion was diminished indicating lower viability.

The relative intracellular pH distributions derived from these cells are shown in Fig. 5. These measurements indicated that the modal pH was progressively lower in all three subgroups, with even greater pH heterogeneity than in the 'open' flasks.

Measurements of the relative DNA denaturation of cells from closed flasks are shown in Fig. 6. Evidently, most cells were still cycling with relatively few in the non-cycling state.

Discussion

Variation in the growth fraction and in the thymidine labeling index occurs in solid tumors because of nutritional and gas gradients that result from inadequacy of the vascular supply [3]. This proliferative heterogeneity should also be associated with metabolic heterogeneity. Hitherto, metabolic heterogeneity was more theoretical than substantive, because biochemical techniques only measured chemical constituents of cells as 'averages' for the cell population, and potential heterogeneity was ignored.

The advent of flow cytometry (FCM) enables measurements to be made on individual cells in large numbers, and is therefore an ideal tool for studying heterogeneous cell features. Nevertheless, most applications of FCM have been of a quantitative nature, but the development of a fast technique for measuring intracellular pH [2] now permits the study of factors that might influence metabolic heterogeneity in a dosimetric sense.

The experiments presented here clearly indicate that the distribution of intracellular pH in a cell population can be influenced by the environment. Relative anoxia and relative nutrient depletion both cause acidification and an increase in pH heterogeneity. This increased metabolic heterogeneity implies that cellular metabolic processes are not rigidly controlled, but are capable of responding to changing environmental conditions. Furthermore, this trend toward metabolic heterogeneity is not revealed by cell cycle analysis using relative DNA content distributions (Figs. 1, 4). In fact, there was a dosimetric reduction in cell cycle heterogeneity with progressive dilution of the culture medium, in both 'open' and 'closed' environments.

However, the appearance of relative cell cycle homogeneity (fewer $S+G_2M$ cells) is deceptive because partial denaturation of DNA *in situ* revealed heterogeneity in chromatin structure that was influenced both by medium dilution and by relative anoxia. Since the cell density was equivalent in both the 'open' and 'closed' environments, it is possible that the greater proportion of cycling G_1 cells in the closed environment

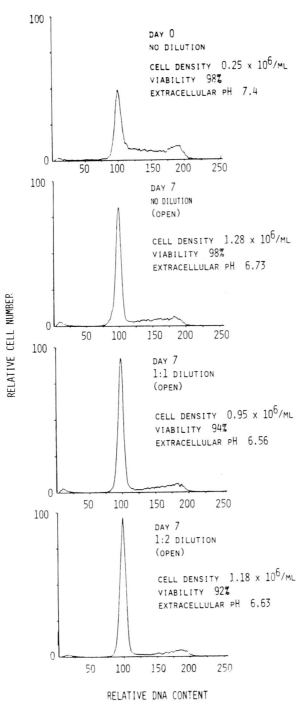

Fig. 4. Comparable DNA content distributions from CA-46 human lymphoma cells grown in vertical closed flasks that limit free gas exchange producing relative anoxia. Cell density was unaffected, but the extracellular pH was slightly lower and the viability was reduced

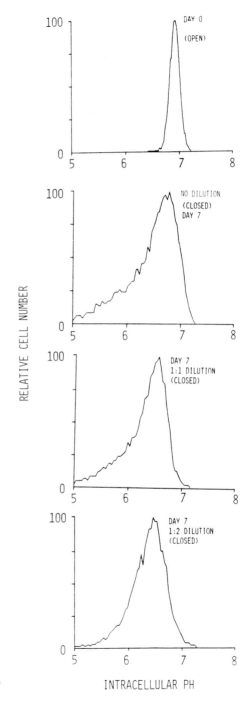

Fig. 5. Intracellular pH distributions from cells grown in closed flasks. Medium depletion and dilution produced greater broadening and acidification of the cell population

RELATIVE DNA DENATURATION

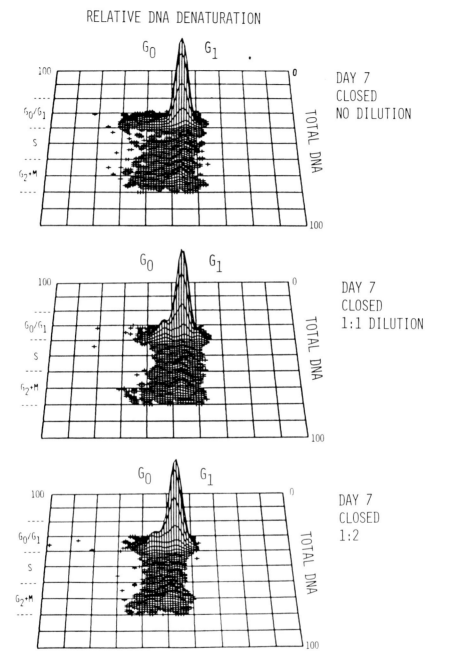

Fig. 6. Acid denaturation of DNA of cellular pH from each of the three groups on day 7. Two cell populations can also be seen, but relative anoxia caused fewer cells to be in G_0 at this stage of cell growth

may be the result of lower cell growth delaying the onset of plateau phase and the consequent establishment of a non-cycling population.

Since the intracellular pH is manifestly more sensitive to the nutritional and gas environment than it is to the extracellular pH, the possibility exists that tumor cell metabolism could be manipulated so as to influence sensitivity to cytotoxic agents. If such metabolic manipulations also reduce cell heterogeneity, the response to cytotoxic therapy should be more uniform, and the chance of cells surviving such treatment will be reduced. This view is supported by the clinical observation that advanced tumors are much more difficult to treat successfully, probably because greater cell numbers result in more cell to cell variation, and because advanced tumors usually have inadequate vascularisation.

References

1. Shackney SE, McCormack GW, Cuchural GJ (1978) Growth rate patterns of solid tumors and their relation to responsiveness to therapy: an analytical review. Ann Intern Med 89:107–121
2. Valet G, Raffael A, Moroder L, Wunsch E, Ruhenstroth-Bauer (1981) Fast intracellular pH determination in single cells by flow cytometry. Naturwissenschaften 68:S 265
3. Tannock IF (1968) The relation between cell proliferation and the vascular system in a transplanted mouse mammary tumor. Brit J Can 22:258–273
4. Darzynkiewicz A, Traganos F, Andreeff M, Sharpless T, Melamed MR (1979) Different sensitivity of chromatin to acid denaturation in quiescent and cycling cells as revealed by flow cytometry. J Histochem Cytochem 27:478
5. Woo KB, Funkhouser WK, Sullivan D, Alabaster O (1980) Analysis of proliferation kinetics of Burkitt's lymphoma cells. Cell Tiss Kin 13:591
6. Crissman HA, Steinkamp JA (1982) Rapid, one step staining procedures for analysis of cellular DNA and protein by single and dual laser flow cytometry. Cytometry 3: No. 2, 84–90

Flow Cytometric Measurement of the Metabolism of Benzo [A] pyrene by Mouse Liver Cells in Culture

James C. Bartholomew, Charles G. Wade, and Kathleen K. Dougherty

Laboratory of Chemical Biodynamics, Lawrence Berkeley Laboratory, University of California, Berkeley, California 94720, USA

Summary

The metabolism of benzo[a]pyrene in individual cells was monitored by flow cytometry. The measurements are based on the alterations that occur in the fluorescence emission spectrum of benzo[a]pyrene when it is converted to various metabolites. Using present instrumentation the technique could easily detect 1×10^6 molecules per cells of benzo [a]pyrene and 1×10^7 molecules per cell of the diol epoxide. The analysis of C3H IOT 1/2 mouse fibroblasts growing in culture indicated that there was heterogeneity in the conversion of the parent compound into diol epoxide derivatives suggesting that some variation in sensitivity to transformation by benzo[a]pyrene may be due to differences in cellular metabolism. The technique allows sensitive detection of metabolites in viable cells, and provides a new approach to the study of factors that influence both metabolism and transformation.

Introduction

Benzo[a]pyrene (BaP) is a polycyclic aromatic hydrocarbon that must be metabolized by the cell before it becomes mutagenic or carcinogenic [1—3]. This metabolism takes place by an inducible enzyme system called aryl hydrocarbon hydroxylase (AHH) [4—6]. Studies have shown that the ability of cells to metabolize BaP is highly variable [7—14]. In mice, much of this variability has been related to genetic factors, and the locus determining the inducibility of AHH has been identified [15]. Boobis and Nebert [16] have described a positive association of this locus with the binding of BaP to DNA and the susceptibility of mice to chemical carcinogenesis.

In humans, the involvement of genetic factors in AHH induction or chemical carcinogenesis has not been clearly established. Okuda et al. [17] have studied AHH in monocytes from monozygotic and dizygotic twins and found considerable variation even when cells were harvested from the same individual at different times. They estimated from this study that from 55 to 70% of the variation observed was genetically controlled. Other studies with human lymphocytes in culture have also shown great individual variability in the metabolism of BaP, and that persons with lung cancer can more active-

*This study was supported by the Division of Biomedical and Environmental Research of the US Department of Energy. CGW was supported by an grant from NIH (Grant HL12528) and from the Robert A. Welch Foundation (Grant F-370)

ly metabolize the carcinogen than persons without lung cancer [18, 19]. These studies suggest the possibility that individuals may possess different susceptibilities to carcinogenesis by polycyclic aromatic hydrocarbons depending on their ability to metabolize the compounds to derivates active in binding to cellular components. Other factors besides genetic such as position in the cell cycle and differentiation state of the tissue [20] have been shown to influence carcinogen metabolism, and so metabolism may even vary from cell to cell within a tissue.

Because of the importance of cellular metabolism in defining the activity of chemicals like BaP in carcinogenesis, and because of the variability of this metabolic potential, careful analysis of the metabolism of such chemicals becomes a necessary part of any risk assessment. With such a complex metabolic profile as depicted in Fig. 1 for BaP, the detailed study of the quantities of individual metabolites in different types of tissues or even in different individuals has been difficult. Normally, such an analysis involves the extraction of metabolites from the cells followed by quantification of the compounds by techniques such as HPLC. However, many of the metabolites bind covalently to cellular components making quantitative extraction not possible with present procedures. Extraction techniques are also time consuming and necessarily lead to an averaging of the metabolites over all the cells in a tissue preparation.

We have been developing methods that do not involve extraction for obtaining metabolic information from individual cells exposed to BaP. The approach is to utilize the changes that occur in the fluorescence emission spectra of cells as they convert BaP into metabolites. We have made these measurements in a flow cytometer and have obtained information about population heterogeneity of BaP uptake and metabolism on a single cell level.

Materials and Methods

Cell Culture

The C3H 10T 1/2 cells are mouse fibroblasts capabile of metabolizing BaP [21, 22]. The medium used to grow the cells was Eagle's minimal medium [23] (GIBCO, Grand Island, N.Y.) containing 10% donor calf serum (Flow Laboratories, Rockville, MD) and 10 μg/ml insulin (Schwartz/Mann, Orangeburg, N.Y.). All cells were judged free of mycoplasma by incorporation of H^3-thymidine (20.1 Ci/mM; New England Nuclear, Boston, MA) into the nucleus of the cells and not the cytoplasm [24]. Stock cultures were maintained by subculturing the cells twice weekly at a cell density of 1 x 10^4/ cm^2. All cell cultures were actively growing at the time of addition of test compounds.

Chemical Additions

The BaP and metabolites of BaP were added to the culture medium from a stock solution (1.0 mg/ml) in DMSO (Matheson, Coleman, and Bell, Los Angeles, CA). The final

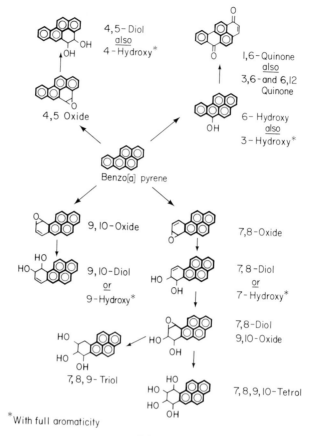

4,5- Diol
also
4 - Hydroxy*

1,6-Quinone
also
3,6- and 6,12
Quinone

4,5 Oxide

6- Hydroxy
also
3- Hydroxy*

Benzo[a] pyrene

9, 10-Oxide

7,8-Oxide

9,10-Diol
or
9-Hydroxy*

7, 8-Diol
or
7- Hydroxy*

7,8-Diol
9,10-Oxide

7, 8, 9- Triol

7,8,9,10-Tetrol

*With full aromaticity

Fig. 1. Metabolites of benzo[a]pyrene

concentration of DMSO was 0.5% in the controls and test cultures. BaP was obtained from the Aldrich Chemical Company, Inc., Milwaukee, Wis., and the metabolites were obtained through the NCI Standard Chemical Carcinogen Reference Repository.

The spectra of all compounds were determined in a Perkin-Elmer MPF-3A (Norwalk, CT) spectrofluorometer. The compounds were added to cells and allowed to equilibrate for 1 h prior to harvesting the cells. The cells were resuspended in isotonic buffer at 1×10^6 per ml and the spectra were measured. All spectra were recorded with slit widths of 4 nm.

Flow Cytometry

The flow cytometer used in these experiments has been described by Hawkes and Bartholomew [25]. Cells traversed the flow chamber at approximately 500 cells/sec in a stream that intersected the beam of an argon ion laser (Model 171 Spectra Physics, Inc. San Jose, Ca) tuned to 351.1 and 363.8 nm. The typical power output at these wavelengths for these experiments was approximately 100 mW. The fluorescent light prior to reaching the photomultipliers (PMTs) was split using a Corning 0–51glass filter (Corning, Corning, N.Y.) to reflect approximately 10% of the beam to one PMT and 90% to a second PMT tube. Band pass filters (Baird-Atomic, Bedford, Mass.) having approximately 10 nm bandwidths were selected to analyze for signals characteristic of BaP or its metabolites. In the two parameter measurements, BaP fluorescence was recorded with the PMT receiving only 10% of the emitted light since its concentration in these experiments was in excess over the metabolites. In all experiments, the cells to be analyzed were harvested by trypsinization and passed directly through the flow cytometer without further processing. The data from each analysis was stored, plotted and analyzed by use of VAX 780 computer (Digital Equipment Corp., Maynard, Mass.).

Radioactivity Measurements

$14C$-BaP (59 Ci/mole) was obtained from Amersham Corporation (Arlington Heights, Il.) and unlabelled BaP from Aldrich Chemical Company (Milwaukee, Wis.). $3H$-BaP diol epoxide (1.3 Ci/mmole) was kindly provided by Dr Ken Straub. Samples were counted in Aquassure liquid scintillation cocktail (New England Nuclear, Boston, Mass.).

Results

Fluorescence Monitoring of BaP Metabolism

AHH metabolism of BaP results in the production of many oxidized products only some of which are shown in Fig. 1. The metabolites most active in binding to cellular macromolecules are the epoxides. With BaP, the epoxide that binds most efficiently to DNA and may be the ultimate carcinogenic form is the 7, 8, 9, 10-tetrahydro-7, 8-diol-9, 10-epoxy-benzo[a]-pyrene (BaP diol-epoxide) [26–30].

The metabolism of BaP by cells generally produces compounds that are still fluorescent, but have alterations in their spectra when compared to that of the parent compound. Figure 2 presents some of the spectra of different metabolites present as individual compounds in C3H 10T 1/2 mouse fibroblasts. The spectra were obtained by exposing 2×10^6 cells growing in culture to 1 μg/ml of the authentic metabolite. The derivatives of BaP which have added a single hydroxyl group without disturbing the aromaticity are characterized by an emission spectrum shifted to longer wavelengths (⟩ 427 nm) relative to BaP. Whereas; the diol or diol epoxide derivatives gave emission spectra shifted to shorter wavelengths (390 and 380 nm respectively).

In Figure 3, the emission spectra of C3H 10T 1/2 cells and medium which have been treated with BaP for 1 to 48 h are shown. The 1 h incubation is long enough for the cells to accumulate BaP to equilibrium, but not long enough to induce the enzymes required for metabolism [20]. At 1 h, the spectra of both the cells and the medium looks very similar to that of unmetabolized BaP. After 24 h incubation, the BaP has

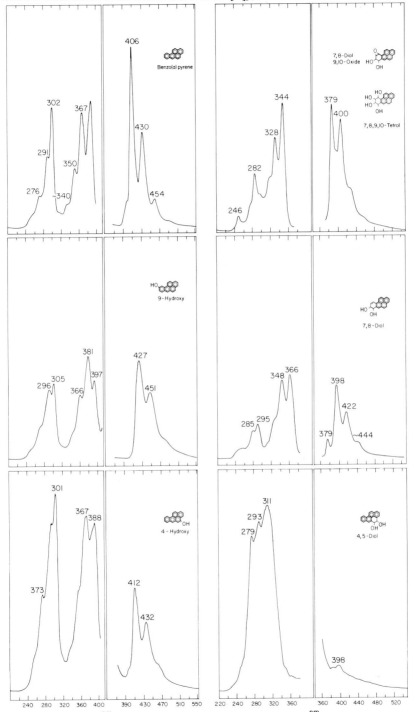

Fig. 2. Fluorescence spectra of benzo[a]pyrene metabolites. Growing C3H 10T 1/2 mouse fibroblast cells (2×10^6 were exposed to 1.0 ug/ml authentic metabolite for 1 h prior to harvesting the cells by trypsinization. The metabolites were each dissolved in DMSO prior to addition. The spectra were taken as described in the text

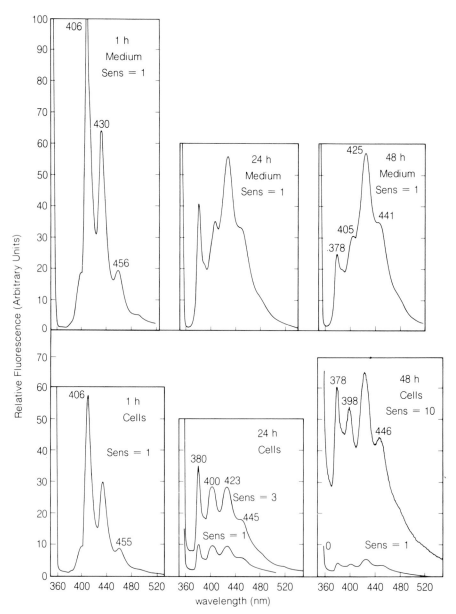

Fig. 3. C3H 10T 1/2 metabolism of benzo[a]pyrene. Growing C3H 10T 1/2 mouse fibroblast cells (2x10⁶ were exposed to 1.0 ug/ml benzo[a]pyrene for the times indicated. At the end of the incubation times the medium was removed and saved. The cells were washed and trypsinized from the dishes, and the spectra were measured by exciting at 360 nm and recording the emission as described in the text

been metabolized and the resulting spectra indicate the presence of phenols and BaP diol epoxides in both the cells and the medium. It should be pointed out that based on spectral characteristics, it is not possible to distinguish BaP diol epoxide from the tetrol which is derived from the BaP diol epoxide. The medium contains more of all the metabolites than the cells, and the ratio of signal characteristics of phenols to that of BaP diol epoxides is higher in the medium than in the cells. This suggests, not suprisingly, that cells bind the BaP diol epoxide more efficiently than the phenol. By 48 h there is very little evidence of signal from unmetabolized BaP either in the medium or in the cells. In the medium, the major signal comes from the phenols, but a significant BaP diol epoxide signal was present. The cells clearly contain compounds fluorescing with the characteristics of BaP diol epoxide, and also contain a significant phenol signal. Although the fluorescence spectra give only a qualitative indication of the types of metabolites present, the general features of the metabolism have been confirmed by HPLC analysis of metabolites extracted from both the medium and the cells (data not shown).

Flow Cytometry of Cells Treated with BaP and BaP Diol Epoxide

Adaptation of the fluorescence measurement techniques seen with a large population of cells to the single cell level can be done by using a flow cytometer. Figure 4a shows the fluorescence emission histogram of BaP in cells analyzed by flow cytometry after 1 hr exposure to BaP and measuring the fluorescence intensity with a band pass filter centered at 410 nm. The fluorescent signal was correlated in this experiment with the average amount of BaP per cell as determined by scintillation counting of an aliquot of cells labeled with ^{3}H-BaP of a known specific activity. The average amount of BaP per cell present in this experiment was approximately 1×10^6 molecules/cell. In the same manner, cells were treated with BaP diol epoxide and the histogram of fluorescence intensity/cell at 380 nm was recorded (Fig. 4b). The level of BaP diol epoxide in this sample was approximately 1×10^7 molecules/cell. With the present experimental system, the sensitivity was limited by background signal from untreated cells. Much of this signal may be due to endogenous fluorescence from NAD(P)H [31]. The shape of the fluorescence distributions for either BaP or BaP diol epoxide did not change appreciably as the concentration of either compound was raised through 2 orders of magnitude from that used to obtain the data in Fig. 4. Both the mean and the mode of the distributions was a linear function of the average amount of compound per cell as measured by radioactivity until the cells became saturated with hydrocarbons. Using this procedure we determined that saturation with these compounds depended on the amount and type of serum in the medium as well as the cells being studied (data not presented).

In general, during metabolism of BaP both BaP and its metabolites are present. To compare the relationship between the measurement of different relative concentrations of BaP and BaP diol epoxide in cells we compared the ratio of fluorescence at 380 and 410 nm with the known average concentration of compound per cell calculated from determining the amount of radioactivity present in an aliquot of cells. Figure 4 c and d presents the fluorescence ratios obtained from two different imput ratios of BaP

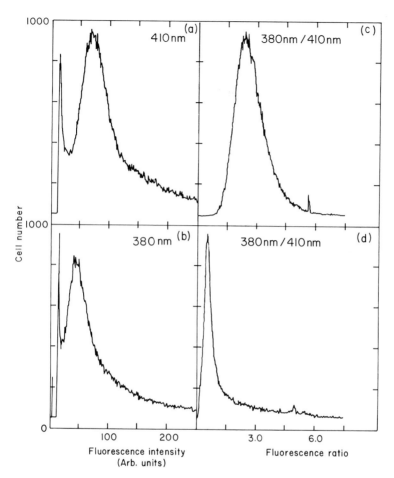

Fig. 4a–d. Detection of benzo[a]pyrene and benzo[a]pyrene diol epoxide in individual cells. Cells were exposed for 1 h to known amounts of radioactive benzo[a]pyrene (**a**), benzo[a]pyrene diol epoxide (**b**), or mixtures of the two compounds (**c, d**). The cells were removed from the dishes by trypsinization and the amount of each compound was determined in a known number of cells by liquid scintillation counting

and BaP diol epoxide. Figure 5 shows that the relationship between fluorescence ratio and molar ratio is not linear over the range of concentrations tested. The fluorescence ratio is lower than expected from the radioactivity measurements when the level of BaP diol epoxide is high relative to BaP. This may be due to the secondary emission bands of BaP diol epoxide contributing to the signal intensity at 410 nm with these high molar ratios. Of course, the secondary emission signals could be subtracted from the 410 nm signal based on the contribution expected from the measured 380 nm intensity.

As seen in the spectra presented in Fig. 3, most of the metabolism that occurs by 24 h can be identified as shifting the fluorescence emission peak at 410 nm to either 380 nm or 427 nm characteristic of metabolism towards either the BaP diol epoxide

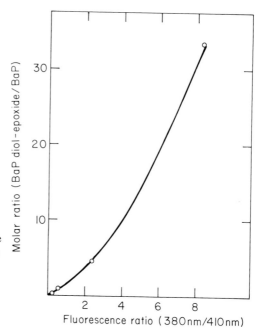

Fig. 5. Standardization of the fluorescence ratio measurements. Cells were exposed to different amounts of benzo[a]pyrene diol epoxide as described for Fig. 4, and the average amount of each compound in the cells was determined by radioactive measurements

Molar ratio (BaP diol-epoxide/BaP)

Fluorescence ratio (380nm/410nm)

or phenols respectively. Measuring the ratio of these two signals on a per cell basis gives an indication whether the population is homogeneous in the type of metabolic pattern it follows. Figure 6 shows the results from analyzing a population of C3H 10T 1/2 cells determining this ratio after they had been treated with BaP for 1 and 24 h. At 1 h the 380 nm/427 nm signal was very low reflecting the low amount of BaP diol epoxide present in the cells at this time. The ratio increases dramatically after 24 h of incubation reflecting the metabolism that has occurred to both BaP diol epoxide and the phenol derivatives. There is heterogeneity in the measurements with some cells in the population having a very high ratio indicating extensive metabolism to the BaP diol epoxide with little production of phenols.

When two parameter measurement are made on cells exposed to BaP using flow cytometry the relative proportion of metabolites going down the "Diol Epoxide" pathway verses the "Phenol" pathway can be determined in single cells. In these experiments we compared the fluorescence intensity at 380 nm with that at 427 nm. As seen in Fig. 3, emission at these wavelengths contribute most significantly to the spectra of cells which have metabolized BaP for 24 h. The bivariate distributions obtained from these cells is shown in Fig. 7. This figure is a composite of samples analyzed 1 h and 24 h after addition of BaP to C3H 10T 1/2 cells. At 1 h most of the signal along the log 427 nm axis comes from the secondary emission signal of the BaP. The 380 nm signal is low and very dispersed and comes from endogenous fluorescence in the population. The lower intensity distribution below that from the 1 h sample is from cells not treated with BaP which were mixed with the BaP treated cells after harvesting. After 24 h the log 427 nm signal is primarily from the major emission from phenols (see Fig. 3). After the 24 h incubation period, the log 380 nm signal increased considerably in-

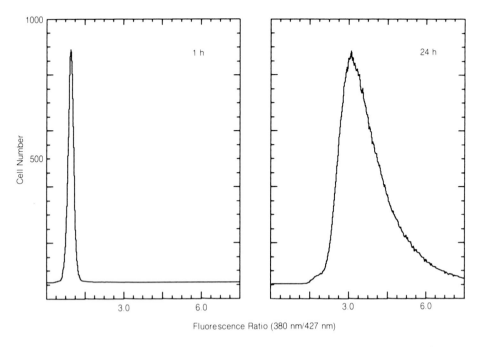

Fig. 6. Ratio of 380 nm/427 nm fluorescence from C3H 10T 1/2 cells metabolizing benzo[a]pyrene

dicating metabolism through the BaP diol epoxides. The distribution along the log 380 nm axis is more dispersed than along the log 427 nm axis suggesting more heterogeneity of metabolism to diol epoxide compounds per cell than to phenols.

Discussion

The techniques described demonstrate that fluorescence spectral analysis of a polycyclic aromatic hydrocarbon like benzo[a]pyrene as it is being metabolized by cells in culture can give information regarding the qualitative nature of the metabolism. Precise quantitative determination of the metabolites is not possible without considerably more knowledge about the multitudinous interactions of the metabolites with all the cellular components. However, both the spectra and the fluorescence histograms do contain indications of the quantities of the metabolites present. The procedure was used here to study BaP metabolism, but can be used with any other compound that changes its spectral properties upon metabolism, or which can be marked with a fluorescent tag.

The application of flow cytometry to the detection of classes of metabolites in individual cells is possible with the use of narrow band pass filters centered in the spectral range that different types of metabolites fluoresce. The sensitivity of the system, while not maximized in our hands, is comparable to that obtained when analyzing metabolites by standard extraction and HPLC analysis.

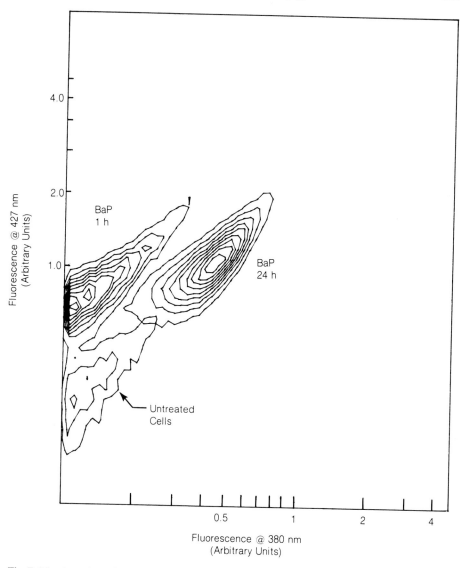

Fig. 7. Bivariate plots of log 380 nm signal vs 427 nm signal in cells exposed to benzo[a]pyrene. This figure is a composite of data from cells analyzed after 24 h or 1 h exposure to BaP. The cells were harvested as described in the text. The initial concentration of benzo[a]pyrene was 1 ug/ml. Prior to analyzing the 1 h sample cells which were not exposed to benzo[a]pyrene were added to the sample and their location in indicated on the plot

A major advantage of the flow technique over extraction methods is that individual cells are analyzed and so population heterogeneity can be determined under different experimental conditions. For example, it is conceivable to couple the metabolite measurement with DNA content determination and relate metabolism of BaP to the cell cycle position. Also, such experimental variables as the relationship of medium components to the heterogeneity of BaP uptake and metabolism can be determined. Since the cells are not fixed for the analysis, the technique may allow metabolic parameters to be related directly to transformation frequencies. Cells with high ratios of BaP diol epoxide to other metabolites could be sorted to determine whether they also have a high transformation frequence when related to the rest of the population. Such direct correlations are not possible with classical methods of analysis for metabolites nor with immunological reagents being used to detect carcinogens [32–34]. It should be pointed out that the concentration of BaP used in these experiments is similar to that used in standard transformation studies.

Direct analysis of the metabolites of BaP per cell by flow cytometry has some unique applications to studies of population heterogeneity and its relation to experimental parameters. The analysis is probably not as sensitive as immunological detection of metabolites, nor as specific as HPLC analysis of extracted samples. However, the sensitivity and selectivity coupled with the viability of the cells after analysis is high enough to contemplate many unique experiments.

References

1. Heidelberger C (1975) Chemical carcinogens. In: Snell EE (ed) Ann. rev. biochem, Vol. 44, Annual Reviews, Palo Alto, CA, pp 79–121
2. Gelboin HV (1967) Carcinogens, enzyme induction, and gene action. In: Haddow A, Weinhouse S (eds) Advances in cancer research, Vol 10, Academic Press, Inc., New York, pp 1–81
3. Diamond L, Baird WM (1977) Chemical carcinogens in vitro. In: Rothblat GH, Cristofalo VJ (eds) Nutrition and metabolism of cells in culture, Vol 3, Academic Press, Inc., New York, ppd 421–470
4. Diamond L, Gelboin HV (1969) Alpha-naphthoflone: An inhibitor of hydrocarbon cytoxicity and microsomal hydroxylase. Science 166:1023–1025
5. Gelboin HV, Huberman E, Sachs L (1969) Enzymatic hydroxylation of benzopyrene and its relationship to cytotoxicity. Proc Natl Acad Sci USA 64:1188–1194
6. Gelboin HV, Wiebel FJ (1967) Studies on the mechanism of aryl hydrocarbon hydroxylase induction and its role in cytotoxicity and tumorigenecity. Ann NY Acad Sci 179:529–547
7. Busbee DL, Shaw CR, Cantrell ET (1972) Aryl hydrocarbon hydroxylase induction in human leukocytes. Science 178:315–316
8. Cantrell ET, Warr GA, Busbee DL, Martin RR (1973) Induction of Aryl hydrocarbon hydroxylase in human pulmonary alveolar macrophages by cigarette smoking. J Clin Invest 52:1881–1884
9. Levin W, Conney AH, Alvors AP, Markath I, Kappos A (1972) Induction of benzo[a]pyrene hydroxylase in human skin. Science 176:419–420
10. Kouri RE, Ratrie H, Whitmire CE (1973) Evidence of a genetic relationship between susceptibility to 3-methylcholanthrene-induced subcutaneous tumors and inducibility of aryl hydrocarbon hydroxylase. J Natl Cancer Inst 51:197–200
11. Nebert DW, Gelboin HV (1969) The in vivo and in vitro induction of aryl hydrocarbon hydroxylase in mammalian cells of different species, tissues, strains, and developmental and hormonal states. Arch Biochem Biophys 134:76–89
12. Yamusaki H, Huberman E (1977) Metabolism of the carcinogenic hydrocarbon benzo(a)pyrene in human fibroblast and epithelial cells. II. Differences in metabolism to water-soluble products and aryl hydrocarbon hydroxylase activity. Int J Cancer 19:378–382

13. Huberman E, Sachs L (1973) Metabolism of the carcinogenic hydrocarbon benzo(a)pyrene in human fibrobalst and epithelial cells. Int J Cancer 11:412–418

14. Whitlock JP, Jr, Gelboin HB, Coon HG (1976) Variation in aryl hydrocarbon (benzo[a]pyrene) and predominantly diploid rat liver cells in culture. J Cell Biol 70:217–225

15. Thorgeirsson SS, Nebert DW (1977) The Ah locus and the metabolism of chemical carcinogens and other foreign compounds. In: Klein G, Weinhouse S (eds) Adv. cancer res., Academic Press, New York, pp 149–193

16. Boobis AR, Nebert DW (1977) Genetic differences in the metabolism of carcinogens and in the binding of benzo[a]pyrene metabolites to DNA. In: Weber G (ed) Adv enz regul 25:339–362

17. Okuda T, Vesell ES, Plotkin E, Tarone R, Bast RC, Gelboin HV (1977) Interindividual and intra-individual variations in aryl hydrocarbon hydroxylase in monocytes from monozygotic and dizygotic twins. Cancer Res 37:3904–3911

18. McLemore TL, Martin RR, Busbee DL, Richie RC, Springer RR, Toppell KL, Cantrell ET (1977) Aryl hydrocarbon hydroxylase activity in pulmonary macrophages and lymphocytes from lung cancer and non cancer patients. Cancer Res 37:1175–1181

19. Kellerman G, Luyten-Kellerman M, Shaw CR (1973) Genetic variation of aryl hydrocarbon hydroxylase in human lymphocytes. Am J Human Genet 25:327–331

20. Becker JF, Bartholomew JC (1979) Aryl hydrocarbon hydroxylase induction in mouse liver cells – relationship to position in the cell cycle. Chem Biol Inter 26:257–266

21. Gehly EB, Fahl WE, Jefcoate CR, Heidelberger C (1979) The metabolism of benzo(a)pyrene by cytochrome P-450 in transformable and nontransformable C3H mouse fibroblasts. J Biol Chem 254:5041–5048

22. Gehly EB, Heidelberger C (1982) Metabolic activation of benzo(a)pyrene by transformable and nontransformable C3H mouse fibroblasts in culture. Cancer Res 42:2696–2704

23. Eagle H (1969) Amino acid metabolism in mammalian cell cultures. Science 130:432

24. Culp LA, Black PH (1972) Release of macromolecules from Balb/c mouse cell lines treated with chelating agents. Biochem 11:2161–2171

25. Hawkes SP, Bartholomew JC (1977) Quantitative determination of transformed cells in a mixed population by simultaneous fluorescence analysis of cell surface and DNA in individual cells. Proc Natl Acad Sci USA 74:1626–1630

26. Meehan T, Straub K, Calvin M (1976) Elucidation of hydrocarbon structure in an enzyme-catalyzed benzo(a)pyrene-poly(G) covalent complex. Proc Nat Acad USA 73:1437–1441

27. Huberman E, Sachs L, Yang SK, Gelboin HV (1976) Identification of mutagenic metabolites of benzo[a]pyrene in mammalian cells. Proc Natl Acad Sci USA 73:607–611

28. Wood AW, Levin W, Lu AYH, Yagi H, Hernandez O, Jerian DM, Conney AH (1976) Metabolism of benzo[a]pyrene and benzo[a]pyrene derivatives to mutagenic products by highly purified hepatic microsomal enzymes. J Biol Chem 251:4882–4890

29. Wislocki PG, Wood AW, Chang RL, Levin W, Yagi H, Hernandez O, Jerina DM, Conney AH (1976) High mutagenicity and toxicity of a diol epoxide derived from benzo[a]pyrene. Biochem Biophys Res Commun 68:1006–1012

30. Wood AW, Wislocki PG, Chang RL, Levin W, Yagi H, Hernandez O, Jerina DM, Conney AH (1976) Mutagenicity and cytotoxicity of benzo(a)pyrene benzo-ring epoxides. Cancer Res 36:3358–3366

31. Thorell B (1981) Flow cytometric analysis of cellular endogenous fluorescence simultaneously with emission from exogenous fluorochromes, light scatter and absorption. Cytometry 2:39–43

32. Hsu IC, Poirier MC, Yuspa SH, Grunberger D, Weinstein IB, Yoken RH, Harris CC (1981) Measurement of Benzo(a)pyrene-DNA adducts by enzyme immunoassay. Cancer Res 41:1091–1095

33. Montesano R, Rajewski MF, Pegg AE, Miller E (1982) Development and possible use of immuno-logical techniques to detect individual exposure to carcinogens: international agency for research on cancer/international programme on chemical satety working group report. Cancer Res 42:5236–5239

34. Seidman M, Mizusawa H, Slor H, Bustin M (1983) Immunological detection of carcinogen-modi-fied DNA fragments after in vivo modification of cellular and viral chromatin. Cancer Res 43:743–748

High-Affinity Monoclonal Antibodies Specific for Deoxynucleosides Structurally Modified by Alkylating Agents: Applications for Immunoanalysis

J. Adamkiewicz, O. Ahrens, and M.F. Rajewsky

Institut für Zellbiologie (Tumorforschung), Universität Essen (GH), Hufelandstrasse 55, D-4300 Essen 1, FRG

Introduction

Mutagenic and carcinogenic chemicals, most cancer chemotherapeutic agents, as well as ionizing radiation and ultraviolet light, cause structural alterations of DNA in the chromatin of target cells (see Town et al. 1973; Lawley 1976; Wang 1976; Grover 1979; Rajewsky 1980; Waring 1981; Singer and Kuśmierek 1982). In the case of alkylating chemical agents, covalent binding generally occurs between nucleophilic centers (electron-rich N and O atoms) in DNA and strongly reactive, electrophilic derivatives, generated from the respective parent compounds either enzymatically ("metabolic activation") or via non-enzymatic decomposition (see Magee 1974; Miller and Miller 1979, 1981; Gelboin 1980). As for most of the DNA alterations caused by various types of radiation, the precise structure of many chemical DNA adducts remains to be elucidated. Nevertheless, several classes of DNA-reactive chemicals have already been very thoroughly studied with regard to their DNA binding products, and in relation to the potential consequences of these DNA adducts in terms of cytotoxicity, mutagenesis, or malignant transformation. Thus, the various types of structural modifications of DNA resulting from the exposure of cells to alkylating N-nitroso compounds (alkylnitrosoureas, alkylnitrosoamines, alkylnitro-nitroso-guanidines) have for the most part been identified (Lawley 1976; Grover 1979; Singer 1979; Rajewsky 1980, 1983).

The exposure of cells to DNA-reactive chemicals *in vivo* or in culture at subtoxic dose levels generally leads to the formation of specific reaction products in DNA at very low frequencies. Therefore, highly sensitive methods are required for their detection and quantitation, for dosimetry in terms of the degree of specific DNA modification in cells, tissues, or individuals, and for the study of particular types of DNA adducts in relation to DNA repair, mutagenesis, or carcinogenesis. The sensitivity of conventional radiochromatographic techniques (Baird 1979; Beranek et al. 1980) is limited by the specific radioactivity of the respective radiolabeled agents and by the relatively large amounts of DNA (cells) required for analysis. As a rule of thumb, these methods will detect one alkylated base in $\sim 10^6$ molecules of the corresponding unmodified base. With the exception of recently developed ^{32}P "post-labeling" techniques (Randerath et al. 1981; Gupta et al. 1982; Lo et al. 1982), radiochromatography requires the use of radioactively labeled, i.e., laboratory-synthesized chemicals. These conditions preclude the detection of specific carcinogen-DNA adducts in (e.g., human) tissues and cells exposed to low doses

Biological Dosimetry. Edited by W.G. Eisert and M.L. Mendelsohn
© Springer-Verlag Berlin Heidelberg 1984

of non-radioactive (e.g., environmental) agents. This situation has been changed by the recent development of immunoanalytical procedures for the detection and quantitation of specific carcinogen-DNA adducts (Rajewsky et al. 1980; Müller and Rajewsky 1981; Poirier 1981; Adamkiewicz et al. 1982; Müller et al. 1982). Immunoglobulins are characterized by an exceptional capability to recognize subtle alterations of molecular structure. Both the specificity and the sensitivity of immunoanalysis either by antibodies purified from conventional antisera or by monoclonal antibodies are, therefore, extremely high. Furthermore, the antigens (i.e., haptens in the case of naturally occurring or carcinogen-modified nucleosides need not be radioactively labeled. Earlier pioneering studies by Erlanger and his associates (Erlanger and Beiser 1964; Beiser et al. 1968; Erlanger 1973; Meredith and Erlanger 1979), Plescia et al. (1964), Plecia (1968), Halloran and Parker (1966), Grabar et al. (1968), and Stollar and Borel (1976), have provided the basis for the development of immunological assays for the quantitation of a number of specific carcinogen-DNA adducts in DNA hydrolysates, notably for deoxynucleosides modified by N-nitroso compounds (Briscoe et al. 1978; Müller and Rajewsky 1978, 1980, 1981; Kyrtopoulos and Swann 1980; Rajewsky et al. 1980, Adamkiewicz et al. 1982; Müller et al. 1982; Saffhill et al. 1982; Van der Laken et al. 1982), N-acetoxy-N-2-acetylaminofluorene (Poirier et al. 1977, 1979; Leng et al. 1978; DeMurcia et al. 1979; Sage et al. 1979; Spodheim-Maurizot et al. 1979; Hsu et al. 1980; Kriek et al. 1982; Van der Laken et al. 1982), benzo(a)pyrene (Poirier et al. 1980; Hsu et al. 1981), and by aflatoxin B_1 (Haugen et al. 1981; Groopman et al. 1982).

As part of a research programme on molecular mechanisms in chemical carcinogenesis (Rajewsky et al. 1977; Rajewsky 1982, 1983), we have developed (and are further expanding) a collection of high-affinity monoclonal antibodies directed against alkyl-deoxynucleosides produced in DNA as a result of the exposure of mammalian cells to N-nitroso carcinogens (Rajewsky et al. 1980; Adamkiewicz et al. 1982; Müller et al. 1982). This collection presently includes monoclonal antibodies specific for O^6-ethyl-2'-deoxyguanosine (O^6-EtdGuo), O^6-butyl-2'-deoxyguanosine (O^6-BudGuo), O^6-isopropyl-2'-deoxyguanosine (O^6-iProdGuo), and O^4-ethyl-2'-deoxythymidine (O^4-EtdThd).

2 Materials and Methods

2.1 Alkyl-Ribonucleosides and Deoxynucleosides, and Radioactive Tracers for Competitive Radioimmunoassay (RIA)

Alkyl-ribonucleosides were used as haptens and coupled to the carrier protein keyhole limpet hemocyanin (KLH; molecular weight $\sim$800 000; $\sim$100 hapten molecules bound per molecule of KLH) for immunization (see Müller et al. 1982). O^6-ethylguanosine (O^6-EtGuo), O^4-ethylthymine riboside (O^4-EtrThd), and O^6-isopropylguanosine (O^6-iProGuo), were synthesized and purified by thin-layer chromatography on silica gel and Sephadex G-10 chromatography (see Adamkiewicz et al. 1982). O^6-n-butylguanosine (O^6-BuGuo) was a gift from Dr. R. Saffhill (Paterson Laboratories, Manchester, U.K.).

O^6-EtdGuo, O^6-BudGuo, O^4-EtdThd, O^6-iProdGuo, and various other alkylated nucleic acid components, were synthesized and purified as described (see Adamkiewicz et al. 1982). For use in competitive RIAs, radioactive tracers with high specific [^{3}H]-activity were synthesized according to Müller and Rajewsky (1978, 1980) and Rajewsky et al. (1980).

2.2 Immunization

Adult female rats of the inbred BDIX strain (Druckrey 1971) were immunized by injection (into about 20 intracutaneous sites) of an emulgated mixture of 50 μg of hapten-KLH conjugate in 50 μl of phosphate buffered saline (PBS), 50 μl of aluminium hydroxide and 100 μl of complete Freund's adjuvant. The animals were boosted by the same procedure after 5 weeks. A second intraperitoneal (i.p.) booster injection was administered 8 weeks later, and spleen cells were collected 3–4 days thereafter.

2.3 Fusion of Spleen Cells with Myeloma Cells, and Culture and Cloning of Monoclonal Antibody-Secreting Hybridoma Cells

As previously described (Rajewsky et al. 1980; Adamkiewicz et al. 1982), cells of the HAT (hypoxanthine/aminopterine/thymidine)-sensitive rat myeloma line Y3-Ag.1.2.3 (Galfré et al. 1979) were fused with spleen cells from immunized BDIX-rats, using polyethylene glycol (PEG 4000). The hybrid cells were cultured in multi-well tissue culture plates (1–2 x 10^6 cells/well) containing Dulbecco's modified Eagle's-HAT medium, with 1mM sodium pyruvate, 5 x 10^{-5}M mercaptoethanol, and 20% fetal bovine serum. Culture supernatants were tested for the presence of specific antibodies by an enzyme linked immunosorbent assay (ELISA). Hybridoma cells from positive cultures were cloned, and once more recloned, without aminopterine in the presence of X-irradiated (40 Gy) BDIX-rat spleen cells as "feeders". Aliquots of the positive clones maintained in culture were injected i.p. into Pristan-pretreated, X-irradiated (4 Gy) BDIX-rats for growth and antibody production in the ascites. Isotype analysis of the monoclonal antibodies was carried out by the Ouchterlony technique (immunoprecipitation in agar), using anti-rat-isotype antisera. Antibody concentrations in cell culture media or ascitic fluid, and the antibody affinity constants for the respective alkyl-deoxynucleosides, were calculated from the data obtained by competitive RIA (Müller 1980). Antibodies were isolated with the aid of specific hapten-immunosorbents (haptens coupled to epoxy-activated Sepharose 6B) at acid or alkaline pH (Müller and Rajewsky 1980).

2.4 Enzymatic Hydrolysis of DNA

DNA isolated from rat tissues by a modified Kirby method (Goth and Rajewsky 1974) or by hydroxylapatite adsorption (Müller and Rajewsky 1980), was enzymatically hydrolyzed to nucleosides with DNase I (EC 3.1.4.5), snake venom phosphodiesterase

(EC 3.1.4.1), and grade I alkaline phosphatase (EC 3.1.3.1), as described (Müller and Rajewsky 1980). Adenosine deaminase (EC 3.5.4.4; 0.3 units/ml) was sometimes used to convert deoxyadenosine (dAdo) to deoxyinosine (dIno) in the DNA hydrolysates prior to analysis (Müller and Rajewsky 1980). After separation by high pressure liquid chromatography (HPLC), 2'-deoxyguanosine (dGuo) and 2'-deoxythymidine (dThd) concentrations in the DNA hydrolysates were determined spectrophotometrically by peak integration. HPLC was also used for the separation of different alkylation products from the same DNA sample, prior to analysis by competitive RIA.

2.5 Competitive Enzyme-Linked Immunosorbent Assay (ELISA) and Competitive Radioimmunoassay (RIA)

The competitive ELISA was performed in 96-well microtiter plates coated with bovine serum albumin conjugates of the respective haptens (see Adamkiewicz et al. 1982). To test cell culture supernatants for the presence of monoclonal antibodies, these were added to duplicate wells (without or with the respective alkyl-deoxynucleoside at a final concentration of 0.15 mM), incubated at 37°C for 90 min, and subsequently washed with PBS containing Triton X-100 (0.05%). Alkaline phosphatase-conjugated goat-anti-rat IgG was then added, and the incubation mixture was removed after another incubation for 90 min at 37°C. The wells were thoroughly washed, and incubated for 1 h at 37°C with a 10 mM solution of p-nitrophenyl phosphate (the alkaline phosphatase substrate). Thereafter, the plates were either screened for positive wells by eye (wells without added alkyldeoxynucleoside, yellow; wells containing the respective alkyldeoxynucleoside, colorless), or the degree of binding of alkaline phosphatase-conjugated antibody to the solid phase was determined spectrophotometrically.

The competitive RIA (Farr 1958) was carried out as described (Müller and Rajewsky 1978, 1980). In a total volume of 100 μl of Tris-buffered saline (with 1% bovine serum albumin [w/v] and 0.1% bovine IgG [w/v]), each sample contained $\sim$2.5 x 10^3 dpm of the [^{3}H]-labeled tracer, and antibody solution diluted to give 50% binding of tracer (in the absence of inhibitor), plus varying amounts of hydrolyzed alkylated DNA, or of other natural or modified deoxynucleosides to be tested for cross-reactivity (inhibitor). The samples were incubated for 2 h at room temperature (equilibrium), and 100 μl of a saturated ammonium sulphate solution (pH 7.0) were then added. After 10 min the samples were centrifuged (3 min at 10.000 x g), and the [^{3}H]-activity in 150 μl of supernatant was measured by liquid scintillation spectrometry. The degree of inhibition of tracer-antibody binding (ITAB) was calculated as described by Müller and Rajewsky (1980).

3 Results and Discussion

Monoclonal hybridoma-secreted antibodies (Köhler and Milstein 1975), contrary to polyclonal antisera raised in animals, are unlimited in quantity and not contaminated with nonspecific antibodies. Therefore, monoclonal antibodies are reagents of the highest possible degree of purity which guarantee reproducible analytical results over extended periods of time. While rather high antibody purity can sometimes also be obtained by extensive affinity purification of specific antibodies from polyclonal antisera (see, e.g., Müller and Rajewsky 1980), such antiserum-derived antibody preparations remain limited in quantity and cannot match monoclonal antibodies with respect to purity.

The detection limit of a competitive RIA for a given alkyl-deoxynucleoside is primarily defined by the affinity constant of the respective antibody. However, even when antibodies are available with very high affinity constants for the alkyl-deoxynucleoside in question, the limit of detection by competitive RIA remains dependent on the specific radioactivity of the tracer which cannot be increased above a given value. Moreover, antibody specificity (i.e. the degree of cross-reactivity with other DNA constituents) becomes a most important factor in the analysis of hydrolysates of DNA containing both natural and modified deoxynucleosides. At a given level of cross-reactivity, the detection limit cannot be lowered, unless prior to the RIA the concentration of cross-reactants can either be reduced, or the cross-reactants can be completely removed, e.g., by separating the alkylation products to be measured from the complete DNA hydrolysate with the aid of HPLC (Adamkiewicz et al. 1982).

The best anti-alkyl-deoxynucleoside monoclonal antibodies of our present collection have antibody affinity constants $\geq 10^{10}$ l/mol (Rajewsky et al. 1980; Adamkiewicz et al. 1982). Table 1 shows some examples of high-affinity monoclonal antibodies directed against deoxyguanosines with different alkyl residues covalently linked to the oxygen-6-atom. When tested by competitive RIA, many of these monoclonal antibodies cross-react with normal DNA constituents only to a very low degree; i.e., they often require a $\rangle 10^7$ times higher amount of the corresponding unaltered deoxynucleosides to inhibit tracer-antibody binding by 50%. However, not all monoclonal antibodies in our collection exhibit this extreme specificity. Therefore, a sufficiently high number of hybridomas should be produced if monoclonal antibodies with maximum affinity and specificity for a given modified deoxynucleoside are desired. At 50% inhibition of tracer-antibody binding in the competitive RIA, our best monoclonal antibodies directed against O^6-EtdGuo will detect 40–60 fmol of this alkylation product in a 100 μl RIA-sample. When a probability grid is used to transform the sigmoidal inhibition curves into straight lines (Müller and Rajewsky 1980), accurate readings are possible at inhibition values $\langle 50\%$. Thus, using anti-(O^6-EtdGuo) monoclonal antibody ER-6 (Rajewsky et al. 1980), reading at 20% inhibition reduces the detection limit for O^6-EtdGuo to $\sim$10 fmol in a 100 μl RIA-sample. This corresponds to an O^6-EtdGuo/dGuo molar ratio as low as $\sim$3x10^{-7} (i.e., the equivalent of $\sim$700 O^6-EtdGuo molecules per diploid mammalian genome) in a hydrolysate of 100 μg of DNA (equivalent to the DNA content of $\sim$1.6 x 10^7 diploid mammalian cells).

One of the most important applications of monoclonal antibodies will obviously be the direct detection of specific DNA modifications in individual cells with the aid of immunostaining procedures. This approach will make it possible to demonstrate the presence, and to monitor the level, of specific DNA adducts (resulting, e.g., from exposure to carcinogens or chemotherapeutic agents) in very small samples of (e.g., human) cells. Moreover, different types of cells at different stages of differentiation

Table 1. Rat x rat hybridoma-secreted, high-affinity monoclonal antibodies specific for different O6-alkyl-deoxyguanosines (examples from a collection of anti-alkyl-deoxynucleoside monoclonal antibodies developed at the Institut für Zellbiologie[Tumorforschung], University of Essen, F.R.G.; Adamkiewicz et al. 1982; J. Adamkiewicz and M.F. Rajewsky, in preparation)

Alkyl-deoxy-nucleoside	Immunogen	Designation of antibody (isotype)	Antibody affinity constant (l/mol)	Radioactive tracer (specific [3H]-activity)	Detection limit of RIA (fmol/100 μl-sample) [a]	Cross-reactivity with the constituents of unmodified DNA[b]
O6-EtdGuo	O6-EtGuo-KLH	ER-6 (IgG2b)	2.0×10^{10}	O6-Et[8,5'-3H]dGuo ($\sim$27 Ci/mmol)	40	2.5×10^{6}
O6-BudGuo	O6-EtGuo-KLH	ER-11 (n.d.)	9.1×10^{9}	O6-n-Bu-[8,5'-3H]dGuo ($\sim$22 Ci/mmol)	60	2.5×10^{6}
O6-iProdGuo	O6-iProGuo-KLH	ER-1005 (IgG2a)	1.2×10^{10}	O6-iPro-[8,5'-3H]dGuo ($\sim$17 Ci/mmol)	50	2.4×10^{6}

[a] Determined at 50% Inhibition of Tracer-Antibody Binding (ITAB) in the competitive RIA

[b] Cross-reactivity determined for a mixture of dGuo, dIno, dCyd, and dThd, at the molar ratios present in an adenosine deaminase-treated hydrolysate of rat DNA. Values given as the multiple of the 50% ITAB value for the respective modified deoxynucleoside, but determined at 10−15% ITAB (because value for 50% ITAB not reached by the ITAB-curve within the concentration range of unmodified deoxynucleosides added to the RIA sample).

n.d.: not determined

and/or development, could be analyzed with respect to their capacity for enzymatic activation of chemicals to the respective DNA-binding derivatives, and for the elimination of specific adducts from DNA (DNA repair). This might, for example, facilitate, the detection of individuals with hereditary defects in DNA repair enzymes (risk groups in terms of carcinogen exposure), and the prediction of individual response to cancer chemotherapy.

So far the results of our attempts to use monoclonal antibodies for the demonstration of carcinogen-DNA adducts in cells by immunostaining have been promising. Thus we have established a standardized procedure for the quantitation of specific alkyl-deoxynucleosides in the nuclear DNA of individual cells by direct immunofluorescence, using tetramethylrhodamine isothiocyanate-labeled monoclonal antibodies and a computer-based image analysis of electronically intensified fluorescence signals (Adamkiewicz et al. 1983; J. Adamkiewicz, O. Ahrens, and M.F. Rajewsky, in preparation). With a fluorescent anti-(O^6-EtdGuo) monoclonal antibody, the present detection limit for O^6-EtdGuo in the nuclei of individual cells previously exposed to an ethylating N-nitroso compound (e.g., N-ethyl-N-nitrosourea) is ~700 O^6-EtdGuo molecules per diploid genome, i.e., similar to the detection limit for the same ethylation product in a hydrolysate of (O^6-EtdGuo)-containing DNA analyzed by competitive RIA (see above). Furthermore, monoclonal antibodies permit the direct localization of specific alkyl-deoxynucleosides in individual DNA molecules by immune electron microscopy (Nehls et al. 1983). Thus O^6-EtdGuo can be localized in double-stranded DNA by the binding of monoclonal antibody ER-6 without requirement for a second antibody carrying an electron-dense label (Nehls et al. 1983, 1984).

Acknowledgements. This work was supported by the Deutsche Forschungsgemeinschaft (SFB 102/ A9) and by the Commission of the European Communities (ENV-544-D[B]).

References

Adamkiewicz J, Drosdziok W, Eberhardt W, Langenberg U, Rajewsky MF (1982) High-affinity monoclonal antibodies specific for DNA components structurally modified by alkylating agents. In: Bridges BA, Butterworth BE, Weinstein IB (eds) Indicators of genotoxic exposure. Banbury Report 13. Cold Spring Harbor Laboratory, Cold Spring Harbor, New York, pp 265–276

Adamkiewicz J, Ahrens O, Huh N, Rajewsky MF (1983) Quantitation of alkyl-deoxynucleosides in the DNA of individual cells by high-affinity monoclonal antibodies and electronically intensified, direct immunofluorescence. J Cancer Res Clin Oncol 105:A15

Baird WM (1979) The use of radioactive carcinogens to detect DNA modification. In: Grover PL (ed) Chemical carcinogens and DNA. CRC Press, Boca Raton, pp 59–83

Beiser SM, Tanenbaum SW, Erlanger BF (1968) Purine- and pyrimidine-protein conjugates. Meth Enzymol 12:889–893

Beranek DT, Weis CC, Swenson DH (1980) A comprehensive quantitative analysis of methylated and ethylated DNA using high pressure liquid chromatography. Carcinogenesis 1:595–606

Briscoe WT, Spizizen J, Tan EM (1978) Immunological detection of O^6-methylguanine in alkylated DNA. Biochemistry 17:1896–1901

DeMurcia G, Lang MCE, Freund AM, Fuchs RPP, Daune MP, Sage E, Leng M (1979) Electron microscopic visualization of N-acetoxy-N-2-acetylaminofluorene binding sites in Col E-1 DNA by means of specific antibodies. Proc Natl Acad Sci USA 76:6076–6080

Druckrey H (1971) Genotypes and phenotypes of ten inbred strains of BD-rats. Arzneim Forsch 21: 1274–1278

Erlanger BF (1973) Principles and methods for the preparation of drug-protein conjugates for immunological studies. Pharmacol Rev 25:271–280

Erlanger BF, Beiser SM (1964) Antibodies specific for ribonucleosides and ribonucleotides and their reaction with DNA. Proc Natl Acad Sci USA 52:68–74

Farr RS (1958) A quantitative immunological measure of the primary interaction between I BSA and antibody. J Infect Dis 103:239–262

Galfré G, Milstein C, Wright B (1979) Rat x rat hybrid myelomas and a monoclonal anti-Fd portion of mouse IgG. Nature 277:131–133

Gelboin HV (1980) Benzo[a]pyrene metabolism, activation, and carcinogenesis: Role and regulation of mixed-function oxidases and related enzymes. Physiol Rev 60:1107–1166

Goth R, Rajewsky MF (1974) Molecular and cellular mechanisms associated with pulse-carcinogenesis in the rat nervous system by ethylnitrosourea: ethylation of nucleic acids and elimination rates of ethylated bases from the DNA of different tissues. Z Krebsforsch 82:37–64

Grabar P, Avrameas S, Taudou B, Salomon JC (1968) Formation and isolation of antibodies specific for nucleosides. In: Plescia OJ, Braun W (eds) Nucleic acids in immunology. Springer, Berlin New York, pp 79–87

Groopman JD, Haugen Å, Goodrich GR, Wogan GN, Harris CC (1982) Quantitation of aflatoxin B_1-modified DNA using monoclonal antibodies. Cancer Res 42:3120–3124

Grover PL (ed) (1979) Chemical carcinogens and DNA, Vol I and II. CRC Press, Boca Raton

Gupta RC, Reddy MV, Randerath K (1982) ^{32}P-postlabeling analysis of non-radioactive aromatic carcinogen-DNA adducts. Carcinogenesis 3:1081–1092

Halloran MJ, Parker CW (1966) The preparation of nucleotide-protein conjugates: Carbodiimides as coupling agents. J Immunol 96:373–378

Haugen Å, Groopman JD, Hsu IC, Goodrich GR, Wogan JN, Harris CC (1981) Monoclonal antibody to aflatoxin B_1-modified DNA detected by enzyme immunoassay. Proc Natl Acad Sci USA 78: 4124–4127

Hsu IC, Poirier MC, Yuspa SH, Yolken RH, Harris CC (1980) Ultrasensitive enzymatic radioimmunoassay (USERIA) detects femtomoles of acetylaminofluorene-DNA adducts. Carcinogenesis 1:455–458

Hsu IC, Poirier MC, Yuspa SH, Grunberger D, Weinstein IB, Yolken RH, Harris CC (1981) Measurement of benzo(a)pyrene-DNA adducts by enzyme immunoassay and radioimmunoassay. Cancer Res 41:1091–1095

Köhler G, Milstein C (1975) Continuous cultures of fused cells secreting antibody of predefined specificity. Nature 256:495–497

Kriek E, van der Laken CJ, Welling M, Nagel J (1982) Immunological detection and quantification of the reaction products of 2-acetylaminofluorene with guanine in DNA. In: Bartsch H, Armstrong B (eds) Host factors in human carcinogenesis. IARC Scientific Publications No 39. International Agency for Research on Cancer, Lyon, pp 541–549

Kyrtopoulos SA, Swann PF (1980) The use of radioimmunoassay to study the formation and disappearance of O^6-methylguanine in mouse liver satellite and main-band DNA following dimethylnitrosamine administration. J Cancer Res Clin Oncol 98:127–138

Lawley PD (1976) Carcinogenesis by alkylating agents. In: Searle CE (ed) Chemical carcinogens. ACS Monograph No 173. American Chemical Society, Washington, D.C., pp 83–244

Leng M, Sage E, Fuchs RPP, Daune MP (1978) Antibodies to DNA modified by the carcinogen N-acetoxy-N-2-acetylaminofluorene. FEBS Lett 92:207–210

Lo KM, Franklin WA, Lippke JA, Henner WD, Haseltine WA (1982) New methods for the detection of DNA damage to human cellular DNA by environmental carcinogens and anti-tumor drugs. In: Bridges BA, Butterworth BE, Weinstein IB (eds) Indicators of genotoxic exposure, Banbury Report 13. Cold Spring Harbor Laboratory, Cold Spring Harbor, New York, pp 253–263

Magee PN (1974) Activation and inactivation of chemical carcinogens and mutagens in the mammal. In: Campbell PN, Dickens F (eds) Essays in biochemistry, Vol 10. Academic Press, London, pp 105–136

Meredith RD, Erlanger BF (1979) Isolation and characterization of rabbit anti-m^{7}G-5'-P antibodies of high apparent affinity. Nucleic Acids Res 6:2179–2191

Miller EC, Miller JA (1981) In search of ultimate chemical carcinogens and their reactions with cellular macromolecules. In: Fortner JG, Rhoads JE (eds) Accomplishments in cancer research 1980. Lippincott JB, Philadelphia, pp 63–98

Miller JA, Miller EC (1979) Perspectives on the metabolism of chemical carcinogens. In: Emmelot P, Kriek E (eds) Environmental carcinogenesis. Elsevier/North-Holland Biomedical Press, Amsterdam, pp 25–50

Müller R (1980) Calculation of average antibody affinity in anti-hapten sera from data obtained by competitive radioimmunoassay. J Immunol Meth 34:345–352

Müller R, Rajewsky MF (1978) Sensitive radioimmunoassay for detection of O^6-ethyldeoxyguanosine in DNA exposed to the carcinogen ethylnitrosourea in vivo or in vitro. Z Naturforsch 33c: 897–901

Müller R, Rajewsky MF (1980) Immunological quantification by high affinity antibodies of O^6-ethyldeoxyguanosine in DNA exposed to N-ethyl-N-nitrosourea. Cancer Res 40:887–896

Müller R, Rajewsky MF (1981) Antibodies specific for DNA components structurally modified by chemical carcinogens. J Cancer Res Clin Oncol 102:99–113

Müller R, Adamkiewicz J, Rajewsky MF (1982) Immunological detection and quantification of carcinogen-modified DNA components. In: Bartsch H, Armstrong B (eds) Host factors in human carcinogenesis. IARC Scientific Publications No 39. International Agency for Research on Cancer, Lyon, pp 463–479

Nehls P, Spiess E, Rajewsky MF (1983) Visualization of O^6-ethylguanine in ethylnitrosourea-treated DNA by immuno electron microscopy. J Cancer Res Clin Oncol 105:A23

Nehls P, Rajewsky MF, Spiess E, Werner D (1984) Highly sensitive sites for guanine-O^6-ethylation in rat brain DNA exposed to N-ethyl-N-nitrosourea in vivo. EMBO J 3: in press

Plescia OJ (1968) Preparation and assay of nucleic acids as antigens. Meth Enzymol 12:893–899

Plescia OJ, Braun W, Palczuk NC (1964) Production of antibodies to denatured deoxyribonucleic acid (DNA). Proc Natl Acad Sci USA 52:279–285

Poirier MC (1981) Antibodies to carcinogen-DNA adducts. J Natl Cancer Inst 67:515–519

Poirier MC, Yuspa SH, Weinstein IB, Blobstein S (1977) Detection of carcinogen-DNA adducts by radioimmunoassay. Nature 270:186–188

Poirier MC, Dubin MA, Yuspa SH (1979) Formation and removal of specific acetylaminofluorene-DNA adducts in mouse and human cells measured by radioimmunoassay. Cancer Res 39:1377–1381

Poirier MC, Santella R, Weinstein IB, Grunberger D, Yuspa SH (1980) Quantitation of benzo(a)pyrene-deoxyguanosine adducts by radioimmunoassay. Cancer Res 40:412–416

Rajewsky MF (1980) Specificity of DNA damage in chemical carcinogenesis. In: Montesano R, Bartsch H, Tomatis L (eds) Molecular and cellular aspects of carcinogen screening tests. IARC Scientific Publications No 27. International Agency for Research on Cancer, Lyon, pp 41–54

Rajewsky MF (1982) Pulse-carcinogenesis by ethylnitrosourea in the developing rat nervous system: molecular and cellular mechanisms. In: Nicolini C (ed) Chemical carcinogenesis. Plenum Press, New York, pp 363–379

Rajewsky MF (1983) Structural modifications and repair of DNA in neurooncogenesis by N-ethyl-N-nitrosourea. Rec Res Cancer Res 84:63–76

Rajewsky MF, Augenlicht LH, Biessmann H, Goth R, Hülser DF, Laerum OD, Lomakina LYa (1977) Nervous system-specific carcinogenesis by ethylnitrosourea in the rat: molecular and cellular mechanisms. In: Hiatt HH, Watson JD, Winsten JA (eds) Origins of human cancer. Cold Spring Harbor Laboratory, Cold Spring Harbor, New York, pp 709–726

Rajewsky MF, Müller R, Adamkiewicz J, Drosdziok W (1980) Immunological detection and quantification of DNA components structurally modified by alkylating carcinogens (ethylnitrosourea). In: Pullman B, Ts'o POP, Gelboin H (eds) Carcinogenesis: Fundamental mechanisms and environmental effects. Reidel, Dordrecht Boston London, pp 207–218

Randerath K, Reddy MV, Gupta RC (1981) ^{32}P-labeling test for DNA damage. Próc Natl Acad Sci USA 78:6126–6129

Saffhill R, Strickland PT, Boyle J (1982) Sensitive radioimmunoassays for O^6-N-butyldeoxyguanosine, O^2-N-butylthymidine and O^4-N-butylthymidine. Carcinogenesis 3:547–552

Sage E, Fuchs RPP, Leng M (1979) Reactivity of the antibodies to DNA modified by the carcinogen N-acetoxy-N-aminofluorene. Biochemistry 18:1328–1332

Singer B (1979) N-nitroso alkylating agents: formation and persistence of alkyl derivatives in mammalian nucleic acids as contributing factors in carcinogenesis. J Natl Cancer Inst 62:1329–1339

Singer B, Kuśmierek JT (1982) Chemical mutagenesis. Ann Rev Biochem 52:655–693

Spodheim-Maurizot M, Saint-Ruf G, Leng M (1979) Conformational changes induced in DNA by in vitro reaction with N-hydroxy-N-2-aminofluorene. Nucleic Acids Res 6:1683–1694

Stollar BD, Borel Y (1976) Nucleoside specificity in the carrier IgG-dependent induction of tolerance. J Immunol 117:1308–1313

Town CD, Smith KC, Kaplan HS (1973) Repair of X-ray damage to bacterial DNA. Curr Top Rad Res Q 8:351–399

Van der Laken CJ, Hagenaars AM, Hermsen G, Kriek E, Kuipers AJ, Nagel J, Scherer E, Welling M (1982) Measurement of O^6-ethyldeoxyguanosine and N-(deoxyguanosine-8-yl)-N-acetyl-2-aminofluorene in DNA by high-sensitive enzyme immunoassays. Carcinogenesis 3:569–572

Wang S-Y (1976) Photochemistry and photobiology of nucleic acids. Academic Press, New York

Waring MJ (1981) DNA modification and cancer. Ann Rev Biochem 50:159–192

Subject Index

1,4-diacetoxy-2,3-dicyano-benzol (ADB) 300
2,3-dicyano-hydroquinone (DCH) 300
20-methylcholanthrene 46
20-methylcholanthrens 44
4-nitroquinoline-1-oxide 46
5-fluorouracil 44
5-fluorouracil 46
6-mercaptopurine 44, 46
8-azaguanine (8-AG) 153

aberration frequencies 77
– loss 4
– – unstable 4
– yield 15, 23
– dicentric 3, 4, 6, 11
aberrations 12, 25, 171
–, chromosomal 25
–, duplicated 17
–, heterogenous 26, 29
–, stable type 79
–, unstable type 79
abnormalities 12
–, genetic 12
A-bomb dosimetry system 77
abortion
–, habitual 89
–, spontaneous 89
acentric chromatin 21
– elements 15
– fragments 26, 27, 67, 68
acentrics 17, 20, 21, 79
acid DNA/RNA denaturation 299, 301, 304
– stability 243
acridine orange (AO) 100, 156, 220, 273, 301
acriflavine-Feulgen 118
acrosome 86
adducts 146
aflatoxin B1 326
agarose coated 118
agents
–, alkylating 63, 64, 150, 325
–, chemotherapeutic 325
–, mutagenic 44
–, non-mutagenic 44

alkyldeoxynucleosides O6-ethyl-2'-deoxyguanosine
    (O6-EtdGuo) 326
alkylnitro-nitroso-guanidines 325
alkylnitrosoamines 325
alkylnitrosoureas 325
alkylribonucleosides 326
alloantigen 233
alloenzyme 190
allogeneic stimulation 233
ames test 143, 217
amino acid substitution 162
anaphase 171
– bridge 16
andrological methods 85
aneuploid cell lines 37
aneuploids 135
aneuploidy 88
–, autosomal 89
– gaps 94
ANLL 62
–, promyelocytic 62
anoxia 305
antibodies 161, 162, 186, 235, 326
aryl hydrocarbon hydroxylase (AHH) 311
automatic counts 257
autoradiograph 153, 155, 256
autoradiographic analysis 94, 244, 255
azaguanine 150

background factor 72
– frequency 15
– signals 67, 71
badge 8
banding techniques 61
BaP diol epoxide 314, 318
base substituiton 143
beam splitter
– –, bifringent 237
– –, polarizing 237, 238
B cells 229
benzo(a)pyrene 43, 44, 46, 121, 311, 313,
    326
biochemical pathways 299
biologic function 203

biological dose estimates
– – sensitivity   17
– dosimetry   38, 127, 141, 216, 240,
   143
– effects   51
– indicator system   15
– and physical methods   8
– vectors   203
blood cells
– –, human mononuclear   240
– forming cells   188
– group determination   166
bone marrow
– – cells   40, 43, 183, 219
– – –, rat   40
– – grafts   196
– – murine   39, 40, 45, 232
– – syndrome   183, 195
brain syndrome   183
break repair time   7
bromdeoxyuridine (BrdU)   4, 15, 18, 144
burst promoting activity   193

C3H mice   219, 220
calibration
– in vitro   15, 20
Cancer therapy   255, 331
carbon tetrachloride   102
carcinogen damage   178
– exposure   331
– metabolism   312
carcinogenic   285, 311
– potential   92
carcinogen(s)   40, 85, 93, 171, 325
cauda epidydimides   112
causative agent   64
cell(s)
– classes   229, 295
– cycle   51, 53, 217, 255, 273, 301, 312
– – kinetics   219, 220, 273, 275
–, cycling   61, 301
– debris   68
– division   41
– –, first   4, 18
– –, – in vitro   79
– growth   240
–, Hela   45
– heterogeneity   299
– inactivation   133
– interphase   275
–, kinetic analysis   220
–, L-cells   41
– length   240
– maturation kinetics   102

– metabolism   309
–, mitotic   275, 279
–, mononuclear   220
– morphology   216
–, mouse liver   311
–, natural killer   250
–, non-dividing   274, 279
–, non-cycling (G_0)   301
–, null   250
– number   301
–, physiological status   244
–, polyploid   279
– populations   216
–, preneoplastic   285
–, prepartory conditions   244
– proliferation   17, 18, 173
– separation   184, 194
– sorting   161
– survival   51, 55, 258
–, synchronized   275
–, T-helper   250
–, T-suppressor   250
–, tetraploid   285
– transformation   51
– viability   301
– volume distribution   265
cellular clonogenic capacity   51
– components   312
– constituents   235
–, DNA   240
– membrane   236
centritole   86
centromere   25, 30, 68
– staining   67, 73
centromeric dips   31
CFU-S   195
chemical carcinogenesis   311, 326
– mutagens   43, 135
– noxae   135
– treatment   219
chemotherapy   103, 104
chinese hamster
– – cells   25, 45
– –, M3-1 cell   31, 32
– – ovary (CHO) cells   229
chromatid   73
chromatin   41, 205, 216, 325
– alteration   99
– changes   210
– clumping   285, 287
– condensation   99, 104
–, diffuse   99
– distribution   203, 285, 292
– features   246
–, isostained   205

chromatin matrix   243, 247
− stability   244
− stained   203
− structure   105, 243, 301, 305
− studies   105
− texture   203, 295
chromomycin A3   25, 26
chromosomal aberrations   37, 44, 194
− abnormalities   61, 62, 89, 117
− analysis of human sperm   93
− translocations   117
chromosome aberration   11, 51, 265
− − analysis   12
− − data   20
− − dosimetry   3, 9
− − frequency   11, 32, 55
− −, heterogenous   25
− −, homogenous   25
− −, inherited   26
− − yield   17
− aberrations   4, 11, 51−53, 55, 77, 79, 265
− −, exchange-type   79
− −, numerical   61
− −, radiation-induced   15
− analysis   61
− −, centromere   25, 30, 68
− −, conventional   31
− aneuploidy   123
− changes   61
− −, non-random   61
− −, structural   15
−, Chinese hamster   25
− classification   25
− damage   3, 51, 52, 67, 73, 171
− −, artificial   57
− dosimetry   3, 9
− examinations
− −, fixed cells   11
− −, hair sheaths   11
− −, skin   11
− fragments   171, 172
− frequencies (dicentric)   32
− histogram   53, 55
−, human   25, 67
−, kangaroo rat   25
− lesions   7
−, mouse   25
−, muntjac   25
− number   117
− patterns   61
− polymorphism (constitutional)   72
− preparations (non-banded)   79
− profile   30
− staining   69, 72
− suspensions (monodisperse)   51, 73

chromosome DNA flow histograms   52, 55
chronic myelogeneous leukemia human   26
− myeloproliferative disorder   63
chronically exposed to radiation   8
circulating lymphocytes   216
clastogenic agents   26
− events   41
clastogens   4, 52
clinical management   10
clonal proliferation   63
clones   52, 53
clonogenic capacity   51, 52, 53, 256
− cells   259
clonogenicity   55
cobalt-60 gamma   7
colcemid   31, 41, 68, 69
colchicine   43, 44, 46, 229
colon carcinoma cells   25, 26
colony formation   184
− morphology   144
colony forming cells   259
color hue   296
committed progenitor cells   193
compartment   11
computer analysis   295
− generated histogram   52
− graphic   295
− sorting   171
Concanavalin A (Con A)   204, 232
condensation   99
confidence limits   10, 27
contour plot   27, 225
conventional staining   17
counterstain   73
count, sperm   105
critically accident   5
cross-linking agent   150
crystal violet   171
culture conditions   7, 18
−, quasi-continous   274
−, times   22
cultures, cell lines   40
−, lymphocyte   4
culturing period   4
cyclophosphamide 40, 44, 46, 203, 210, 214
cytogenetic analysis   4, 127
− damage   37, 44
− dosimetry   5, 20, 23
− effects   1
− examination   78
− methods, conventional   3, 51
− samples   78
− scoring, conventional 171
− study   77
cytoplasmic droplet   86

cytoplasmic membrane 229
– proteins 249
cytostatic agents 41
cytotoxic action 127
– agents 309
– effect 158
– precursor 233
– therapy 299
cytotoxicity 325

damage, chemical 243
–, radiation 243
DAPI 40
data aquisition 225, 236, 239, 240
debris frequency 123
decision space 216
decondensation 153
–, proteolytic 120
degree of depolarization 235, 239
– – polarization 239
deletion 15, 143
denaturation (in situ) 106
densitometric system 173
density gradient separation 195
deoxynucleosides 325
depolarization, fluorescence 236
depolymerization 243
depurination 243
dicentric chromosomes 4, 10–12, 15–17,
    17–21, 25, 26, 57, 67, 68, 79, 94, 172
– frequency 33
– yield 5, 20
diethylnitrosamine (DENA) 106, 107, 285
differentation 295
– state 312
digitized images 204
dimethylsuberimidate 162
diploid sperm 117
discriminant function 212
dispersion, DNA content 37
disulfide bonding 99
division–delay 276
– mechanism 273
dizygotic twins 311
DNA–adducts 142, 325
– content 172, 236
– – dispersion 37, 41
– – errors 111
– – histograms 51, 54, 219
– – of individual chromosomes 51
– denaturation 301
DNA-distribution 273
– distribution analysis 40, 276

DNA-hydrolysates 326
– modifications 329
– recombination 146
– repair 325
– – enzymes 331
– specific fluorescent dyes 25, 229, 266
– strand breaks 52
– synthesis 205
DNA/RNA measurements 219
DNA enzymatic hydrolysis 327
dose dependence 43, 51
– dependent alterations 285
– dependent delay 279
– effect curve
– – –, acute gamma 10
– – –, in vitro 9
– – relationship 4, 7, 9–11, 13, 16, 19, 33,
    43, 51, 104, 115, 121, 132, 146, 220, 285
– – –, linear-quadratic 16, 17
– estimation 3, 4, 10, 23, 77
– –, individual 10
– –, biological 15
– –, cytogenetic 15
– rates 5
– –, acute 7
– –, chronic 7
– response curve 6, 33, 115, 121, 220
dose-RBE relations 20
dosimeter 184
–, thermoluminescent 15
dosimetric applications 90
– sensitivity 111
dosimetry, chromosome 3, 7, 15
–, cytogenetic 3, 15
double focus arrangement 236
– y-body test 93
doubling dose 115
drug permeability 229
– resistance 144
– resistant cell 299
– – colonies 144
– sensitivity 299
– treated 212
drugs 183
dual laser analysis 167
dye, adenine-thymine specific 26
–, DNA specific 25, 229, 266
–, guanine-cytosine specific 26
dye uptake 230
dysfunction 37
–, epididymal 91
–, testicular 91

Ehrlich ascites cells (EAT)   45, 273
electrophilic derivatives   325
embryo death   100
embryonal cell carcinoma   103
embryonic gonad   89
emission spectra   314
entropy of the nuclear image   287
environmental agents   37, 61, 63
– harzards   43
– insults   203
– modifying agents   99
– pathology   295
– pollutants   285
– results   203, 212
enzymatic activity   299
enzyme-linked immunosorbent assay
   (ELISA)   328
enzyme inducible   311
enzymes   184
eosin   112
epidemiological studies   63
epi-illumination   236
epoxides   314
equivalent whole body dose   8
erythrocytes   183
–, chicken   40
–, frog   40
erythroid progenitor cells   193
erythropoietin   193
ethidium bromide   68, 107, 127, 274
exchange events   80
exposure   4, 23
–, homogenous   10
–, low level   235
extinction scanning   219

fast interval processor   150
fatty-acid chains   236
femal germ cells   94
fertility   105
fertilization   90, 117
Feulgen method   118, 150, 153, 174,
   203, 204, 210, 243, 285
fibroblasts   243
ficoll hypaque gradient   68
film badge   4, 15
first division cells   17
FITC fluorescence   73
flagellum   86
flow chamber   239
– cytogenetics   67
– cytometer   156, 230, 300
– cytometric analysis   51, 53, 111, 265

flow cytometry   25, 51, 69, 99, 104, 106, 128,
   161, 219, 273, 311, 313
– –, dual beam   25, 224, 236
– –, multiparameter   220
– –, slit-scan   25, 237
– karyogram   53
– karyotype, bivariate   27
flow-cytograms   38
fluorescence distribution   71
– emission   69, 219, 240
– – spectrum   311
– excitation, epi-illumination   238
– plus Giemsa (FPG) staining   4, 12, 17
– polarization   235, 237, 240
– spectra   230
fluorescent-debris   33
– staining   178
formaldehyde   219, 224
forward mutation   144
fourier analysis   53, 54
– power spectrum   53, 55
fragmentation   123
frame encoding   297
– shift   143
frames   296
Friend leukemia virus   204, 210
fusion   89

G_0/G_1-phase   37, 221, 279, 305
G_1/S block   274
G_2+M-phase   37, 221
G_2-phase   274
gallocyanin   112
gamatogenesis   85
gamma-radiation   18, 221
gas gradient   299
gene activity   244
– expression   143
– loss   161, 166
– mutation, induced   146
Gene-Tox comittee   92, 93
genetic–counselling   4
– damage   26
– –, heritable   93
– –, integrated   166
– disease   26
– hazard   146
– information   141
– insult   161
genome mutations   135
genomic DNA   28
genotoxic-agents   149
– chemicals   219

genotoxic-effects   161
– treatments   38
germ cell(s)   129, 133
– ejaculated   99
– generation   87
– mutagenicity   92
– pool   94
–, testicular   99
germinal-effects   90
– epithelium   87
– mutation assay   145
Giemsa stain   7, 78
glacial acetic acid   204
glycophorin A   161
gonad dose   127, 135
grain-counting   257
granularity   205
granulocytes   187
grayness   174
growth fraction   299, 305

haematopoiesis   183
haptens   326
harlequin-staining   18
harzard analysis   141
heat treatment   236
HeLa cells   45
hematologic disorders   61
hematopoietic effects   181
hemoglobin C   162
– inherited variant   164
– S   162
– sites   164
– variant   161
hemopoietic system   219
hepatocytes   285
hepatocarcinogenesis   285
hereditary defects   331
heterogeneity, metabolic   299, 301
HGPRT-variants   149, 158
Hiroshima   77
histidine dependence   143
histocompatibility complex   233
histogram thresholding   174
HLA determinants   166
Hodgkin's disease   104
Hoechst 33258   25, 26, 67, 68, 274
Hoechst 33342   229
human hemoglobin variants   161
– malignancies   61
human-sperm/hamster-egg fertilization system
    94
hybrid cells   327
hydrodynamic focussing   222, 238

hypersensitive individuals   141
hypocenter (bombing)   78
hypogranular morphological variant   62
hypotonic treatment   53, 69
hypoxia   88

image analysis   111, 112, 144, 243, 331
– digitizer   171
– processing   174
immunoanalysis   325
immunofluorescence   161, 205
immunoglobulins   326
–, fluorescein-conjugated   69
immunologic effects   181
immunostaining   329
incubator   259
index of dispersion   38
– sort   118
indicator cells   203
induction of aberration   16
– of dicentrics   20
infertility   91
insecticides   64
insertions   79
instituiton, health   7
–, research education   7
insults, biologic   203
–, chemical   203
–, physical   203
integrated optical density   150
intercalated dye   240
intercalation   143
inter-city difference   80
interphase cells   171
– death   11
interstitial deletion   62
intestinal syndrome   183
intracellular pH   299
inversions   79
ionizing radiation   41, 51, 93, 127, 135, 174,
    177, 183, 188, 219, 274, 325
irradiated fraction   11
irradiation   135, 203, 265
– dose   203
isopropyl-N(3-chlorophenyl)-carbamate   46

Japanese houses   78

karyotype   5
– analysis   80, 117
karyotypic instability   27

keyhole limpet hemocyanin   326
kinetic effects   253
kinetics-directed therapy   255
kinetochore staining   69
Klinefelter's syndrome   88

labeling index   150, 154, 255
labelled nucleus   154
Lag-phase   244
L-cells (mouse)   41
LDH-X test   145
lectins   232
Lesch-Nyhan syndrome   149
lesions, carcinogenic   161
–, heritable   161
LET   5, 7, 17
lethal damage   274
leukemia   26, 61
–, acute   61, 64
– – lymphoblastic   63
– – monoblastic   62
– – myelogenous   63
– – non-lymphoblastic   61
–, chronic myelocytic   61
–, secundary   64
–, smoldering   63
–, subclassification   64
leukocytes human (in vitro)   45
–, circulating   4
light microscope   171
– scattering multiangle   240
linear discriminant analysis   287
linear-quadratic relations   17
Lineweaver-Burk plot   230
lipid bilayer   236
– vesicles   236
lipopolysaccharide   204
LLNL system   77
locus test   145
low dose effects   10
– – level   67
low-LET irradiation   265
lupus erythematous   234
lymph node   231
lymphocyte aberration   11
– cultures   18
lymphocytes   12, 17, 22, 40, 67, 203,
    216, 265, 311
–, circulating   4, 266
–, cultured   79
– – peripheral   67
–, human   17, 22, 40, 171
–, interphase   231

lymphocytes, interphase death   23
–, mouse   40, 232
–, peripheral blood   17, 21, 67, 203, 233,
    265
–, PHA-treated   9, 144
–, splenic   203
–, subpopulations   231
lymphoma cells   300, 302

macrocolonies   256
macromolecular structures   235
– synthesis   99
macrophages   233, 265
Mahalonobis distance   113
male germ cells   99
malignancies human   26
malignant disease   12
– transformation   325
mammalian cells   55, 229, 273
– spermatozoa   41
manual-scoring   74
manual grain counts   256
marker chromosome   25, 26, 29
markermolecule   235
maturation   62
mature sperm   99, 100
meiotic division   130
– nondisjunction   94
– stages   135
melting profiles   236
membrane–markers   205
–, outer cellular   236
– permeability   229, 230
– structures   185
messenger molecules   197
metabolic activation   142, 146
– effects   253
– heterogeneity   299
– inhibitor   230
– profile   312
metabolism   8, 9, 311
–, BaP   312
metaphase cells   5, 45, 67, 79
– chromosomes   171
– finder   3, 9, 13
– spreads   4, 31, 52
– –, second cycle   7
metaphases, first division   21
–, harlequin   18
–, stained   18
methods, morphological   243
methyl alcohol   204
methylcholanthrene   43

MgCl-treatment   240
microcolony   259
microcultures   15
micronuclei   15, 17, 20–22, 44, 171,
    177, 281
micronucleus test   45
microprocessor, Z-80   237, 239
microscope photometer   205
– scoring   52, 69, 150, 184
microscopic slides   244
microsomes   144
microtubule inhibitor   274
Microtus monatanus   123
– oregoni   123
misrepair   143
mithramycin   127
mitochondria   107
mitogen   149
mitomycin-C   121, 150
mitosis   41, 53, 273
– irregular   17
mitotic apparatus   37, 41
– cells   25, 51
– delay   11, 18
– index   69, 276
mixed lymphocytes reaction   233
mobility rotational   235, 240
model calculation   52
modifications, structural   325
monocentrics   79
monoclonal antibodies   142, 144, 325, 331
monocytes   183, 311
monolayer culture   31
mononuclear cells   153, 158, 219
monosomies   27, 61
monozygotic twins   311
morphologic effects   253
– errors   111
morphological abnormalities   92
– differences   217
morphology sperm test   93
morphometric analysis   204
mouse bone marrow   40, 45, 269
– L-cells   40
– L-fibroblasts   244
– lymphocytes   40
– mammary tumour   299
– spermatocytes   46
– spermatozoa   45
– thymocytes   40
M-phase   274, 279
Muellerian system   89
multivariate analysis   216, 285
murine leukemia virus   203
– peritoneal cells   265

murine spleen cells   231
mutagen   178
– screening   44
mutagenesis   141, 219, 325
–, chemical   142
– testing   143
mutagenic   311
– actions   127, 135
– agents   11, 37, 41, 67, 85, 93, 143, 171,
    325
mutagenicity testing   38, 43, 135
mutagens, chemical   171
–, physical   171
mutants   229
mutated cells   12
mutation   52
–, radiation-induced   135
– rate   146
– studies   155
mutational changes   161
myeloblastic   62

Nagasaki A-bomb survivors   77
neutron component   77
neutrons   5, 18–20
N-nitroso compounds   325, 326
nocodazole   274
non-cycling population   309
non-dividing cells   281
non-ionizing radiation   274
non-proliferating cells (Q cells)   255
non-random changes   64
nondisjunction   117
nonviable cells   107
noxious agents   85
nuclear chromatin   243, 288, 295
– DNA   106
– histones   99, 100
– membrane   286
– morphology   285
– organisation   7
– power industry workers   11
– properties   173
– proteins   249
– size shape   172
– texture   172
nucleolus   286
nucleophilic centers   325
nucleus   289
null mutation   166
nutrient depletion   305
nutrition gradient   299

oligospermia 88
oocytes 94, 127
optical density 174
– –, integrated 286
orchitis
–, aspermatogenic 89
–, isoimmune 89
orienting flow cytometry 121
ORNL system 77
ovarian germ cells 94
overdose cases 4
over-exposure 12, 15, 135
oxygen supply 51

pachytene 129
partial body exposure 13, 23
particle stream velocity 239
peripheral blood chromosomes 99
– – lymphocytes 3
peritoneal fluid 265
persons at risk 8
pertubation 274
petroleum products 64
pH distribution 299
– heterogeneity 305
PHA (phytohemaglutinin) 149, 153,
    176, 204, 232
pharmacological behavior 142
phenocopies 144, 168
phenol 317
Philadelphia chromosome 61
physical dosimetry 15, 51
phytohaemaglutinin (PHA) 149, 153, 176,
    204, 232
plateau phase 309
plating effiency 188
pluripotent haematopietic stem cells
    183, 192
poisson distribution 10
– statistics 5
polarization, degree 237
–, plane 236, 237
polarized light 235
polychlorinated biphenyls 285
polyclonal horse antibodies 161
polycyclic hydrocarbons 43
poly-L-lysine 204
polymorphonuclear white blood cells 187
polyploid sperm 117
polyploidization, dose dependent 273
population heterogeneity 312
– at risk 11
– survey 11
possible causative agents 64

post-irradiation mitoses 15
precursor cells 100, 256
– uptake 94
precursors 63, 183
preleukemia 61, 62
preparation procedure 53
preparations, G-banded 80
procarbazine 145
progenitor cells 183
proliferating cells 173, 255
proliferative capacity 51, 52
– heterogeneity 305
promoting agents 285
promutagen 142
propidium iodide (PI) 25, 31, 240, 265, 301
protamines 99, 100
protein content 236
– matrix 244
– structure 299
proximal break 62
purine analogues 149
– biosynthesis 149
pyrene 46

Q-cell recruitment 258
quinacrine mustard 93

radiation 219
– accident 8
– damage 52, 55
– dose 51, 56, 67
– effects 55, 129
– –, cytotoxic 129
– –, mutagenic 129
– exposure 15, 111, 117, 171
– –, accidental 128
–, gamma 3, 57
– induced 171, 273
– – chromosome aberration 127
– – damage 93
– lesion 273
–, neutron 3, 7
– qualities 15–17, 20, 21
– responses 265
– sensitivity 273, 282
–, UV 142
– workers (repair defective) 22
–, X-ray 5, 7
radiation-damaged cells 295
radiation-exposure limits 128
radicals 142
radioactive decay 8

radioactive thymidine   135
radiobiological effects   9
radiographers   11
radioimmunoassay (RIA)   328
radioisotope exposure   78
radiological protection   3
radionuclides   8
radioprotection   191
radioprotective effect   196
radiosensitivity   52, 184, 195, 258
radiotherapy   78
rat myeloma line   327
reciprocal translocations   79
recuplex accident   5
red blood cells   161
– cell ghosts   162
refractive indices   237
refractory anemia   63
regulatory factors   193
relative biological effectiveness (RBE)   7, 20
repair   146, 274
– defectives   146
– kinetics   282
– mechanisms   142
reproductive abnormalities   99
– capacity   55
– death   51, 258
– disorders   85
– effects   83
resistant clone   230
restriction enzyme   145
reticulocytes   183
retinoblastoma (human)   26
reverse mutation   143
reversible alterations   235
revertability   143
revertant phenotype   143
rheumatoid arthritis   234
rhodamine   107, 123
ring chromosomes   57, 79
RNA synthesis   153
rotational freedom   236

S-phase cells   129, 221, 273, 305
salmonella typhimurium   143
scanning   245
Schiff's reagent   155, 204
scleroderma patient   69
searching phase   9
secondary antibody labeling   168
segmentation   154
semen bioassay   85
– quality   89
– sample   90, 104

semen sample   90, 104
– volume   90
seminal contamination   90
– cytology   91
seminiferous epithelium   89
– tubule   87
seminoma   103
sensitivity   55
Sertoli cells   86, 88
sex chromosomes   62, 122
shape abnormalities   111
shielding factors   78, 80, 81
sickle hemoglobin   145
sister chromatid exchange   219
size distribution   75, 156
skewness   289
skin burns   8
slide analysis   67
– chamber   259
slit-scan flow cytometry   32, 116, 224
smoldering leukemia   62
solid tumors   258
somatic cell mutation   161, 164
specimen handling   4
sperm   86, 99, 111
– analysis   85, 111
– chromatin differentation   99
–, chromosome constitution   93
– – concentration   90, 106
– count   91
– functionality   90
– head morphology   92, 99, 105, 112
– maturation   99
– morphology   91, 92, 111
– –, chemical effects   92
– –, radiation effects   92
– motility   90, 106, 107
–, nuclear DNA   105
–, nucleus   99
– shape abnormalities   92
– viability   106
spermatid   99, 100, 128, 132
–, diploid   127
spermatocytes   128, 135
spermatogenesis   85, 86, 91, 93, 99, 111, 117, 127
–, abnormal   88
–, duration   87
spermatogenic damage   90
– dysfunction   92
– process kinetics   86
spermatogonia   86, 99, 127, 128
spermatozoa   86
spermiogenesis   99, 121
spermiohistogenesis   128

spindle fiber formation 177
spleen 231
– cells 219
– colonies 188
– colony assay 198
splenocytes 203, 206
spontaneous leukemia 64
staining characteristics 271
– –, conventional 19, 20
–, FPG (fluorescence plus Giemsa) 4, 19, 21
– properties 295
– techniques 184
statistical uncertainty 10
stem activating factor 193
– cells 183
– cultures 187
strand breakage 143
structural alterations 235, 325
– irreversible 240
– reversible 240
subfertility 88, 100
subpopulations 158, 265
survival 274
survivors atomic bomb 77
synergistic effect 219

T cell activation 233
T cells 229
T helper cells 233
T-lymphocytes 15
T suppressor cells 233
T65D system 77
target cells 184
teratogens 85
teratoma 103
test
–, Ames 143, 217
–, chromosomal aberration 217
testicular biopsies 100
– carcinoma 103
– cell kinetics 101
– cells 135
– dysfunction 104
– radiation 117
– steroidogenesis 88
testis 88
tetraploid cells 286
tetraploidy 89
textural features 292
texture 174, 203, 216
– of chromatin 203
therapeutic strategies 255
thermal denaturation 105
thresholding technique 112

thymine dimers 142
thymocytes 190
–, mouse 40
time-lapse cinemicrography 259
– photography 144
tissue metabolism 88
– sections 285
toxic agents 64
– effets 293
– exposure 112, 161
transformation 311
– frequencies 322
transit time 4
translocations 26, 27, 62
–, reciprocal 62
transplantable tumors 52
transplantation 188
trenimon 43, 44, 46
treshold 119
triploidy 89
trisomies 26, 27, 61
trisomy 8, 61
tritiated thymidine 149, 153
tritium gas 9
Triton-X 100 53
trypsinization 243
tubular degeneration 88
tumor cell colonies 255
– cells, L 1210 mouse 240
Turner's syndrome 88, 89
TV camera 171, 173
– scanning 285

ultraviolet light 45, 325
uptake, BaP 312
urethane 44, 46

V79 Chinese hamster cells 52, 55
V79 chromosomes 54
variant frequency 150, 154
vas deferens sperm 102
velocity sedimentation 195
viability 300
video image 257
vinblastine 53
viral infection 203
visual scoring 117
vole 94

white cell count 4
whole body exposure 7

whole body ratiation   67, 205
worker
–, chemical industry   11
–, radiation   5, 11

x-ray irradiation   18, 55, 57, 203, 258

Y-chromosomal nondisjunction   93

zero dose estimates   8

Biophysics

Editors: **W. Hoppe, W. Lohmann, H. Markl, H. Ziegler**
With contributions by numerous experts

1983. 852 figures. XXIV, 941 pages
ISBN 3-540-12083-1

Contents: The Structure of Cells (Prokaryotes, Eukaryotes). – The Chemical Structure of Biologically Important Macromolecules. – Structure Determination of Biomolecules by Physical Methods. – Intra- and Intermolecular Interactions. – Mechanisms of Energy Transfer. – Radiation Biophysics. – Isotope Methods Applied in Biology. – Energetic and Statistical Relations. – Enzymes as Biological Catalysts. – The Biological Function of Nucleic Acids. – Thermodynamics and Kinetics of Self-assembly. – Membranes. – Photobiophysics. – Biomechanics. – Neurobiophysics. – Cybernetics. – Evolution. – Appendix. – Subject Index.

Biophysics is the English translation of the completely revised second German edition of this book. A wealth of information, new sections and chapters, and extra figures make **Biophysics** an indispensable text for advanced students and lecturers of physics, chemistry, biology, and medicine. Indeed every scientist interested in biophysical questions will find much in this book essential for a deeper understanding of modern biophysical research.

From the reviews:

"… Most of these contributions are very informative reviews of 10–20 pages written for undergraduate students or scientists who wish to broaden the scope of their knowledge. They provide a convenient source of information for many facets of biophysics and cannot be found in such a concise way in other textbooks…"*Nature*

"… this book will surely reach the highest standard in its category…" *Bioelectrochemistry and Bioenergetics*

Springer-Verlag
Berlin
Heidelberg
New York
Tokyo

Radiation and Environmental Biophysics

ISSN 0301-634X   Title No. 411

Managing Editor: U. Hagen, Neuherberg

Editors: J. J. Broerse, Rijswijk; F. Dunn, Urbana, IL; H. Fritz-Niggli, Zurich; A. S. Garay, College Station, TX; R. Goutier, Liège; U. Hagen, Neuherberg; H. Jung, Hamburg; A. Kellerer, Würzburg; J. T. Lett, Fort Collins, CO; H. Lieth, Osnabrück; B. D. Michael, Northwood Middlesex; S. M. Michaelson, Rochester, NY; H. Muth, Homburg, Saar; P. Oftedal, Blindern, Oslo; S. Okada, Tokyo; R. B. Painter, San Francisco; H. Pauly, Erlangen; H. H. Rossi, New York; H. P. Schwan, Philadelphia; K. Wagener, Jülich.

The journal is devoted to fundamentals as well as to applications. Its range of interest includes:

- biophysics of ionizing radiations (including radiations chemistry, radiation biology, radiation genetics; incorporation of radionuclides; biophysics as the basis of medical radiology, nuclear medicine and radiation protection)
- biophysics of nonionizing radiations: ultraviolet, visible and infrared light (including laser light), microwaves, radio waves, sound and ultrasound
- biophysical aspects of environmental and space research, i. e., physical parameters used in the description of ecosystems, mathematical treatments of the environment, mechanism and kinetics of flows of matter and energy in the biosphere
- biological effects of such physical factors of the environment as temperature, pressure, gravitational forces, electricity and magnetism.

The treatment of these themes may include both theoretical and experimental material and can embrace complex radiobiological phenomena as well as health physics and environmental protection.. Special emphasis is given to fundamental questions. The journal accepts also important papers which are not concerned directly with effects of radiation or environmental problems but with the scientific basis for their understanding, e. g. biophysics of genetic and mutations, mathematical models of biological or environmental systems and the mathematical treatment of cell kinetics.

Springer-Verlag
Berlin
Heidelberg
New York
Tokyo

For subscription information or sample copy write to Springer-Verlag, Journal Promotion Department, P.O.Box 105 280, D-6900 Heidelberg, FRG

Biological Dosimetry

Among the various types of dosimetry, biological dosimetry relies on selective, quantitative biological responses to measure dose. It is particularly valuable in situations where physical or chemical dose is unknown (as in accidents), where dose is difficult to standardize (as in irregular scheduling), or where individual biological susceptibility is likely to vary (as in repair defectives or pharmacologic variants). The rapid recent development of quantitative cytometry has provided a new multidisciplinary, potentially powerful analytical tool for biological dosimetry. These cytometric techniques offer the promise of objective, precise, sensitive, cellular and subcellular classification coupled with highly rapid, statistically sound rates of analysis. They are ideally suited to the measurement of biological dosimetric responses in many mammalian systems.

ISBN 3-540-12790-9
ISBN 0-387-12790-9